I,robot
Information Sheet

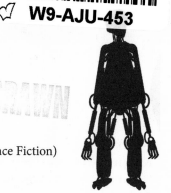

Publication Date: September 20, 2008

Author: Howard S. Smith
Illustrator: Kathy Harestad
Title: *I,robot*
Genre: Technothriller (Speculative Fiction, Science Fiction)
Intended Audience: 14 years and up

Publisher and Imprint:
Robot Binaries & Press
704 Spadina Avenue #134
Toronto, Ontario M5S 2S7 Canada
howard@robotpress.net • Toll-Free Fax: 1-866-693-0120 • www.robotpress.net

Price: US $17.95
Pages: 408   Trade Paperback 6x9 trim size
ISBN: 978-1-894689-06-9
Library and Archives Canada Cataloguing in Publication:
    PS8637.M5625I7 2008   C813'.6
Illustrated: yes – maps and technical diagrams
Bibliography: yes

Distributor: Atlas Books, Ashland, Ohio • www.atlasbooksdistribution.com
Wholesalers: Ingram, Baker&Taylor
Canadian Distributor: Scholarly Book Services, Toronto • www.sbookscan.com

Brief Synopsis:

Technothriller. Japanese detective stumbles onto deployment of military robots. Cutting-edge technology. Fast read.

Commentary on Title:

The title of the book is intentionally similar to Isaac Asimov's classic *I, Robot* but of different meaning. In Asimov's stories, the robot-based plots develop interesting twists because the robots must follow their rigid rules known as the three laws of robotics. Some sixty years later, this book updates Asimov's work with a realistic technology for the robots, as well as a realistic driving force – military need – for their emergence. This is one reason the title is used, but the other one is that the main character of this book, Haruto, is in fact a human 'robot' who must follow his own internal rigid rules. And by following these rules an interesting twist in the plot arises, and ultimately leads to the emergence of vast numbers of robots in our world, thereby bringing Asimov's vision to fruition.

Author biography:

Dr. Howard S. Smith is an MIT-trained engineer with an interest in artificial intelligence — the supermarket self-checkout machines are *all* based on his work — and natural intelligence — evolution of the brain.

MAJOR NATIONAL MARKETING AND PUBLICITY CAMPAIGN:

• National Media Attention
• National Review Attention
• National Print Advertising
• Online Promotion to news, culture, entertainment, robotics,
    AI and science fiction sites
• E-mail and viral video campaign
• Widespread ARC distribution
• Targeted Mailings
• Off-the-Book-Page Attention
• In-store signage: 7-foot and 3-foot robot posters
• In-store signage: Cover blowups

I, robot
ATLAS BOOKS FALL 2008 CATALOG

• To place orders in the U.S., please contact Ingram or
Baker & Taylor, or call Atlas Books Distribution, 1-800-BookLog.

• To obtain a copy of the Atlas Books Catalog, go to
www.atlasbooksdistribution.com/booksellers.htm.

• To place orders in Canada, please contact Scholarly Book Services
in Toronto at 1-800-847-9736, Fax 1-800-220-9895
or go to www.sbookscan.com.

• To place orders elsewhere, please contact Robot Binaries & Press
at info@robotpress.net.

ForeWord Reviews *of* Good Books Independently Published

Review in ForeWord Magazine, March/April 2008:

"…Innovative, fast-paced, and extraordinarily well-written, it's honestly better than both of those stories [Isaac Asimov's book or Will Smith's movie] put together…

The premise could easily have been pulled from tomorrow's headlines. Japan is threatened by an increasingly hostile North Korean regime that sinks its ships, captures its sailors, and even fires missiles through its airspace, yet finds itself constitutionally prohibited from developing offensive weapons that many in the military feel are necessary for proper defense. Besieged by terrorists and disdained by most of the rest of the world, Israel is also in desperate straits. Losing a slow war of attrition with their numerous enemies, the Israelis make a secret deal to trade tactical nuclear weapons and technology to the Japanese in exchange for a horde of artificially intelligent, combat-trained robots. Police inspector Haruto Suzuki is assigned to investigate the death of the owner of an electronics company that makes parts for the androids, and gets caught up in the action.

Suzuki is fascinating. Obsessive-compulsive, he possesses the drive necessary to accomplish almost anything, yet this single-mindedness encumbers his life. He is overly conscientious and obsessed with rules. When officers in his unit accept free meals from local merchants, for instance, he turns them in, ruining their careers while unintentionally accelerating his own. He makes keibhu, full inspector, by age thirty-five, an extraordinary achievement. He ruthlessly hones his body, excelling in karate as well. This trait saves his life more than once throughout his globe-spanning adventures. When he meets Mara, a beautiful Israeli woman, he discovers the one thing he'd been missing all his life—true happiness. The challenge is that he is torn between his obsession to follow the "rules" and the consequences that such actions would bring upon his newfound love. This relationship is convincingly written, truly romantic, and not contrived in the least.

Smith, an MIT-trained engineer, really did his homework. Cutting-edge technology is explained in ways that make it readily accessible to the lay-person. Harestad's charts and illustrations help clarify things even further. Everything from nuclear technology to advanced robotics and artificial intelligence is artfully described, believable, and surprisingly exciting. The author even describes how a nuclear test detonation could realistically be hidden from satellite surveillance. There's a thirteen-page bibliography at the end for readers more interested in the technology.

... I, robot is a mesmerizing read with memorable characters, great dialogue, believable technology, and wonderful action."

— ForeWord Magazine

Quoted from review provided prior to publication of March/April issue of ForeWord Magazine

Howard S. Smith's

I, robot

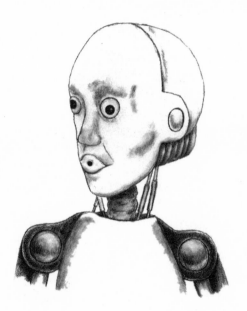

THIS IS AN ORIGINAL STORY CREATED BY
HOWARD S. SMITH.

WHETHER OR NOT YOU'VE READ THE ASIMOV BOOK OR SEEN THE 2004 FILM,
YOU'LL ENJOY THIS NOVEL.

Howard S. Smith's

I, robot

PUBLISHED BY ROBOT BINARIES & PRESS CORP
704 Spadina Avenue, #134, Toronto, Ontario M5S 2S7
www.RobotPress.net

Illustrations, book and cover design by Kathy Harestad www.kathyart.com

'Born Classified': United States Atomic Energy Act of 1946
All information contained herein concerning nuclear weapons has been obtained from
public sources. Notwithstanding, any organization capable of building nuclear weapons
would require far more extensive and precise information than is provided herein.

Library and Archives Canada Cataloguing in Publication
Smith, Howard S., 1958-
        I, robot / Howard S. Smith. — 1st ed., trade pbk
Includes bibliographical references.
ISBN 978-1-894689-06-9 (pbk.)
        I. Title.

PS8637.M5625I7 2008   C813'.6   C2008-900252-0

PRINTED IN CANADA

www.RobotPress.net

Howard S. Smith's

I, robot

ILLUSTRATIONS BY KATHY HARESTAD

Robot
Binaries
&Press

www.RobotPress.net

Yo no naka wa mikka minu ma ni sakura kana

Ōshima Ryōta (1718-1787)

"This is the way of the world:
three days pass, you gaze up —
the blossom has fallen."

International Units: What's a kilometer?

If you are unfamiliar with metric units, please look at the chart at the end of the book.

If you cannot understand the meaning of a foreign or technical word from its use in a phrase, please consult the glossary at the end of the book.

Japanese names

Family name and then *given name*
Example: *Suzuki Haruto* — Haruto is the given name, Suzuki is the family name

Japanese-English Short Glossary

Ohayo gozaimasu — good morning

Konnichiwa — good day

Konbanwa — good evening

Jinzouningen — artificial human, used in Japanese science fiction to refer to robots and androids

Howard S. Smith's

I, robot

INTRODUCTION

II **I**n the 1970s... Israel perfects its nuclear weapons, even patenting the laser technology that enriches uranium isotopes without the need for bulky centrifuges:

United States Patent [19]

Szöke et al.

[11] **4,035,638**

[45] **July 12, 1977**

[54] **ISOTOPE SEPARATION**

[76] Inventors: **Abraham Szöke,** 22 Harakafot Street, Kfar Shmaryahu; **Isaiah Nebenzahl,** 10 Nachshon Street, Haifa, both of Israel

[21] Appl. No.: **563,139**

[22] Filed: **Mar. 28, 1975**

[30] **Foreign Application Priority Data**

Mar. 29, 1974 Israel 44529

[51] **Int. Cl.²** **G01N 27/28; H01J 39/34**

[52] **U.S. Cl.** .... **250/251; 250/284;** 250/423 P

[58] **Field of Search** .... 250/423 P, 284, 283, 250/282, 281, 251

[56] **References Cited**

U.S. PATENT DOCUMENTS

3,558,877   1/1971   Pressman 250/284
3,772,519   11/1973   Levy .... 250/423 P

Primary Examiner—Alfred E. Smith
Assistant Examiner—T. N. Grigsby
Attorney, Agent, or Firm—Hubbell, Cohen, Stiefel & Gross

[57] **ABSTRACT**

A process for the separation of one isotope from a mixture of isotopes and an apparatus for carrying out said process. The process comprises generating a beam of the atoms or molecules, introducing the beam into a laser cavity in such manner that the direction of the laser beam is substantially perpendicular to the direction of motion of the beam, irradiating the beam with a radiation adapted to being about the excitation only of the isotopic species to be separated, reflecting the laser beam back and forth between at least a pair of opposed sides of the laser cavity, such as between two mirrors opposite and parallel to each other which are located on these sides, for deflecting the desired isotopic species, and collecting said species.

17 Claims, 3 Drawing Figures

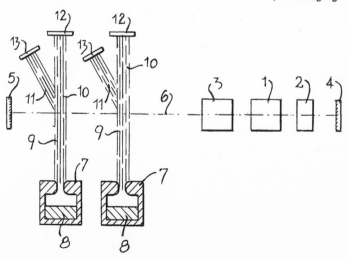

"**In 1971...** ten years before the IBM PC exists — the Japan Robot Association forms. In June 24, 2002, Time Magazine reports, "Japanese engineers are creating a race of obedient machines."

"**By 2006,** android-like Japanese robots are able to work in practical applications.

Actual Construction Robot — Kawada Industries, Japan, circa 2006.
Height - 154 cm. Weight - 58 kg.

"In 2010... the world order starts to unravel, just a bit, but enough: The United States, superpower of the world, is bogged down in Iraq. Iran shouts that it is now part of the nuclear club and a power to be reckoned with. Japan watches in fear as North Korea unveils its nuclear weapons. The Chinese Army, and economy, continue to grow. And Israel realizes it may not have many friends left.

"A few years later...

Pacific Ocean 35°N, 160°E
Japan Maritime Self-Defense Force (Navy) Destroyer Ashigara
June 10 2AM Tokyo Time (June 9 17:00 Zulu)

CHAPTER001

"Missile launched** south of Pyongyang!"

"Stand ready." Captain Watanabe scanned the water with his binoculars. Little light came from the cloudy skies. He could no longer make out the North Korean frigate at the horizon.

The control bridge fell silent. Four subordinate officers, plastered in their seats, staring at green flatscreens in front of them ... waiting.

Watanabe gripped his binoculars tighter. What was happening to this world? Korea — North Korea! — had nuclear missiles.

"Sir, Yokosuka announces the missile is over-flying Japan now... Over Tokyo... now. Altitude eighty thousand meters and climbing. Vector is on course."

The only sounds on the bridge were the soft hum of the electronics and the breathing of the men. The labored silence stretched on, with the officers staring at their screens, and Captain Watanabe gripping his binoculars and wiping his brow.

"On radar now," Master Petty Officer Hirashi shouted. "Computer is waiting for data to calculate vector."

The background of Hirashi's screen started flashing. "Sir, vector estimate is twelve thousand meters short!"

Watanabe dropped his binoculars. The left lens shattered on the metal floor and broke the silence of the room for good.

Hirashi's radar screen changed to a flashing orange background. "Sir, increase to fourteen thousand meters short and two thousand meters wide of expected impact."

"Position and time!"

"Impact at six thousand meters southeast of our position in ninety-four seconds, Sir!"

"Damn them!" Watanabe yelled at the floor. The North Koreans knew that the *Ashigara* was there observing.

Watanabe started thinking. After the nuclear warhead exploded a swell would come at them from the southeast. To avoid capsizing he ought to steer the ship *into* the wave. But that was to make way toward the explosion and the radiation. Steering one hundred eighty degrees away from the wave would take the ship farther from the blast and fallout, and prevent the wave from hitting it broadside, but a stern is not a bow. If the wave were large enough, when it hit the rear of the ship, the stern would rise, push the bow underwater and capsize his boat.

Damn those Koreans. Damn those nuclear weapons.

He'd take his chances with the sea and get as far as possible from the nuke. The decision had only taken seconds. "Helmsman, make your direction northwest at full propeller."

"Yes, Sir."

The roar of the four Ishikawajima turbines coming up to full speed with one-hundred-thousand horsepower filled the ship.

"Transfer ballast from bow tanks into stern tanks and initiate full anti-broaching. Seal hatches. Blast shields down. Electromagnetic pulse protection on. Infirmary to stand ready."

"Yes, Sir!" The officers typed madly into their keyboards. In a few seconds sirens and computer-generated voices sounded throughout the ship, followed by the clanging of blast shields and hatches.

"Northwest at full propeller now, Sir."

"Stern buoyancy?"

"Decreasing, Sir." Hirashi looked at the monitors again. "But not enough time to transfer full ballast... Impact in forty seconds, Sir."

"Position?"

"Estimate impact southeast fifty-eight hundred meters from current coordinates, Sir."

Watanabe opened his wallet and looked at the photos of his wife and twin girls. There was nothing else he could do.

"Ten..." Hirashi said. "Five...four...three...two...one...Impa—" A bright light pushed through the cracks at the edges of the blast shields.

"Brace for impact!" Watanabe shouted. Then an eerie silence.

A loud blast shattered the quiet. Then the massive wave slammed into the ship and over its decks. The *Ashigara's* bow went down under, and the ship listed to port.

Watanabe grabbed onto a pole to keep from toppling over. Damn it, stop, stop! The ship kept listing.

Watanabe's heart was beating faster. With each beat, the port side of the ship approached the waterline. Ballast and loose cargo slid portside, driving the ship further to its side. Time slowed and the ship hung there — its tilted deck only meters from the sea, and creeping closer.

Watanabe pulled against the pole, somehow imagining he could right the ship, his hands so tight they were white. Stop. Please don't end this way. He wedged his feet against the now vertical floor. With every muscle in his body, he pulled against that pole, straining and groaning.

The ship edged closer and closer to the waterline. Chairs and flatscreens broke loose of their attachments and dropped onto the port wall.

And then, as if his will and his pulling had had an effect, the ship slowly started to right itself.

"Damage report," Watanabe shouted.

Hirashi climbed back in his seat. "Propulsion good. Electrical good. Ship's seals good. Negative radiation intake."

"Relay telemetry data to Yokosuka."

"Already done, Sir."

"What does ship's computer say about the blast data?"

"Yield estimated at nine point two kilotons, Sir."

Watanabe wiped the sweat from his eyes. His left hand was still shaking. They'd been lucky. He looked at the laminated map of Japan affixed to the starboard wall and frowned. Maybe next time they would not be so fortunate.

CHAPTER002

"**J** *inzouningen!*"

Haruto was startled for a second, then pushed past the young policeman standing by the doorway.

Suzuki Haruto, Police Inspector, was tall and slim. Usual dark blue trousers, white shirt with starched collar, blue tie, and loose dark blue jacket. Every hair was in place, even his short mustache was combed.

Haruto sat down on the left side of the table between the graying Superintendent-General and the young secretary taking minutes on her laptop.

The Superintendent-General thumbed through the contents of the folder. "Policeman Fujii, please come here."

The policeman standing by the doorway moved to the front of the conference table and stood at attention again.

"Policeman Fujii, you have confessed to corruption. Will you be making a voluntary request this morning to transfer to Traffic Section, or will you be voluntarily resigning?"

Fujii's face reddened, his neck veins bulged. "I never confessed. I was protesting… against Inspector Suzuki's *ridiculous* allegations… that having a meal with a business owner means that seven policemen in our station should have their lives destroyed. You can't destroy careers like this!"

The Superintendent-General turned to Haruto.

"When I was a young policeman, others too did not think it necessary for policemen to follow the same rules the rest of society follows," Haruto said. "Small misconducts became big ones. Finally, we had the rape of a young woman detainee *in* the Kikuya-bashi station *by* a police officer. If we don't take the rules seriously, it could happen again." Haruto looked at everyone sitting at the table. "Rules are there for a reason. Policeman Fujii broke the rules."

The Superintendent looked at Fujii. "Voluntary request for transfer to Traffic Section is accepted."

"No! This isn't fair! All I did was a have a meal with —"

"Voluntary *resignation* is accepted then."

CHAPTER003

The yellow neon bars of the Kiiro Tower Hotel loomed in the distance. Haruto signaled and then veered into the left lane, cutting off the car behind him.

The Nissan's horn blared for a good ten seconds but it didn't concern Haruto. What was happening to this world? Work should have been better at the new police station but it seemed even worse. Should have taken the posting at the Keishicho administration. Boring maybe, but at least there would have been peace.

Why did they hate him so? He did his job. He always did his job. *Jinzouningen*. He hated that term. His wife knew that. Why did she have to call him that when she left him?

Haruto signaled, jerked the steering wheel to the left and veered off Meiji-dori onto the Metropolitan Expressway. Another horn sounded.

He was a good cop. *Keibu-ho* before he was thirty — almost unheard of in the force. And now *Keibhu*, a full ranking Police Inspector, at thirty-five. And he was a good husband. Not once had he missed a birthday, an anniversary, a special occasion. How could anyone fault him?

The Kiiro was about to whiz by. Haruto yanked his car into the hotel's semi-circular driveway. He jumped out, tapped the door two times before closing it, and walked toward the lobby.

A policeman — a *junsa* — stood by room 3602. When he saw Haruto come toward him, he bowed. "Inspector, the body is inside. The medical examiner is there too."

Haruto entered the suite. He saw the coroner, dressed in an ordinary business suit, turning over the body.

"Stop!" Haruto yelled. "What are you doing?"

The coroner, a sixty-ish man, jumped back. "Who are you?"

"Suzuki Haruto — *Keibhu*. Where is the forensics team? You're contaminating the crime scene with your hair and DNA. What for? To finish your work a few minutes early?"

"Ah, the famous Inspector Suzuki. You manage to take down half of your station in shame yet you still remain Inspector, continuing with *your* career. Am I now committed to the same fate?"

"No," Haruto said, and beckoned to the policeman standing at the door. The *junsa* came running to Haruto and bowed.

"I want a full forensics team here immediately."

The *junsa* bowed again, lifted his police walkie-talkie, and started screaming commands into it.

"Suzuki-san, no one cares about this murder, assuming that's what it is." The coroner pointed to the body — a slightly overweight Caucasian man, about fifty-five years old in a pinstripe suit, lying on the carpet. A large bruise on his right temple, but there was no blood anywhere.

"Suzuki-san — where are the reporters?" the coroner said. "Normally the *junsa* would be driving off the hordes of photographers."

"I assume their attention is on last night's nuclear missile."

The coroner nodded. "In shame we could've given in to trade extortion, but instead we take the other path of shame and we let nuclear bombs fly over our heads."

"But it detonated on an old Korean test ship halfway to Hawaii. It will have no effect on our lives."

"Ahh... *jinzouningen* suits you well, Suzuki-san. No effect on our lives? Next time where do you think it will detonate?"

"Negotiations with North Korea have been successful in the past, and they can be successful again in the future. I have faith in my government. They know the best path for us. And the path for me now is to do my job. Who is this dead man? Why was he killed?"

"It's obvious he's a Western tourist or businessman," the coroner said. "A thief who preys on hotel guests probably knocked on his door, was allowed to enter, punched him in the head, and robbed him."

Haruto stared at the dead man. "Perhaps you're right."

The *junsa* at the door knocked once to alert Haruto and then entered along with a middle-aged white woman.

"I — am — Inspector — Suzuki," Haruto said in slow English.

"*Konbanwa, Keibhu*-san," the woman replied without a trace of accent.

"Who are you?" Haruto asked in Japanese.

"I am Co'en Ayaka." The woman began to cry, but pulled a tissue from her purse, rubbed her eyes and regained her composure quickly. Haruto admired the very non-Western control. "That is my husband. When did he die?"

"I would estimate two days ago," the coroner said. "We'll have to do more tests."

"What business was your husband in?" Haruto asked. "How long were you staying in this hotel?"

"My husband owns an electronics export business. Nothing worth

being killed over. I don't even know what he's doing here in this hotel..."
The woman started weeping.

Haruto took her arm and helped her sit down in the room's upholstered chair. "I need to ask you more questions. I am sorry."

The woman wiped the tears from her eyes and nodded.

"Where are you from? Where did you learn to speak Japanese so well?"

The blond-haired woman paused for a moment to hold back her tears and then smiled faintly at him. "I was born in Japan. I went to school here, just like you."

CHAPTER004

Nelson Bennett tried to open his eyes. The phone rang again and again. He grabbed for the blur of the telephone receiver.

"Who the bloody hell is there? It's the middle of the night!"

"Sorry to disturb you. I need a large favor from you, old chap."

"Is that you, Mohammed?" No one else he knew spoke with an upper-crust affectation that was nearly self-parodying.

"Yes. Bit of an urgent printing job required… I just sent the files to you. Could you have a look? I'll hold on."

The printer lifted himself out of bed and rubbed his eyes for a few moments. He stumbled to the laptop on the other side of the room and clicked three times on the screen menus. *Need 7000 placards and 200 banners ASAP*, the e-mail read. He clicked on the attachment and two-tone colors filled the screen.

"Mohammed — we can get to work on it today. I can deliver it to your company tomorrow."

"Actually, I need it for this afternoon. Sorry for the rush."

"I'll have to stop all my other jobs. It'll cost you… fifty-percent rush fee."

"Fine. Just get the work to my office by noon."

Qana, Lebanon
June 10   6:30 AM Local Time (03:30 Zulu)

The half-dozen men, fit and in their early thirties, wearing jeans, t-shirts and sneakers, filed past the rock carvings outside the village and down the path to the large cave.

An older man, Ibrahim, sat on the cave's floor with his laptop open, surrounded by four large canvas sacks. As the men entered the cave they murmured "*Salaam alaykum*" and sat down on the floor.

"*Wa alaykum as-salaam*," Ibrahim said, the rays from the rising sun outside the cave reflecting off the side of his face. "Come here, let me show you something."

The older man reached into one of the canvas bags and pulled out a gray metal ring just over twenty centimeters in diameter, with four fins welded on it at equally spaced distances.

"These fins slip onto the lower section of the two-twenty-millimeter missiles." He pushed a fin and it moved slightly.

"How do we control the fins?" one of the younger men asked.

Ibrahim smiled and held up his finger. "You don't — the computer does."

Ibrahim plugged the USB cable from his laptop into a socket tucked under a rocket fin. He pointed to a colorful relief map of northern Israel on the laptop. With his hand, he beckoned one of the younger men over.

"Do you know your targets well?"

The young man nodded.

"Show us," Ibrahim said. "Put the cursor over the oil refinery outside of Hayfā."

The young man moved the cursor over the Haifa oil refinery.

"Now click it," the older man said.

A red dot started flashing on the map over the oil refinery. In about eight seconds, a text box popped up on the computer screen: *Coordinates successfully loaded. GPS computer functional.*

Ibrahim pointed to the first canvas bag. "In this bag are another five sets of fins, already programmed for the oil refinery. Attach the fins to six rockets, and at noon launch them from the village. The fins' GPS computer will move the fins and direct the rocket to the programmed target."

The younger men gasped. Then one of them said, "This will cut off half their petrol supply. What if they respond?"

The older man laughed. "They're not stupid enough to hit this village again."

"But what if they do?"

"That's what these other sacks are for. And I have plenty more — from our friends to the east." Ibrahim gave three more of the men each a canvas sack. "If there is a response, then you will each fire your missiles with these programmed fins."

Ibrahim turned to the two men without sacks. "Are the tank traps set up on the east and west funnels?"

The two younger men nodded.

"How many of the tandem anti-tank missiles do you have in launchers?"

"Five hundred and forty," and, "Three hundred and two," the men responded.

"Good. That should be enough. And if not, I can get you more."

CHAPTER005

Avrim, a clean cut, boyish-looking man in a blue Air Force uniform with *Tat Aluf* (Brigadier General) insignia, looked through the glass panels at the soldiers in front of their computer screens. All seemed peaceful at the moment. He walked over to his desk and sat in front of his own computer. Suddenly two short high-decibel shrills filled the building.

Shit, another launch. Twenty, thirty, some days even forty times a day. The rockets rarely hit anything, but there were so many of them that accidents happened. People got tired and stopped using the bomb shelters. He thought of the Druze mother and child who were hit yesterday and frowned.

Behind the glass panels the five-by-five meter overhead display lit up with a map of southern Lebanon and northern Israel. A single green dashed line came out of the village of Qana and edged toward Israel. Ten seconds later a second green dashed line appeared on the screen, also coming out of the village of Qana. Two missiles.

Avrim opened his online agenda. He noticed a conflict with two meetings tomorrow morning. He started composing an e-mail when three loud shrills filled the building.

Avrim jumped up. At the top of the overhead display screen, the green dashes were arcing from Qana over the Israeli border. In large red letters below blinked the message: *Est. target: Haifa oil refinery.*

Army Major Kirzner, heavyset, thirtyish with buzz-cut hair, came running into the General's office. "Looks bad, Sir. Do you want to try to shoot them down?"

"The oil refinery has its own Trophy system. It'll blast away at any incoming missile. The Patriots won't work against this. Let's wait."

Avrim's right eye started twitching. He rubbed it against his shirtsleeve but the twitch only got worse and the eye started to tear. The red letters flashed faster: *Est. target: Haifa oil refinery 20 seconds.* With each blink of the General's eye, the number changed. *19. 18. 17.* Avrim breathed in and closed his eyes. He opened them again. *12. 11. 10.* The display split vertically. The blinking message and countdown remained on the left side. The right side showed a live video feed of the oil refinery.

9. 8. 7. Avrim stopped breathing but forced his eyes open. *6. 5. 4.* His heart pounded harder. *3. 2. 1.* One of the green dotted lines turned into a solid red one stretching from Qana to the oil refinery, and on the right side, the video feed showed an explosion a few meters outside the refinery's chain link fence. Avrim smiled and breathed in deeply.

"See, I told you. Best thing to do is wait. They never hit anything."

"Yes, Sir."

"The Patriots cost a million dollars each and do a lousy job against these low-altitude incoming missiles."

"Yes, Sir."

"You see —" Avrim stopped his lecture and gasped. The second green dotted line had become a solid red one, also tracing a path from Qana to the oil refinery. On the right side of the screen, there was the live video feed — flames growing higher and higher, as the refinery became engulfed in dark smoke.

"Sir, what should we do?" Kirzner asked.

TWEEP-TWEEP. Two short high-decibel shrills filled the building again. Another set of green dashed lines were leaving Qana on the screen, travelling toward Israel. Another two short loud shrills pierced the room. A second set of green dashed lines. *TWEEP-TWEEP, TWEEP-TWEEP.* Four sets of green dashed lines from Qana were arcing their way to the Israeli border.

Avrim felt his heart pound. The twitch started once more, and vision in his right eye became blurred. "Hit the origin. Make this stop."

Kirzner rushed to the phone on the General's desk. "Send coordinates of origin to all overhead F-35's and to all artillery pieces in range. Ten-minute barrage."

Within thirty seconds, an F-35 dropped its payload of 500-pound bombs onto the launching origin — which was actually the yard of a high school in Qana. On the large display board a red circle blinked around Qana. Another thirty seconds later, the artillery platoon at Zarit, Israel started lobbing 155mm shells at the origin. Ten seconds later, another F-35 dropped its payload of 500-pound bombs on the origin. Five seconds after this, artillery shells from the platoon at Eilon, Israel, streamed in.

The second F-35 swept low, returned to the origin, and sent video feed to the large display board — no village could be seen, only small fires and thick, black smoke pouring out.

Shit. Avrim's right eye started twitching again.

CHAPTER006

Bobbies encircled the Foreign Office and locked arms against the wave of protestors.

The Chief Inspector looked out at the sea of demonstrators stretching past Downing Street up Whitehall. Many of the women had their heads covered and many of the men wore short beards. Thousands of placards and banners waved in the crowd. *Israel kills children. Israel guilty of war crimes. Death to Israel.* He saw a few that said *Death to America* also.

Leaders with power megaphones speaking fluent British English came to the front of the crowd. "Death to Jews. Death to Israel. Death to America. Death to Secretary Matthews." The crowd took up the chant. "Death to Jews. Death to Israel," the crowd of many thousand yelled back with enthusiasm.

Uttering a death threat against a sitting Secretary. We should arrest them. Then at the park corner of the Foreign Office, a phalanx of protestors suddenly pushed through the police officers and ran up the stairs. Bloody hell, the Secretary might actually be in the building right now.

The Chief Inspector grabbed the walkie-talkie off his belt. "Water cannons to the park corner immediately. All inner ring officers, do not allow penetration of the building. Permission to fire in the air. Exercise caution."

The inner ring of police officers removed their service revolvers, pointed them skyward and sent off a loud volley of shots. A moment later a six-wheeled truck arrived at the building's park corner and fired a stream of water at the protestors, driving them back.

On the seventh floor of the building, Foreign Secretary Matthews parted the curtains slightly, looked at the crowd below, and shivered.

3PM

The Foreign Office Press Room looked deserted. Secretary Matthews frowned.

"Not to concern yourself, Ma'am," her cheerful young aide said. "We have CNN, BBC and a few of the newspapers here. The others simply couldn't get through the line of protestors outside."

"The shots into the empty pressroom will look bad," the Secretary said.

"I spoke to the cameramen already. Everybody is taking forward view shots only."

The Secretary smiled. "Good work. It's an important briefing. We don't want more trouble overseas... more terrorism here at home. Shitty little country... always causing troubles."

"Not to worry, Ma'am. I think the people appreciate your efforts. Your polls were up four points last week."

The aide went up to the podium. "The Secretary of State for Foreign and Commonwealth Affairs, The Right Honorable Nancy Matthews, will be making a statement."

The room became quiet.

"Good afternoon, ladies and gentlemen of the press," the Secretary said. "The British government condemns the Israeli actions in Lebanon this morning. The Israelis attacked Qana in 1996, killing 102 innocent people. They attacked Qana again in 2006, killing 28 innocent people. Again, this morning they attacked Qana, killing many civilians. This is a deliberate war crime. In response to this crime, the British government is hereby implementing full sanctions on Israel. All trade is suspended. All cultural events are suspended. Under war crimes legislation, any Israeli entering Britain will be arrested, unless he or she can show no ties to the Israeli military or government. Thank you."

CHAPTER007

Avrim watched the CNN feed on the large display screen. Shivers went down his spine as the British Foreign Minister said, "Any Israeli arriving in our country will be arrested." *Any Jew will be arrested.* He remembered the stories his grandparents told him of Germany. Shit. How in one day had he done all this to his beloved Israel?

Kirzner ran into his office. "General, the Defense Minister himself is on the way over."

"They saw the CNN segment?"

"No, Sir. I don't think it's that. The blaze destroyed Haifa's cracking towers. It'll take over a year to rebuild. Ashod is already down for maintenance. We just lost all our refined petroleum."

TWEEP-TWEEP. Two loud shrills suddenly filled the building. Then two more, *TWEEP-TWEEP.* And over and over again, *TWEEP-TWEEP, TWEEP-TWEEP.* Avrim looked at the overhead display. A dozen green dashed lines were coming from all over southern Lebanon, arcing their way to the Israeli border.

Avrim grasped his sternum and winced. He lost his footing and fell to the floor, rubbing his lower chest.

"I'll call the medics, Sir. Permission to assume command?"

Avrim shook his head. "No, please, get the bottle of pills on my desk and a glass of water."

Avrim got up on one knee and raised himself. He popped four Gaviscon tablets into his mouth, chewed and washed them down with water. Relief came as the antacid coated his esophagus.

On the large screen, the dozen green dotted lines had already left Lebanon and were now in northern Israel.

"Sir, we need to do something," Kirzner said.

"They usually don't hit anything," Avrim said, his voice trembling. "I can't order bombardment of all the points of origin. You saw what happened this morning."

Avrim stopped breathing as one of the green dotted lines became a solid red line, extending from just outside of Chlhine, Lebanon to Akko, Israel. His face turned pale as he held his breath, waiting for the live video feed.

Four seconds later, the large display split into two. On the right side of the screen was a grainy picture of a burning warehouse. Below the image was a caption in large red letters: *Sat. live feed: Alev Electronics Industry.*

A few moments later, another one of the green dotted lines became solid red. And in a few more seconds the right side of the screen changed showing another burning building with another caption, this time: *Sat. live feed: Rambam Hospital.*

Like the turning lights of a discotheque, the large display screen became a mad rush of colored lines and rapidly changing video clips, almost all of them showing bright orange flames against a background of dark smoke.

"Sir, we need to do something," Kirzner repeated.

Avrim did not respond.

"There are eight points of origin of the missiles. We have enough artillery pieces in place to take every one of them out in the next minute, Sir."

Avrim stared at the burning buildings on the large display screen.

"Sir, permission to give the order?"

Avrim shook his head. "What if all those missiles were launched from nursery schools? What will the world say?"

"Unlikely, Sir. Request permission to give the order."

"No, we need boots on the ground. There's no other way."

"Ground invasion, Sir?"

"Yes. Order for penetration of border. Wrap around points of origin and destroy."

The Major's eyes opened wide. "That's going to need a... a lot of men, Sir."

"Deploy Golani Brigade — Battalions 12, 13 and 14. Immediately. In and out."

"Yes, Sir. What route?" Kirzner pointed to the printed map of northern Israel and most of Lebanon on the General's office wall.

Avrim didn't respond, but kept staring at the burning images on the video screen.

"Sir!" Kirzner said.

Avrim turned his head and looked at the Major pointing to the printed map.

"We have United Nations' troops along the border here, and we have French troops along the border here. Unless we're going to engage those troops, there's only this spot on the east or this spot on the west to enter Lebanon. Which entry vector, Sir?"

"The ground commander will decide. Commence operations immediately."

"Yes, Sir." Kirzner yanked the phone off the General's desk and blurted out a rapid-fire list of orders.

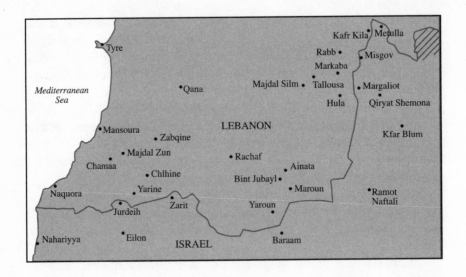

Chlhine, Lebanon
June 10   6 PM Local Time (15:00 Zulu)

CHAPTER008

Tank after tank churning up clouds of brown dust streamed into position in a large circle surrounding the village of Chlhine. Once they were in place, the fast-moving Merkava-5 tanks moved further in, closing the circle. When the Merkavas reached the first outlying buildings, a large set of light brown Achzarit armored personnel carriers rushed in between the tanks, locking their tracks at the last minute. Almost simultaneously, their hatches opened and four hundred commandos with darkened faces and assault rifles poured out. In staggered formation the men pushed forward toward the center of the village, with soldiers peeling off from the others to take command of buildings as they encountered them.

The streets of the village were deserted, but eyes pressed against the windows of the stone houses and peered out. In four minutes, Gideon Battalion reached the town square. A spot-welded frame of a rocket launcher was lying in the dirt. A young commando yelled orders at two even younger soldiers, and they lifted the metal frame.

Avrim's back muscles relaxed as he watched the video feed of the soldiers returning to the tanks with the rocket launcher frame and four suspicious young prisoners from the village. It would turn out okay. A few more days. The troops would mop up the mess there and return to Israel.

Kirzner knocked on the General's open office door, and entered with a graying, short man in a civilian suit, the Minister of Defense. The Minister walked over to the glass panels and extended his hand.

Avrim smiled and took it.

"Busy day, Avrim…," the Minister said.

"Sir, I believe we have regained control again. In a few days we'll remove half of their stockpile of rockets in the south."

"What's your plan?"

Avrim pointed to the large video screen. "The tanks are moving now to Chamaa. There were four rocket launches from there today. Within an hour it will be cleaned up."

"And then?"

"At Chamaa the battalions will split. One force to the south to Naquora. Another to —"

The Minister put up his hand. "I thought French peacekeeping forces were in Naquora. Are you going to engage them?"

"No, Sir. French are just south of Naquora. Makes it all the more convenient for them to close their eyes and allow the missile launches, doesn't it?"

"Good. Please continue with your plan."

"As I said, one force south to Naquora, with another split north with one force to Mansoura and another cross-country to Zabqine."

"Sir!" Kirzner said.

"One moment," Avrim said. "You're probably wondering about Rachaf. We had multiple launches from there as well. We'll continue west from Zabqine. I expect arrival by —"

"Sir!" The Major pointed to the live video feed on the large screen.

One tank and two armored personal carriers were in flames. The Minister's jaw dropped as he watched a screaming young soldier with burning clothes pull himself out of a carrier and run into the bush.

The successive white flashes of anti-tank tandem-missile launches illuminated one side of the display screen followed by almost simultaneous brighter flashes on the other side of the screen. Dozens and dozens of launches overwhelmed the tanks' Trophy anti-missile systems — for every missile Trophy destroyed with its explosive blasts, two came in. Each tandem-missile's first charge set off a tank's reactive (explosive) armor, then a tiny fraction of a second later, the larger explosive charge in the rear of the missile cut through the unprotected metal of the tank like a knife through hot butter.

Flash after flash of light rolled off the large screen, punctuated by burning soldiers jumping out of their vehicles.

Avrim stood frozen with his face pressed against the glass panel, rubbing his lower chest.

"General, how many men do you have there?" the Minister asked.

"Three battalions. About four thousand men."

"Oh my God… Do something!"

Do what? Bomb his own troops? Send more troops in? Avrim stared at the screen, feeling the acid come up from his stomach again.

CHAPTER009

aruto gasped as he breathed in the formaldehyde-tinged air in the autopsy room. He walked over to the coroner. "I don't know how you can stand the fumes here."

The coroner turned away from the body on the table and smiled. "How can the fumes bother a *jinzouningen*?"

Haruto ignored him. "What did he die of?"

"I'm almost there… give me another minute."

The coroner turned back to the partially dissected body and made three fast cuts with the scalpel. He pulled out the heart, about the size of a closed fist, and now a dull whitish red, and studied it.

"No areas of ischemia. Inspector, I have yet to finish the toxicology and microscopic studies, but I think the cause is the obvious one. Someone hit him in the head."

"Could he have slipped? Any possibility of accidental death?"

"I don't think so." The coroner pointed to the right side of the head. "A force to the temple fractured the thin temporal bone and tore the middle meningeal artery. It was a fast blow and caused most of its damage underneath the skin."

Haruto felt his breakfast coming up into his throat. He bowed to the coroner and left the autopsy room at a fast walk. He'd never liked that place. He inhaled the fresh air of the hallway.

A few steps down the corridor was Forensics Laboratory. Haruto tapped silently on the tile outside the doorframe, two times to the left, two times to the right, and then went in. The young technician panicked when she saw Haruto and dropped her newspaper.

"Suzuki-san, I was just reading about the missile that flew over our heads." She pointed to the bright flash lighting up the ocean, splattered in full color across the front page of the newspaper. "What do you think of this?"

"I think I should do my job and you should do your job. What have you found on my case?"

"Inspector, you know we need more time for the DNA and hair analysis."

Haruto smiled. "That's why you started last night, right?"

The technician hit a few buttons on her keyboard and started to read from the flatscreen in front of her. "Eleven different fingerprints found in the hotel room. We're in the process of tracking them all down. Probably most are guests and cleaning staff. Two dozen prints and partial prints recovered from the money and cards in the wallet. Some tissue underneath the right fingernails — I'll have results later. Lots of hairs and fibers, like usual. You know I don't do the analysis and need to wait for the results."

"Can I have the victim's wallet?"

"Some robbery — you could retire with the cash in this wallet."

The technician gave Haruto a bulging brown leather billfold. Haruto counted the notes — four million yen. That was enough money to buy a car. What type of thief would leave this behind?

Haruto popped out the identification cards in the right inner sleeve of the wallet. Driver's license — Co'en Satoki. A VISA credit card. A business card — Co'en Electronics, Roppongi, Shibuy-ku. Another business card — Toshifumi Haruka, President, Autonomous Products, division of Mikiyasu Industries, Sumida River Road. Haruto kept the cards and returned the wallet.

Haruto left the building, backed out his light brown Honda and pulled onto the expressway. In about thirty minutes he turned onto Roppongi Boulevard and rolled down to the three-story concrete warehouse. A red neon sign was mounted above the first floor — *Co'en Electronics Trading Company — Specialty Electronics — Wholesale Only.* The parking lot looked deserted.

Haruto went up to the door. He pulled hard but it didn't budge. He then saw the piece of paper taped on the inside of the glass — *Closed for 1 Week for Mourning due to Death of Founder.*

For a moment, Haruto thought he saw some motion in the glass of the door. He peered inside but only saw a darkened hallway leading to offices and a set of cubicles. He glanced behind him — just the empty parking lot.

Well, perhaps he could learn something from... Haruto looked at the other business card... Toshifumi Haruka, President, Autonomous Products. He got in his car and pulled onto the road.

The tall, thirty-something Korean came out of the bushes and watched Haruto drive away. He ran across the street and put his car into gear.

The Korean watched Haruto's Honda turn north onto the expressway, slowed his car for a few seconds, then he too went onto the expressway's north ramp. For ten minutes the Korean stayed about five car lengths behind Haruto. Then he saw Haruto turn onto Sumida Road. The

Korean's eyebrows lifted. This man in the brown Honda must be involved with Co'en, too.

The Korean drove past the Sumida Road exit. Surprise was good.

Haruto drove down Sumida Road but couldn't find a sign for Mikiyasu Industries or for Autonomous Products. He pulled to the side of the road to a stand selling boxed lunches. "Where can I find Mikiyasu Industries?"

The old man behind the open counter shook his head.

"How about Autonomous Products Company?"

The old man shook his head.

"You don't sell lunches to workers at these companies?"

"Maybe. Never see those names," the old man replied.

"Where are those 'maybe' workers from?"

The old man pointed to the large fenced complex, about twenty meters in from Sumida Road and occupying the area leading to the river. Four-meter high chain-link fence topped with razor wire surrounded a dozen dull warehouse-like buildings, each five times the size of a soccer field. A single wide road led out of the complex, past a gate in the fence, and onto Sumida Road.

"Thank you."

Haruto drove over to the gate, and saw the intercom to the left. He got out of his car and started walking toward the fence.

The foot came out of nowhere.

Before Haruto's brain could register the blur his eyes saw or the whooshing sound his ears heard, his latissimus dorsi muscles jerked his shoulders back, and his left biceps jerked his arm in front of his head and slapped the foot away from his face.

Haruto looked at the Korean man in front of him. For the moment he spent considering why this man was trying to hurt him, he lost focus. The Korean's fist smashed unblocked into his rib cage.

An electric pain shot through Haruto's chest, but instead of thinking about it, Haruto's mind focused on the man in front of him. Before Haruto even took in a breath, his leg smashed into his opponent's left kidney in a violent *mawashi geri* kick. Then Haruto's right arm shot out in what looked like an *oi-tsuki* strike, but at the last second his right wrist dorsiflexed back at a speed faster than the eye could see and smashed into the Korean's nose, breaking the bone in two places and pulverizing the cartilage. Blood gushed from the Korean's face.

The Korean did not press on his nose to stop the bleeding or even scream out in pain. He rolled to the ground, grabbed a handful of dirt and launched it at Haruto's face.

In the second it took Haruto to wipe the dirt out of his eyes, the Korean rolled back up and dashed to the small crest on the side of the road. Haruto saw him run over the hill and out of sight.

Haruto pressed his ribs and shook his head. He walked to the gate and pushed the intercom button. Before he could speak his cellphone rang.

"Inspector Suzuki here."

"Suzuki-san," the technician said. "The initial DNA analysis of the tissue found under your victim's fingernails came back. Does not match any offenders on record. Korean ethnic group."

Haruto grimaced and rubbed his rib cage again.

At the same time, a voice came from the intercom — "Can I help you? Why are you here?"

"I am Inspector Suzuki, Metropolitan Police. I need to see Toshifumi Haruka. Immediately."

CHAPTER010

S **till rubbing his right** ribcage, Haruto paced the small stone walk in front of the intercom, his head down. DNA evidence of a Korean struggling with the victim — probably the one to kill Co'en with a blow to the temple. Now a Korean attacks him, a full-ranking *Keibhu*!

And why did Co'en have Toshifumi's business card? It might have nothing to do with the murder. Sometimes these clues went nowhere and the case was as cold as the stones he was walking on. What was Co'en doing in that hotel room if he lived in Tokyo? He looked out at the large warehouses over the fence. What was going on here? Something about this case. He couldn't put his finger on it.

"Suzuki-san…" a cheerful voice called out from the gate.

Haruto jumped back.

"Suzuki-san, did I surprise you?" a young woman said. "Please come here." She patted the passenger seat of the open electric cart she was driving.

Haruto sat down and the cart glided down the road into the bleak entrance yard. Haruto followed the young woman into the first warehouse. The reception area was empty. The room had broken linoleum tiles, peeling walls, and held three dozen flimsy plastic chairs, a wooden counter, and an old glass CRT computer monitor. Haruto followed the woman through the room to swinging metal industrial doors at the far end.

As Haruto stepped through the doorway, he squinted for a moment against the bright lights reflecting off the stainless steel and glass décor of the second room, as modern and polished as the other had been seedy. On one wall, a large backlit plexiglass sign proclaimed *Autonomous Products — a division of Mikiyasu Industries*. On the other side of the room stood a bank of six cylindrical, glistening elevator tubes.

Haruto looked up. No second story was apparent.

The young woman laughed. "We will be going down." She waved Haruto into the open doors of the third cylinder and pressed the *16* button of the twenty buttons on the control panel.

"What does Autonomous Products do?" Haruto asked.

"We are an important division of Mikiyasu Industries."

"That is most honorable. What product do you make here?"

"We make assembly equipment for Mikiyasu Industries. Our equipment is the most automated and of the highest quality."

"Can you give me an example of your equipment?"

"Suzuki-san, forgive my ignorance. You would have to ask a marketing employee," the woman said as the elevator doors slid open. "This way please."

Haruto followed the woman down a wide white concrete corridor that looked like it went on forever. After walking about half a kilometer, they turned to the left. Off-white carpet and oak-paneled walls now replaced the bare concrete. The woman stopped at the open doorway and motioned to Haruto to enter.

The woman bowed. "Inspector Suzuki, I present to you Toshifumi Haruka, President of Autonomous Products, a division of Mikiyasu Industries."

Toshifumi had a skinny face with a large mustache, small eyes, and salt and pepper hair. A fading bruise under the left eye. He smiled weakly and bowed. "I am pleased to make your acquaintance, Inspector. Please come in."

Haruto walked into the cavernous office. There were no windows, but a mural of Mikiyasu-manufactured bulldozers, trucks and ships decorated one wall. On the opposite wall were three large oil portraits — Western men in business suits on either side of a portrait of a much younger Japanese man in academic regalia. At the far end of the office was a brushed aluminum and glass desk. Two white leather sofas forming an L were on a multi-colored modern art rug in the front of the office.

The woman bowed. Toshifumi closed the door behind her, then spun on Haruto, "Are you out of your mind? We are *never* to meet here. I should go to the police and put you and that damn Korean gang in jail."

Haruto looked at Toshifumi in puzzlement. He took out his badge and ID card, and flashed open the holder. "Toshifumi-san, I would be most happy to put all concerned in jail."

Toshifumi froze. The blood quickly drained out of his head, his face turning pale.

"I... see," he said after a moment. "Please, Inspector Suzuki, won't you please sit."

Haruto looked at Toshifumi without expression and sat down on the white couch. He remained silent, curious as to how Toshifumi would extricate himself.

"Suzuki-san..." Toshifumi's professional persona was locked firmly into place once more. "Like all large companies, we are sometimes subject to the nonsense blackmail by the crime syndicates. I apologize for my outburst and my error. As you know, it is my responsibility to handle these situations delicately, to avoid any bad publicity for the company."

"Toshifumi-san, thank you for your explanation. I am familiar with these *yakuza*. But is it not unusual for the President to be handling these small matters? Is it not better to have one of your managers deal with this 'nonsense'?"

Toshifumi nodded profusely. "You are correct, Inspector. It was an error in my judgment."

Haruto smiled. "Toshifumi-san, how did you get that bruise on your face?"

"I was careless and fell at home the other day."

"Toshifumi-san, do you know Co'en Satoki?"

Toshifumi shook his head. Haruto saw the perspiration on his forehead. Good. The man was off balance and vulnerable.

"Co'en-san was murdered the other day. He had your business card in his wallet. How do you explain this?"

"Co'en-san was murdered?"

"Yes, he was murdered."

"Many persons have my business card, Inspector. I meet thousands of persons every year in the course of my business."

"What is your business? What does Autonomous Products manufacture?"

"We make automated assembly equipment for our parent company. We make the robots that weld and paint all of Mikiyasu's cars and trucks, for example."

"Toshifumi-san, may I look around your factory?"

"I am sorry, Inspector. Strict regulations from my parent company. *Everyone* must apply in writing for authorization to enter the factory."

Haruto dropped his smile and opened his eyes wide. "I am the police. There has been a murder. I have every right to look around."

Toshifumi started trembling. "Inspector, I would like to accommodate your request. However, my parent company has invested two *trillion* yen in this facility, which I have been given the honor to run and to protect. What if an imposter came here and stole our secrets?"

"I am investigating a murder. We found your business card on the victim. I will get a warrant and return."

"Mikiyasu is very powerful, Inspector. You may not be able to get your warrant. But you have my full cooperation. For the last week I have been in one meeting after another. Would you like to speak to my assistant? She can tell you where I was at whatever day or time you are interested in."

Waste of time. It was far too easy to fabricate an agenda.

Haruto got up, looked around the office again, and walked over to the portraits. *John J. Hopfield* — *Professor* — *Caltech & Princeton.* **Haku Seiko** — *Professor* — *University of Tokyo.* **Rodney A. Brooks** — *Professor* — *MIT.*

"Who are these men?"

"Autonomous Products honors their discoveries by these portraits," Toshifumi replied, his voice cracking.

Haruto took out a small black notebook from his pocket and jotted down their names.

"Thank you, Toshifumi-san," Haruto said and opened the office door.

The young woman was waiting there. She smiled and bowed. "I am pleased to show you out, Inspector."

CHAPTER〇11

The Mossad meeting room had lightly stained oak-paneled walls. On the front wall was a large flat-panel display, currently only showing a blue screen. On the rear wall in large raised Roman and Hebrew letters: *Where no counsel is, the people fall, but in the multitude of counselors there is safety. — Proverbs XI, 14*

The Prime Minister started tapping his index finger on the large leather-clad meeting table. The rest of the room fell silent at the simple gesture. He looked about fifty-five years old, an energetic fifty-five. "What a disaster in Lebanon." He continued tapping his index finger on the table, harder and faster. "Two thousand casualties to take out *fourteen* enemy? What's happening to this world? What happens when there's the next real war against us? What happens when Iran decides to send in a hundred thousand men through Syria?"

"There's an investigation under way," Amir, Chief of General Staff, said. The graying General stood up. "I accept full responsibility, Prime Minister, with my career, if necessary."

The Prime Minister slammed his fist down on the table. "Good. Very good. So all my Generals wish to resign. What good does that do me? It's not a problem of training, or a problem of discipline, or a problem of intelligence. The country is slowly dying."

"Stop exaggerating," said Uri, Director of Mossad. He was about sixty years old and wore an open, white shirt.

"Uri, forget about yesterday for a moment. Twenty-five rockets a day are hitting us. Casualties are low, yes, but it's destroying the economy. Once investment goes down it will take decades to build up confidence again. Look —" The Prime Minister started gesticulating with his hands. "Right now over half of the computers in the world are powered by processor chips made in Israel. But this morning, IT Chips announced they're shutting their plant near Haifa — that's four thousand jobs lost."

"So we improve our intelligence and hit them again," Uri said.

Amir shook his head. "Intelligence isn't going to help us, Uri. We already know what's going on. In the past, nobody wanted a war they couldn't win, so we had peace. It's different now. They know it. They're emboldened. Now it's al-Haleeb and Syria and their allies. Soon it will be Jordan and Egypt and the rest of the Middle East."

"Look—" the Prime Minister said. "The nuclear threat worked for years and kept us safe. Now it won't be enough anymore. We have nuclear bombs, but *they* have nuclear bombs and dirty bombs, so the effect is cancelled. First Pakistan, then Iran, and God knows who's next. Only a matter of time until al-Haleeb gets a device."

Uri stood up and raised his hands in protest. "That's ridiculous. Al-Haleeb will never get a device. The nuclear situation isn't good, but it's contained."

"Forget that for a moment. Our tanks and planes kept us safe for years. But we can't use them against an enemy mixed in with a civilian population. We've lost the war of world opinion, and the results to the country have been devastating. Over the next twenty years, we'll die a slow death as our economy crumbles, the threat of our deterrence crumbles, and our people lose confidence. As a last blow we'll be invaded, and it's the Holocaust again."

Amir shook his head. "I can't predict what will happen in twenty years, but right now we're still strong. Give me the order and there won't be a Lebanon or a Syria left in two weeks."

"And maybe there won't be an Israel either, when Iran retaliates with its nuclear missiles."

"Gentlemen," Uri said. "Mossad has been working on this for the last two years — *before* the disaster of yesterday. The response to the slow war of attrition they are waging on us is a *counter* war of attrition, but one that is to our advantage. If we were willing to accept the casualties, we could have already started this strategy. And I don't fault you, Prime Minister. You would no longer be Prime Minister if a hundred men a day were dying. I accept that. *But...*" Uri put his finger up in the air. "We cannot let this situation continue. It's like the boiling of frogs."

The Prime Minister and Amir looked at Uri in puzzlement.

"You put some frogs in a deep pot of water — nothing much happens," Uri said. "You put the pot on the stove and turn up the heat, slowly...very slowly. As long as the water heats up bit by bit, the frogs don't jump out of the pot. It isn't until the water becomes too hot that the frogs realize something is the matter, but at that point the hot water incapacitates them and they can't jump out of the deep pot, and they die.

"We're the frogs. We have to act now. Either start hitting back on a small scale in Lebanon immediately and accept the casualties and the political fallout, or follow the alternative plan."

The Prime Minister nodded. "Where's Cohen? Have you heard from him?"

Uri shook his head. "No communication now for two days. He was supposed to have met with the contact at Mikiyasu Industries yesterday. They did about ninety percent of the development, and have access to everything. It was a clean deal. Cost, above and beyond the purchase price, was about two million dollars to pay off Cohen's contact's gambling debts, another billion dollars in unrelated purchases — mostly construction equipment and trucks that would help the economy anyway."

"He's Ollum's cousin," the Prime Minister said. "A screw up, like Ollum, I should have known better."

"Ollum, our ex-Prime Minister, has a cousin who is *from* Japan?" Amir asked.

"Ollum's family was in China for a long time," Uri said. "The rabbi in his grandfather's shtetl in Poland told the whole town to pack up and move to China. The Jews in the other nearby villages all died in the Holocaust, but this town listened to the rabbi and moved to Shanghai, and they survived. Many came to Israel in the fifties, but some of them stayed on in China while others went to live in Japan after the war."

The Prime Minister put up his hands. "Cohen hasn't reported for two days. It's time to become practical. Uri, will Mikiyasu deal directly with us?"

"They're not allowed to, unless all we really want to buy are excavators and dump trucks. That's why we set up the operation with Cohen."

The Prime Minister started tapping his fingers again. "Two thousand dead yesterday. We have to start becoming practical… Amir, you saw the American version. What's your opinion?"

"No better than our own. Their models have a larger machine gun. They both work, but aren't even close to the level we need."

"What about the Taiwanese?" the Prime Minister said.

"Like the Japanese, they're afraid of the four-million-strong Chinese Army," Amir said. "However, Japan started their program thirty years ago. The added experience shows. Plus the Taiwanese want more than money."

"The same thing our Japanese friend in the waiting room wants, right?" the Prime Minister said.

"Yes."

"I'm totally against this," Uri said. "Give us a chance to establish contact with Cohen again. We still have a few days before the ship gets here and Cohen has to pay off in full. It's a good, safe deal. The deal with the Japanese government is crazy! How's the U.S. going to react? How's the rest of the world going to react?"

"Do we really need this technology?" the Prime Minister asked.

Uri nodded without saying anything. The men all sat there silently.

The Prime Minister thrust up his shoulders and head. "We will survive. We always have. Let's show the gentleman in and do the deal."

CHAPTER012

Haruto scanned the lobby display for Professor Haku's name. He crammed into an elevator with a dozen teenaged students and got out on Five. A forty-year-old woman in jeans a bit too tight for her belly and wearing a bit too much makeup was on the phone behind the reception desk. Large metallic letters above her read *University of Tokyo Artificial Intelligence Laboratory*.

The receptionist kept chatting away on the phone. Haruto tapped the reception desk and coughed loudly.

"How can I help you?" the woman said.

"I would like to see Haku Seiko."

"I'm sorry, he's gone on sabbatical."

"When will he be back?"

"I don't know."

Haruto glared at her. "When did Professor Haku leave for sabbatical?"

"I don't know. I've never met the man, and I've been here three years already."

Haruto took out his small black leather notebook. "Can you give me his home address?"

The woman shook her head. "We're not allowed to. Professors' home addresses are private."

Haruto reached into his breast pocket and took out his police identification card. "I am Inspector Suzuki here on police business, and I request the information."

The receptionist quickly became serious. "One minute Suzuki-san," she said and picked up the intercom.

Through a glass door behind the receptionist, Haruto saw a white-haired man walk toward them.

The elderly man bowed to Haruto. "Inspector, I am Professor Tamaki, Director of the Artificial Intelligence Laboratory. Please, won't you come to my office for a cup of tea?"

"Thank you."

Haruto followed the professor through the glass door and down a corridor also constructed out of glass panes. Rack after rack of computers

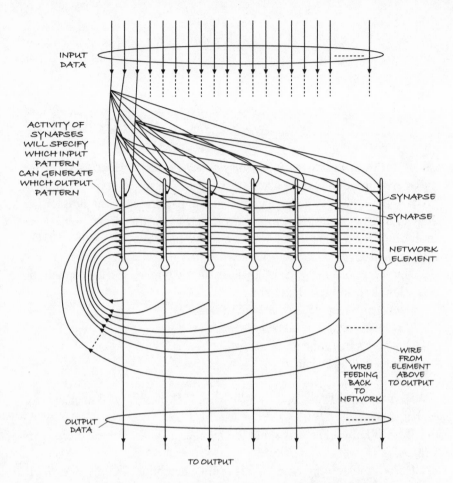

INPUT
DATA

ACTIVITY OF
SYNAPSES
WILL SPECIFY
WHICH INPUT
PATTERN
CAN GENERATE
WHICH OUTPUT
PATTERN

SYNAPSE

SYNAPSE

NETWORK
ELEMENT

WIRE
FROM
ELEMENT
ABOVE
TO OUTPUT

WIRE
FEEDING
BACK
TO
NETWORK

OUTPUT
DATA

TO OUTPUT

HOPFIELD FEEDBACK AUTO-RECOGNITION CIRCUITRY (SIMPLIFIED DEPICTION)

lined up in rooms behind the panes. A dozen twenty-something jeans-clad students sat behind clusters of large flat-panel displays, or chatted at white boards on the walls.

Tamaki's office also had walls made of glass, and a glass desk held together with a shiny stainless steel frame. How large was the cleaning staff here? Mathematical symbols covered an electronic white board, about two-by-four meters, hanging from one of the glass walls. The professor poured Haruto a cup of green tea and sat down beside him on a black couch in front of the desk.

"Inspector, Haku Seiko has been absent from our group for nearly five years now. I have no forwarding address for him."

"Do you know him well?" Haruto asked.

The old professor's eyebrows rose. "Very well. He did his PhD

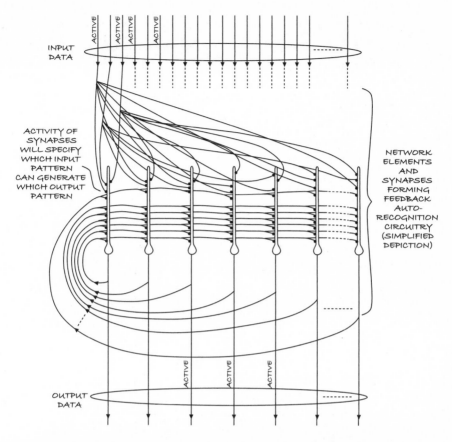

"CORRECT" INPUT DATA (ACTIVE, NONACTIVE, ACTIVE, ACTIVE, NONACTIVE, ACTIVE ...)
TRIGGERS REFLEX OUTPUT DATA (NONACTIVE, NONACTIVE, ACTIVE, ACTIVE, ACTIVE ...).

underneath me about ten years ago. He was a frequent visitor to my house. My whole family knew him well."

"What happened to him?"

Tamaki pointed to the large whiteboard on the glass wall. "Our research, Inspector, is of a theoretical nature. We prove a theorem here, disprove one there. Sometimes, there is some immediate benefit, but usually any value may not be apparent for decades. However, Seiko's work led to immense immediate practical value."

"What did he do?"

"He continued my work."

Haruto nodded in acknowledgement to the elderly man.

"Like us, the Americans have been working hard for the last forty years to find what they call the 'holy grail' of artificial intelligence — a system that could learn and think on its own, much like a man. They did not, and still have not, succeeded. But they came close, and Seiko saw this. He took

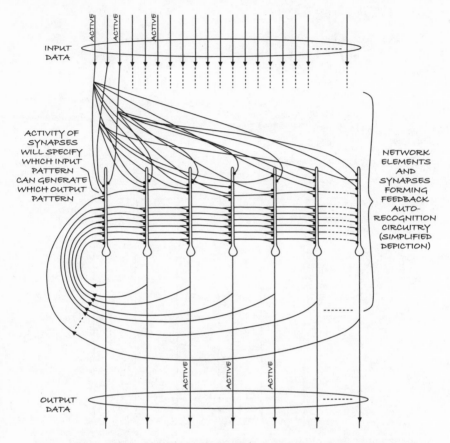

HOPFIELD CIRCUITRY MATCHES INPUT DATA (ACTIVE, NONACTIVE, ACTIVE, *NONACTIVE*, NONACTIVE, ACTIVE ...) SO AS TO TRIGGER THE REFLEX OUTPUT DATA (NONACTIVE, NONACTIVE, ACTIVE, ACTIVE, ACTIVE ...) WHICH IS ACTUALLY ASSOCIATED WITH INPUT DATA (ACTIVE, NONACTIVE, ACTIVE, *ACTIVE*, NONACTIVE, ACTIVE ...).

HOPFIELD CIRCUITRY FORMS A SELF-LEARNING MEMORY UNIT THAT TOLERATES INPUT DISCREPANCIES.

some of their ideas, combined them with my homeostatic system theories, and the result, to all our surprise, was an artificial intelligence system that finally worked."

Haruto flipped to the previous page in his notebook. "Did Seiko use the ideas of Professor Hopfield of Caltech and those of Professor Brooks of MIT?"

"Precisely. But how do you know this? Very little of Seiko's work has ever been published."

"Can you tell me about these ideas, Tamaki-san?"

The professor strode to the large whiteboard and hit a switch. The myriad of symbols and equations disappeared. Tamaki picked up an electronic writing stick and started drawing on the whiteboard.

"This is a Hopfield network. It's a flexible memory unit that learns by itself."

Haruto lifted his hand. "Professor, I'm not sure I understand."

"In an ordinary computer memory circuit, we place the information, a pattern of 1's and 0's, at a particular 'address' and it can be retrieved if we use this same address. In a Hopfield circuit, we don't bother with addresses. If the Hopfield network sees the same input again, then it produces the same particular output again. This is similar to an ordinary memory circuit. However..." The professor's eyes opened wide and he pointed to the whiteboard. "If you give the Hopfield network an input that is not exactly the same as one it has seen before, it will match it automatically to the closest previous input, and give you the output for that previous input."

"What's the advantage?"

"Data from the real world is messy," Tamaki said. "On assembly lines, we go to great lengths to make sure the world is neat and organized, and that's why our assembly-line robots can work. But if something on the assembly line is a degree out of alignment, a human operator, whose brain can easily handle these little deviations, must be there to correct the situation. However, the Hopfield circuit will forgive such a discrepancy, and give you the same output pattern of 1's and 0's, probably the one that is required — all automatically."

"How come more computers don't use such memory?"

"If you want to design a computer to calculate square roots, then you need to use the ordinary memory circuits we've always used. But if you wanted to design a machine that is autonomous, that could function by itself in the real world, you could do a lot of things with the Hopfield memory that you would have a very hard time getting the ordinary memory to do."

"*Autonomous* — you mean like robots that could move around?"

"Yes, the Hopfield memory would be excellent for that. But we've had it since 1982. Why didn't we see robots walking around in 1983? I'll tell you why — it's not that easy to do."

"But Seiko figured out how to do it?"

The professor nodded. "But not by himself. My ideas and the ideas of Brooks were also used."

"What did Professor Brooks do?" Haruto asked.

"Much of the work in artificial intelligence was based on complicated systems of symbols and logic, and much still is. A few years after arriving at MIT, Brooks shook them up with his rejection of this approach to AI. He started building insect robots — small machines that interacted with their surroundings. Here, look at this..."

Tamaki went over to his desk and pulled out a fading glossy reprint, *Elephants Don't Play Chess by Rodney A. Brooks.*

"What did Seiko do with all these ideas?" Haruto asked.

"First, he started designing artificial intelligence systems based on Hopfield's memory circuits. We usually simulate all our work here." Tamaki pointed through the clear glass walls to the many computer screens in the rooms surrounding his office. "However, Seiko said he needed true circuits operating at full speed for his work, so he got a professor in the Electronics Engineering Department to build actual chips where the basic unit was not a logic gate, but a small Hopfield circuit. They were able to cram millions of Hopfield circuits onto each of the chips."

Haruto scribbled furiously in his notebook.

"Then Seiko took much of Brooks' work on insect robots, and plugged into these robots his newly designed AI system based on the Hopfield-circuit chips."

"So Seiko had working robots at this point?" Haruto asked.

"Yes and no. The robots worked, but they didn't do much. A human operator had to guide them through every step of whatever it was they were doing. So Seiko borrowed — and I use the word loosely since he has never given me any credit — my ideas on homeostasis in artificial intelligence."

"What is homeostasis?"

"It is balance," Tamaki said. "The trillions and trillions of chemical processes occurring in our bodies this very moment are controlled by homeostasis in every part of every cell. All these chemical reactions don't run amok — they proceed to some point and then slow down until they are needed again, when they speed up again. I took these ideas from nature, and applied them to artificial intelligence. I showed how a large, complex AI system could self-regulate itself as such."

"Balance seems good," Haruto said. "But what good is a robot that just sits or stands there in balance? How do you get the robot to do things?"

"That's the power of homeostasis," the professor said with a smile. "Homeostasis involves obtaining various goals — either energy to keep the robot's batteries charged, or 'goody tokens' for accomplishing various tasks. The robot self-regulates itself toward these goals."

"Does that actually work?"

"I would say 'yes,' but I am a theoretician. Would it work in the real world? I think it has. Shortly before he left the university, Seiko received a large research grant to build an actual system."

"What did he build?" Haruto asked.

"A set of about twenty 'insect robots' — each about the size of a small footstool, with six legs, looking like giant mechanical cockroaches."

"Where can I see pictures of these insect robots or read about them further?"

The professor shook his head. "Seiko never published his work."

"But is Seiko not a professor at your university still? I thought in order to be a professor you must publish research articles."

"That is very true. However, in Seiko's case, he went from being a Research Associate to a Full Professor without publishing any papers *and* while being on sabbatical all these years. I have no idea how or why."

Haruto rubbed his ear with his pen. He still could not see clearly what this had to do with Co'en Satoki's murder, but there were several possibilities that were starting to make sense. "You actually saw the insect robots. Please tell me about them."

"Making robots in the shape of insects is nothing new," Tamaki said. "What was new was that these insect robots actually worked. Haku could give them a task, such as to look for a girl wearing a red t-shirt, and let the robots go wandering the halls of this building. We would sit down and have a tea, and fifteen minutes later some wide-eyed female student wearing a red t-shirt would follow the robots into our office, laughing, saying the robots ordered her to come here."

Haruto flipped through the pages in his black notebook. "You said a professor in the Electronics Engineering Department built the Hopfield chips for Seiko. Who was this professor? Can I go over to his office to see him now?"

"One minute, Inspector." Tamaki picked up the phone on his desk and hit a few keys. "Can I speak with Nishimatsu Akihiro please?" The professor listened to the phone, and nodded his head a few times before putting down the receiver.

"Inspector, I do not know Professor Nishimatsu personally, but I know the University promoted him last year to a Full Professor. However, his department secretary tells me that Nishimatsu has been on sabbatical for the last four years, and she has never even met the man."

CHAPTER013

The Uenosen Expressway had turned into a large parking lot. The column of cars started up, Haruto rolled a few meters forward, and then braked as the column came to a standstill. Over and over again this traffic dance repeated, and in the next half-hour Haruto's car managed to move about a kilometer. Haruto didn't mind. It was better than going home to an empty apartment.

Haruto turned on the radio. Ayu-sama was singing *Destiny*. Tears welled up in his eyes as he remembered his wedding, holding Michiko so close, this song playing. It was a perfect wedding — everything went just as expected. The marriage was perfect too. He had done everything so well. He had been the perfect husband.

So why did she leave him?

Haruto jerked the steering wheel to the left and took the Kototoi Avenue exit. He turned right at the corner and went a kilometer down the street. He parked his car outside the high-rise apartment building at the corner, tapped two times, closed the car door and went into the lobby.

"Who is there?" an old woman's voice said from the intercom speaker.

"It is I, Suzuki Haruto."

"Please come up, Haruto." The entrance door buzzed open.

Haruto stepped out of the elevator and saw Michiko's mother standing halfway out at the apartment door down the hallway. The old woman bowed to Haruto. "I am leaving to do some shopping. Please come in."

Haruto entered the apartment. Michiko was sitting on the black sofa in the living room. She was in her early thirties, with long straight black hair, large eyes, a small nose, and a narrow face framed by large white earrings. She wore a pair of tight blue jeans with a plain white t-shirt.

Haruto stood rigidly by the vestibule. "How are you, Michiko?"

"Why are you here? What do you want from me?"

Haruto tapped the doorframe twice, walked in and embraced Michiko. She turned her head to the side. He tried to kiss the edge of her lips but his kiss fell on her cheek.

He backed off and sat down in the wide chair facing the sofa. "Why did you leave?"

"I was not happy with my life with you."

"What did I do that was wrong?"

"Nothing. That's the problem. 'Everything by the book,' as you say. Your rule books… I'm tired of following so many rules."

"The rules help us know how to live."

"Did you really need to turn in your entire police station because some of them were getting free meals?"

"They broke the rules."

"They were our friends. Many of their wives were my friends."

"It was police business. It should not concern you."

"What about our sex life?" Michiko said. "Should that concern me?"

Haruto's eyebrows rose. "What about it? Have I not been faithful? Have I not been a good lover?"

"Yes, on Saturday nights. Haruto, tell me when we ever had sex on any other night."

"I was moving up in the ranks. I was trying to organize my life. That's all."

Michiko got up and started pacing the length of the room. "What about a baby, Haruto?"

"We both agreed that once my career was more stable it would be time to start a family. We must make a careful plan and then follow it."

Michiko spun on Haruto and stared at him, her face red. "Plans, rules, rituals, I'm sick of it. I can't be happy with you, Haruto. I don't want to spend my life with a *jinzouningen*. I want a divorce."

CHAPTER014

Haruto snatched the empty parking spot. He tapped his steering wheel twice, jumped out and walked across the street to the multi-story gray building. He didn't bother with the freight elevators but as always took the unfinished concrete stairs to the third floor.

Haruto went into the locker room, where his basket was in its usual place. Arms through the starched white *gi*, tighten the uniform with the worn black belt. He stepped into the empty *dojo*. Shivers went down his spine. Twenty-three years. This place had been a refuge from the cruel world outside its doors for nearly a quarter century.

Haruto put his knees down on the polished gym floor and sat on his ankles. Even after all these years, it still hurt. He breathed in through the nose. *Ichi. Ni. San.* He exhaled through the mouth. He closed his eyes. The other boys teasing him at school. Why hadn't he fit in? He clasped his hands together and let them fall into his lap. In through the nose. *Ichi. Ni. San.* Out through the mouth. Michiko calling him *jinzouningen*. In through the nose...

Soon Haruto no longer felt the pain of his knees against the hard wood floor. He no longer felt the pain of the teasing boys. Or the pain of Michiko's words. As he breathed in and out, his muscles relaxed. Karate was about balance — both with the physical world and the inner emotional one. He was in balance, at peace with himself, at peace with the world.

An elderly man in a bright white karate *gi* tied tightly with a frayed black belt, walked quietly across the *dojo* in bare feet. He faced Haruto, and he too put his knees down on the polished wooden gym floor, sat on his ankles, and started to meditate.

About five minutes later, Haruto opened his eyes slowly and unclasped his hands. He smiled at the sight of *Sensei* and let his arms fall to his sides. Soon the elderly karate master opened his eyes too.

"It is good to see you Haruto-san."

"Thank you, *Sensei*."

"How is your marriage?"

"There are serious problems."

"I am sorry to hear this. How is your work going?"

"I'm working on a difficult case, but I'm pushing hard for a solution."

"Be careful, Haruto." The master smiled gently. "In life sometimes the harder we push, the harder our opponents push back."

"Yes, *Sensei*."

A group of students entered the *dojo*. Most were in their teens or early twenties. They all wore white *gi* karate uniforms but had belts of a variety of colors. A large number of students had a white, yellow or orange belt, while smaller numbers wore green, blue, brown and black belts.

"Will you help me with the class, Haruto?" *Sensei* asked.

Haruto nodded, and he and *Sensei* quickly rose to their feet.

"*Seiretsu!*" the master said.

Almost instantly, the twenty-two students formed a long line in front of the older man. To the left was a student wearing a black belt. Then came the brown belts, a single blue belt, a pair of green belts, the orange and yellow belts, and at the far right, the white belts.

"White and yellow belts — *kumite* with Suzuki-san," *Sensei* said.

Haruto moved to one side of the gym, and about a dozen students with white and yellow belts followed him.

"Who can tell me what *kumite* is about?" Haruto asked.

A muscular seventeen-year-old boy with a yellow belt raised his hand. Haruto nodded at him.

"*Kumite* is the fighting part of karate. In *kumite*, our minds must be focused to achieve full victory. No mercy for our opponents."

Haruto smiled. "Yes, *kumite* is to fight, but it is to fight with balance. To keep your physical balance and your emotional balance. Our opponents here in this club are our friends, so we don't hurt them. And in the real world, perhaps one day your adversary may be your friend, so *kumite* is not about hurting people. It is about keeping your balance."

The students all quickly bowed to Haruto.

"Who are the most dangerous karate fighters in this room right now?" Haruto asked.

A young woman with a white belt giggled and put up her hand. Haruto nodded.

"We are. When we spar with black belts, there is no danger. The black belt can control his punches and kicks, so no one gets hurt. The white belt does not have this control, and is more likely to injure his fighting partner."

"Very good," Haruto said. "Pair up, and five minutes of *kumite* practice. No contact at all to the face. Extremely light contact to the body. If you injure your opponent, you won't have anyone to train with."

The white belts attempted clumsy slow kicks and misplaced punches to their opponents. The yellow belts managed faster moves, but lacked the grace seen in the higher belt colors on the other side of the gym. At the end of five minutes Haruto shouted, "*Yame.*"

The students caught their breath.

Two minutes later Haruto shouted, "*Kumite,*" and the students started to fight again. Fight for five minutes, break for two minutes, over and over again. By the end of the hour, the students were dripping with sweat.

"*Seiretsu!*" the master said.

All the students in the gym lined up by belt color. In perfect unison, the students dropped to their knees and bowed prostrate to the old man.

"*Domo arigato, Sensei,*" the students said as one voice. They popped up on their left knee, stood up and gave one small final bow. The line disbanded, and the students left.

Haruto quickly showered and changed back into his usual street clothes — dark blue trousers, white shirt with starched collar, blue tie, and loose dark blue jacket. Even at night, he wore a tie. He was *Keibhu*, a rank to be honored.

Haruto dashed down the stairs to his car. As he was opening the door of the Honda, the cold barrel of a gun silencer jammed into his neck.

"One little move and you are dead. Do you hear me, policeman?" A Korean-accented, tight, controlled voice. One Haruto did not recognize.

"Yes."

"How are you involved in the robot deal?" the Korean asked.

"I know nothing of any robot deal."

The Korean lifted the barrel to Haruto's head. He put his bandaged nose almost flush with Haruto's, and he stared into Haruto's eyes.

"Do not lie to me." The Korean quickly lifted the gun barrel and pointed it at the streetlight across the road. Even with the silencer the bullet made a loud "thump-whoosh" as it left the gun, followed by the instantaneous shattering of the light. In less than a fraction of a second, the Korean pushed the gun barrel even harder into the side of Haruto's head.

"What are the Israelis giving to Japan for the robots?"

"I don't know."

"Do not lie to me. You are *Keibhu*. You have all the information I have, and even more."

Haruto said nothing.

"We know all about Toshifumi Haruka. He owes hundreds of millions to the Korean syndicate. He was ready to sell out Mikiyasu Industries in a heartbeat. To the Israelis. For what? Tell me, policeman. I have killed before. The Korean jammed the metal silencer barrel even harder into Haruto's head, bruising the scalp. It means *nothing* for me to kill you."

Haruto was walking home from school. The five boys from his class who despised him so. Matching yellow bandanas around their heads. Hit in the mouth. He could still taste the blood. Hit again in the head. Knocked to the sidewalk. Dragged by his legs as his back burned along the ground, to a patch of dirt between the two buildings.

Each boy grabbed a leg or an arm. Then the fifth boy, the largest one, started ripping off Haruto's clothes. His shirt. His pants. His underwear. And then that first punch to the gut. Haruto gasped for air. As he was trying to breathe in, the boy kicked him in the groin. Haruto cried out. The boy kicked him again in the abdomen, the leather shoe bruising the flesh and muscles near the surface and pushing deep into the viscera. The boy pulled back his knee and aimed at Haruto's head. Before the boy released another violent kick, Haruto started retching. Vomit stained with fresh blood flew onto the boy's shoe.

"Ahhh!" the boys yelled together.

"He can't even fight — all he can do is barf his guts," one of the other boys said. Then they started taunting, "*Jinzouningen, jinzouningen.*"

The larger boy grabbed Haruto's clothes and the pack ran off.

Haruto lay there on the dirt, naked, covered with vomit and blood.

"I had nowhere to go," Haruto said.

"What are you talking about?" the Korean yelled, not even trying to muffle his voice now.

Haruto raised his left hand in surrender. His right arm wrapped in a cowardly fashion across his chest and his right hand gripped tightly onto his left shoulder.

The Korean smiled. "Tell me now. What are the Israelis giving to Japan? Give me the details. Weapon systems? Chemicals? Intelligence? Missiles? Do the Israelis have agents in North Korea? Tell me fast, or I will kill you."

"I had nowhere to go," Haruto repeated.

The Korean pushed the silencer barrel even harder into Haruto's head, causing the scalp to ooze with blood. "TELL ME NOW —"

Haruto's right hand was pulling on his left shoulder with five hundred Newtons of isometric muscle force. Haruto released the memory, and then released his hand. The hand crossed the Korean's line of vision too fast to register. Only as Haruto's hand crashed into the Korean's right arm, the one holding the gun, did the Korean react and squeeze down on the trigger. The gun was now fifteen degrees off Haruto's head and the bullet flew out to hit the concrete wall of the industrial building across the road.

As the Korean started bringing down the gun to Haruto's head, pulling on the trigger again, Haruto started the *oi-tsuki* strike with his left hand. Karate was about measured control, but when your life was about to end, you did whatever you could do to go on living. If the *oi-tsuki* punch did not incapacitate the Korean, Haruto would die, and Haruto realized this at a level below conscious thought.

An *oi-tsuki* strike does not begin with the hand or the arm. It begins with the toes. And although it takes place in tens of milliseconds, it is measured in years — year after year of training to alter the neuronal pathways. The collective history of this training is in the pathways that remain.

Haruto's toes pressed down against the road. With a magically fast synchrony of the neurons, the muscles of the lower leg tightened and pressed against the road. And then the upper leg muscles. And then the powerful muscles of the back. And then the shoulder muscles and the triceps muscles. The forearm and finger muscles squeezed the open hand into a tight ball. The forces cut off the blood supply to the large knuckles and they became white.

Like a wounded animal, Haruto cried out a "*kiai*" that boomed through the street, as his muscles exploded.

Millisecond by millisecond, as the Korean brought the gun barrel down to bear on Haruto's head, Haruto's left fist accelerated toward the Korean.

Haruto's knuckle crashed into the Korean's temple. Bone cracked.

The Korean dropped the gun and fell to the road.

Haruto knelt. The Korean was still breathing and had a pulse. Haruto hit speed dial on his cellphone.

"This is Inspector Suzuki. Badge four-oh-seven-seven-four. Suspect down. Request ambulance and forensics team. Ikebukurosen and Shinobazu-dori."

Haruto looked at the Korean again. He was the key to solving the case. He must not die. But soon the Korean's eyes froze into an open, fixed position. Haruto pressed his fingers on the Korean's neck in a vain attempt to find the missing pulse.

A second death. What was happening to this case? The man had died just as Co'en had died, yet Haruto felt no balance. It was hardly a karate victory to be looking at a dead opponent.

CHAPTER015

"**P**roblems with your interrogation techniques tonight, Inspector?" a deep voice said.

Haruto, standing over the body, looked up and saw the coroner step into the lights the forensics team had set up.

"How did he die, Inspector?"

"I punched him."

"You punched him?"

"In the head."

"Are you supposed to be punching suspects in the head, Inspector?"

"When they are holding a gun to *my* head, then yes." Haruto pointed to the 9-millimeter Glock pistol and silencer on the ground.

Haruto walked off down the road. The bright neon lights from Ikebukuro Station and Sunshine City's buildings filled the nighttime horizon with a pink hue. His cellphone warbled.

"Hello. Is this Inspector Suzuki?" a woman's voice said.

"Yes."

"I'm Co'en Ayaka. You gave me your card in the hotel room. I'm sorry to bother you, but I think a man is following me. I fear I may be in danger."

"Can you see him right now?"

"No. I'm in my apartment."

"Stay there. I have the address and am coming over now."

Haruto ran back to the crime scene and sped off in his brown Honda.

CHAPTER016

Haruto pulled into the visitor's driveway of the steel apartment skyscraper. As he walked into the lobby, a woman's scream came from outside. Haruto spun about, and shoved the door open. He could make out in the streetlight a woman's blond hair — Mrs. Co'en — as she struggled against a tall man.

Haruto started running. "Stop! Police!"

The man shoved Mrs. Co'en into the front seat of a silver Lexus.

Haruto continued running toward the Lexus and the blond woman. "Stop! I am the police!"

The man reached into his pocket. A second later a muzzle flash lit up the night as the gunshot broke the silence of the neighborhood.

The bullet did not come even close, but Haruto instinctively rolled onto the ground. Like many police officers, Haruto did not carry a gun. There was no need to — most policemen in Japan would spend their entire careers without ever seeing a shot fired.

Another gunshot rang out. The man ran around his car and opened the driver's door.

Haruto leapt up and sprinted back to his Honda. The Lexus jolted out onto the road and sped down Minami-dori. As Haruto pulled his car onto the road, the silver Lexus turned right on Meiji-dori.

Haruto pressed the accelerator. He looked for his police flasher — he hadn't used it in a long time — and found it underneath the passenger seat. The Honda screeched into a right turn onto Meiji-dori. Haruto held the steering wheel with one hand, planted the light's magnetic base on the roof and flipped it on. The Lexus was still within sight in the distance.

Haruto floored the gas pedal and the Honda sped down Meiji-dori. The colored neon of the Shinjuku retail stores flashed by as Haruto dodged around a car in front of him and moved to the inner lane. The bright lights from Takashimaya Times Square glinted on the silver Lexus.

Haruto kept his foot down on the gas, swerved past another car and fishtailed back into the middle lane. The Lexus had picked up speed and still remained a few hundred meters ahead of him.

In front of Haruto a blue Toyota, a white Nissan and a large truck blocked all the lanes. Haruto did not release the gas pedal but jammed his palm onto the horn. The flashing red light should have been enough!

The Toyota, Nissan and truck finally took notice of the flashing light, and all slowed down in response, blocking the road almost completely. Haruto pounded the horn. He wasn't going to lose the Lexus. He was going to solve this case. Haruto aimed for the gap between the side of the Toyota and the Nissan, kept the pedal down, glanced at the speedometer now reading 150 kilometers per hour. He took a deep breath.

Haruto's Honda jolted violently to the left as its front bumper caught the side panel of the blue Toyota. Haruto kept his hands gripped to the steering wheel, his arms tense, he continued to press on the gas pedal. The Honda's left front wheel came off the ground for an instant as the Toyota's side panel sheered and came up onto the Honda's windshield, blinding Haruto. The Honda shuddered again, the panel went airborne over the windshield, and the Honda broke free.

Haruto floored the accelerator and aimed for the silver Lexus straight ahead of him. He pulled out his cellphone and hit speed dial. "This is Inspector Suzuki. High speed pursuit of silver Lexus — cannot read license plate. South on Meiji-dori. Past Shinjuksen Expressway."

Haruto came closer to the Lexus. He could vaguely make out the driver and Mrs. Co'en in the front seat. The Lexus suddenly swerved off the road toward the Meiji Shrine Gardens, its rear skidding wide for a few seconds before the driver regained control. Haruto hit the brakes, yanked the steering wheel to the right, and followed.

The Lexus barreled down the Gardens' main roadway for twenty seconds, then jerked to the right off the roadway into thousands of irises in full bloom. Haruto took a deep breath and followed the plume of dirt the Lexus was kicking up.

The Lexus pulled out of the flower garden onto terra firma and bolted toward the Meiji-jingu Shrine. It sped through the *ichi-no-torii* — the massive wooden gate of the shrine. Haruto held his breath and followed through into sacred space at 140 kilometers per hour. The Lexus then darted into the forest of trees around the Shrine.

Haruto couldn't see the silver Lexus any longer. He stopped his car. As he stepped out to look around, a bullet crashed through his windshield. Glass fragments sprayed all over him. Haruto jerked back and dove to the ground. Another gunshot rang out and slammed into the rear fender. The Lexus jolted out from between two willow trees.

Haruto jumped back into his Honda and sped after the silver car into the forest.

Haruto jammed the accelerator. He had almost caught up when the Lexus jerked to the left. A large oak tree — dead ahead!

Haruto hit the brake pedal and yanked his steering wheel to the right. The tree came closer and closer.

The passenger door slammed into the oak and tore off. The car vibrated but was still moving. Haruto pressed on the gas again and kept following.

The silver car shot out of the forest onto the Gardens' service road. A few seconds later, it tore through the parking lot of the 1964 Tokyo Olympics Village. Haruto followed, and like the Lexus was soon barreling down a small residential street. The Lexus veered sharply left onto Yamate Avenue, and Haruto followed.

The brown Honda now had no passenger door. Its windshield was shattered. It was missing its front bumper and radiator grille. But it was still running. Haruto pressed down on the gas pedal. The speedometer rose to 160 kilometers per hour, and the nightlights of Tokyo blurred by.

Like a mad cat chasing a mouse, the brown Honda followed the reckless silver Lexus down Yamate-dori, passing the University of Tokyo, and then onto the Shibuyasen Expressway. The Lexus dodged traffic left and right, and so did the Honda, staying on its tail. On a stretch without much traffic, the more powerful Lexus started to pull away from the Honda. Ten seconds later, it came upon a triad of delivery trucks slowing the road, and Haruto caught up. At the last minute, the Lexus veered violently onto the exit ramp at Gaien-Higashi-dori.

Haruto followed the Lexus onto the broad avenue. In a minute, the floodlit Tokyo Tower loomed over the two mad cars. The Lexus bolted off the end of Gaien-Higashi Avenue on to the smaller roads near the Tower.

Eateries lined the small road. Pedestrians were moving about. The Lexus blindly sped down this vulnerable passageway. Haruto started sweating.

A group of diners crossing the road a hundred meters away! For the ever-slightest moment, Haruto hesitated, then slammed on the brakes and pounded his horn. The pedestrians scattered. Haruto pressed the gas again, but no silver car was in sight. Damn, he wasn't going to lose the clue. He wasn't going to let the woman die. Two deaths in this case were enough already.

Haruto accelerated past Hibiya-dori and then, for a split-second, he glimpsed the Lexus, turning left onto the Uenosen Expressway. Haruto slammed the gas pedal to the floor.

The Lexus sped up further. Haruto tried to keep up but the Honda peaked at 160 kilometers per hour. Tokyo Bay, all lit up, went by in a blur on the right. The Lexus gained more ground. Just as Haruto was losing sight of the car again, he saw it veer to the right onto Harumi-dori. Haruto followed, and the two cars barreled across Kachidoki Bridge toward the Harumi Wharf.

Haruto saw the yellow and black wharf markings and slammed on his brakes. A second later, so did the driver of the Lexus. The latter's wheels locked intermittently, but to no avail, and the Lexus smashed through the barrier. The nose of the Lexus tilted down as the car fell free through the air and sliced into the water.

Haruto's heart started pounding as the broken barrier approached. The Honda continued skidding toward the water. It slowed and slowed, but was still moving as it reached the barrier. The front wheels crossed the gap in the wood that the Lexus had just ripped out and the Honda started tilting down toward the water. Finally, the rear wheels dug into the pavement, and the car stopped, its front end dangling over the bay.

The Honda teetered. Then the nose of the car dipped down but didn't seem to be bouncing back up! Haruto threw open the door. The nose started angling almost straight down vertically, the rear wheels lost their purchase, and the car started falling.

Haruto pushed off violently with both legs and caught the edge of the asphalt with his hands as the Honda tumbled below him into the water.

Haruto got to his feet and looked down. His wounded car had already filled completely with water and was sinking. The silver Lexus with intact doors and windshield, was still floating upright. There was not enough light to see if the occupants were alive.

Haruto hit speed dial on his cellphone. "Inspector Suzuki here. Harumi Wharf. Send ambulance, backup and police diver. Suspect's car is in the water."

Haruto heard the woman's screams again, but he couldn't see what was going on. He pushed off his shoes, let his jacket fall to the ground and dove off the pier.

The ocean water was cool, even in June. Haruto swam over to the Lexus. The driver's head was resting on the steering wheel and bleeding. He was Korean, about thirty years old. The passenger compartment was half-filled with water, and the blond-haired woman was screaming and pounding on the window.

Haruto pulled on the passenger door, but it didn't open. Haruto knocked on the passenger window, trying to get the woman's attention. He pointed to the door handle. She just screamed over and over again.

Haruto knocked more forcefully on the passenger window. The woman stopped screaming for a second. Haruto pointed wildly to the door handle. "Open it!"

The woman opened the door, but before she could get her seat belt off, the water poured in, and the Lexus started to submerge.

Haruto took a breath, grabbed onto the door handle, and let the car pull him down too. He pushed on the seat belt button, the shoulder strap sprung back, and Haruto yanked Mrs. Co'en from the sinking car. They quickly floated to the surface, and the woman started coughing. Haruto gripped the back of her dress with one hand and started swimming to the rusty ladder on the side of the wharf.

CHAPTER 017

"**P**lease sit down," Haruto said, motioning to the bench at the edge of the wharf. "Help will be here soon."

"Thank you, Suzuki-san." Mrs. Co'en grabbed him and gave him a large hug.

Haruto did not know what to say. "Please...please...," he finally blurted out. "Please sit down. Help will be here soon."

The woman nodded. She was wearing a blue dress, now soaked, and her blond hair was hanging back in tight messy clumps.

Sirens wailed in the background.

"Are you all right?"

The woman nodded.

"Co'en-sama, why were you with that Korean man?"

"He kidnapped me." She started to cry. "I don't know what's happening, Inspector. Two days ago, my husband was killed and now I am a widow, and I don't even know why. I've received many phone calls from my cousin in Israel over the last few weeks. He wanted to speak to my husband — why, I don't know."

"I am sorry for your difficulties," Haruto said softly.

"I telephoned my cousin in Israel this morning and demanded answers. He told me to go to Harumi Wharf this evening. He said someone would meet me at the entrance to the wharf and explain to me what happened to my husband."

The woman continued speaking. "Just before you came to my building tonight, this Korean man — possibly the one who was following me — forced his way into my apartment and started yelling at me, demanding to know what 'deal' my husband had made. When I couldn't answer him, he told me he knew about the meeting tonight and said I was coming with him to the wharf."

The sirens grew louder. The headlights of a police car swept onto Haruto and Mrs. Co'en, and it braked to a sudden stop, its siren still blaring. Two police officers jumped out of the patrol car, one of them brandishing a long rifle — an unusual sight for Tokyo. A few seconds later an ambulance pulled in, its white stroboscopic lights dancing with the red flashing lights of the police car. Two ambulance attendants ran to open the back door and pulled out a stretcher.

"I'm fine. Please make sure she gets to see a doctor." Haruto turned to Mrs. Co'en. "I will need to speak to you in more detail tomorrow."

The attendants had the drenched woman lie on a stretcher. In a few moments the ambulance departed, siren blazing even louder.

Another half-dozen flashing and wailing patrol cars came down Harumi Avenue, followed by a television news truck — Haruto could make out the TV Tokyo logo on the folded satellite boom. Two more ambulances arrived, with lights flashing, followed by more patrol cars, more news trucks, and in the distance what looked to be a large winch with a battalion of fire trucks.

Haruto stood in the middle of this chaos, his soaked shirt clinging to his chest, his tie dripping, and his trousers letting a stream of water fall onto his bare feet. He picked up his jacket off the wharf, pulled out his police identification and held it up. "I am Inspector Suzuki, in charge of this crime investigation," he shouted, then turned to the police officer holding the rifle. "That isn't necessary. You don't want the newspapers dramatizing the situation. Put it away."

It was too late, however, as the first photographers and news crews on site started snapping pictures and lighting up the wharf with their camera lights.

"Inspector Suzuki." A newswoman jammed a red and orange microphone under his jaw. "Is it true that there have been a string of murders in Tokyo tonight?"

Haruto moved away from the reporter and held up his police identification badge at two other policemen who had arrived. "Please set up a perimeter and secure the crime scene." Haruto pointed at another two policemen. "You, please find the harbormaster. Find out what ships are here tonight."

The battalion of red fire trucks flowed onto the wharf, their sirens blaring and their relentless red stroboscopic lights piercing the night. Three lighting trucks raised their eight-meter telescopic booms and threw on their searchlights, lighting up the wharf brighter than daylight. A winch truck raised its boom, and telescoped it out over the water. The hook and chains at the edge of the boom then lowered into the harbor. Haruto saw a pair of scuba divers climb into the water.

The two policemen Haruto had asked to find the harbormaster came running back. "We can't find the wharf master," one of them said. "But we have a call in to the Coast Guard and into Tokyo Port, and they'll be here shortly."

Then a lone figure pedaled his bicycle into the circus of reporters, fire trucks, and police cars. He had on a police cap, a blue short-sleeved shirt with a police logo sewn on the right sleeve, and wore a thin, blue zippered multi-pocketed vest over the shirt. He held some loose clothing in his arms. The policeman approached Haruto.

"Inspector, it is I, Itou Yoshio." The bicycle policeman pointed to his police radio. "I heard your request for help. I've brought you some dry clothes." He gave Haruto a white polo shirt and jeans.

Haruto bowed. "Thank you, Policeman Itou. I remember you well. Where are you working now?"

"Ginza Koban. I like it better than the station."

"I started my career at the Roppongi Koban. The community police box is a good way for a young officer to gain experience working with people — *jinseikeiken*."

"I am glad to be able to help you, Inspector. Last year at the station, only you had faith in me. I shall always be grateful."

Itou bowed to Haruto and started to walk away.

"Policeman Itou, do you know the Harumi Wharf well?"

Itou turned around. "Yes, Inspector. I patrolled near this area last month. It's close to the Koban, and I still come through the area."

"Where would I find the harbormaster or the wharfmaster?"

"That would be Kauma Ogawa. There's a grilled fish restaurant on one of the side roads here. He'd be there at this time."

"Policeman Itou, please help me with my investigation. I want to speak with Mr. Kauma as soon as possible."

The young policeman bowed, ran to his bicycle and pedaled off.

Haruto saw a group of firefighters lean over the edge of the wharf. The two scuba divers were at the water surface now, with their thumbs pointing up. The winch truck gave off a loud new hum as its geared winch started turning. In a few seconds, the rear of the silver Lexus came out of the water, and dozens of photographers, also now by the edge of the wharf, started snapping pictures.

CHAPTER018

General Otzker's neck veins were bulging as he looked out the starboard window of the conference room. Fire trucks, bright lights, and hundreds of spectators and reporters. This was supposed to be a secret meeting, not the lead story on the local news. "This is outrageous!" He turned to stare at the younger Japanese man.

Colonel Tanaka, about forty years old, slim with gold-rimmed glasses. Like Otzker, the Colonel was in civilian clothes — light blue pants, open white shirt and a blue cotton jacket. Tanaka looked at the fifty-something, Caucasian General. "My team is working on it. The intelligence report I just received said there's a murder investigation going on down there."

"Tanaka, for twenty years I have run testing without a glitch. Do you hear me? Without a glitch, and now you're going to screw me up with some bullshit murder investigation outside my ship, with the reporters snapping away... at my ship!" Otzker walked in a small circle around the end of the conference room. "The test is off. I'm not going to have any part to this."

"General, no test, no robots."

"Fine. No robots. We'll survive. We're not the ones with nuclear missiles zipping over our heads."

"General, we have forty-two *tonnes* of plutonium sitting on the shelf from our power plants. We have already built dozens of fission devices."

"So why don't you tell your superiors that? Go tell them that you have forty-two tonnes of plutonium so *you've* decided *you* don't need the deal. And when they ask you why, you'll tell them that some police investigation took priority. Then they'll ask you how you plan on testing any of your devices without changing your constitution or your relations with the US, or with the entire world for that matter — you *did* sign the Comp Test Ban in '96. Even if your primitive bombs work, they'll give poor yields and they're too heavy to launch. Without tests — lots of them — you can't build advanced bombs that work."

"I still don't understand how you're going to test undetected," Tanaka said, pausing to make eye contact with the older man. "If I fart in the ocean, one of the American acoustic stations will pick it up... and you're going to hide an explosion equal to a million tonnes — *a billion kilograms* — of TNT in the ocean?"

"I told you we would discuss testing methods, as well as bomb designs, once we were underway. I don't see what this has to do with what's going on outside."

Tanaka gazed out and saw the banks of bright lights from the fire department light trucks intersected by a dozen rotating red beams from the police cars and fire trucks. A mob of reporters and photographers and spectators were crushing against the policemen securing the crime scene.

"I apologize, General. There is no need for me to speak to my superiors. I will handle this." Tanaka pulled out his cellphone, hit a speed-dial button and switched to Japanese.

CHAPTER019

Haruto watched as the winch lowered the dangling Lexus to the pavement. Seawater poured out of the car's seams. The Korean driver was still buckled into his seat, head slumped on the steering column. Photographers pushed past the policemen and snapped picture after picture of the car and the firefighters.

"Inspector, this is Kauma Ogawa, Wharfmaster."

Haruto turned to see Itou, the young policeman, standing beside a fifty-ish, short man.

"Kauma-san, have there been any unusual ships at your wharf recently?" Haruto asked. "Any ships from Israel?"

"No ships from Israel, but this ship is most unusual, Inspector." The wharfmaster pointed to a mid-sized passenger cruise liner moored twenty meters down the wharf. "*New Pacific Queen* — Panamanian registration, based out of Guam."

"What's special about that ship?"

"I visit every ship that docks onto my wharf," the wharfmaster said. "That ship looks like an ocean liner but it seems to be for laying telephone cable. There are two huge spools lying on its rear deck. The captain did not want me looking around, but took me one deck up to the ship's theatre, where he offered me refreshments."

"What's so strange about that?" Haruto said.

"Why are those large spools on a liner? Also, do you see the larger passenger liner, the *Holland East* moored just in front of it?"

Haruto nodded.

"The world-famous *Holland East* holds two thousand guests. As you can imagine it has a nice theatre. The theatre of the *New Pacific Queen*, where the captain took me for refreshments, was doubly plush. It doesn't make sense to have this luxury for a smaller, no-name cruise liner."

A loud snap sounded as the relays controlling the banks of lamps on one of the lighting trucks switched off. Then the same happened for the other two trucks, and the crime scene plunged into night again. Haruto, the wharfmaster and Itou turned their heads to see what was happening.

Five seconds later, the street lamps on the wharf as well as all the lights in the buildings along it went dark.

"Excuse me, Inspector — I must see what's happening to my wharf." The wharfmaster ran off down the street toward the Harumi Passenger Terminal.

Itou's police radio screamed to life. "General bulletin. Arrest warrant for Suzuki Haruto, *Keibu*, for the murder of a suspect. Any officer encountering Inspector Suzuki is to arrest him immediately."

The classroom had always smelled of bleach, and its linoleum floor was shiny enough to see your reflection. Haruto was frozen with a mixture of fear, shock, and yes, awe, as the class clown went up to the teacher's desk and ripped the attendance record in half. At that age, one did not think of the stupidity of the act but rather the courage to do it.

The teacher came in a few minutes later — Mr. Juku, a harsh, large teacher, just the way his name sounded. "Who did this?"

"Haruto did!" the other boys shouted. "Haruto ran up to your desk and ripped your papers."

"Haruto, come to the blackboard right now," Mr. Juku said.

He should have said something. Should have defended himself. But the teacher said to come to the blackboard. He always did what the teacher said.

"Haruto, for the next *month*, for an hour every day, while the class is working on their art projects, you will come up to this section of the blackboard, and you will write *I am guilty of bad actions* over and over again. You can start *now*."

Each line he wrote hurt. Haruto could still feel the tears, fighting so hard not to release them. The class laughing at him. Day after day that month, line after line.

"Inspector — I must arrest you!" Itou reached into his utility vest and took out a nylon hand restraints band.

"*Seiretsu!*" the master had said. Haruto had lined up, wearing his white belt, at the right of the line with the other white belts. *Sensei* was not so old then. No gray hair. A strong face with only the odd wrinkle.

The master sent the colored belts to one corner of the gym to practice *kumite*. He stood in front of the white belts. "Over the last few months you have learned the basic punches, kicks and blocks. Whether you are a black belt or a white belt, it is the same moves — *oi-tsuki* punch, *mae-geri* kick, *age-uke* block. We now start the real study of karate — how to put these basic moves together." The master looked at him. "Haruto, come here."

"Yes, *Sensei*." Haruto came out of the line to face the master.

"You will attack me now. Fighting stance."

Haruto's body turned at an angle so only his thin side presented a target to his opponent rather than the width of his abdomen and chest. Legs were bent and spread, with about three-quarters of the weight of his body on the rear leg. Arms were bent and up. His left hand was in a tight fist in front of his face, and his right hand was in a fist, further back at chest level. Eyes were focused on the opponent.

The master turned and looked at the line of white belts. "If Haruto attacks me with an *oi-tsuki* how do I defend myself?"

A teenage girl put up her hand. The *sensei* nodded at her.

"*Age-uke* — rising block with the hand," she said.

The master shook his head. A young man in his twenties put up his hand. *Sensei* nodded at him.

"*Uchi-ude-uke* — inside forearm block."

The master shook his head, and turned to Haruto. "Attack," he yelled.

Haruto executed as perfect an *oi-tsuki* as a white belt could be expected to perform. Balance with legs was kept. Head did not bob up and down. His front left hand thrust out toward the master's chest.

Before the *oi-tsuki* was complete, Haruto felt the sting on his nose. He could not see the master's hand, but he had been hit, hard enough to be stunned for a second, and stop his *oi-tsuki*. It felt like his nose would start bleeding any moment.

"This is not a self-defense class," *Sensei* said. This is a *karate* class. The best defense is a punch or a kick. Block when you have to, but do not become preoccupied with defending yourself. Attack, attack, attack!"

Haruto rubbed his nose to make the pain go away.

Itou held out the hand restraints. "There is an arrest warrant broadcast for you, Inspector."

"Policeman Itou, those charges against me are without merit. I must solve my case. Will you help me?"

Itou looked at Haruto in silence for a moment. "Yes, Suzuki-san. But the next police officer will not. You *will* be arrested."

"Do you know this poem?" Haruto asked.

"*This is the way of the world:*
three days pass, you gaze up —
the blossom has fallen."

"Yes, of course. It is the poem of the cherry blossom, the poem of renewal."

"It is also the poem of the *samurai*. Not knowing when they might die, they accepted a brief life, but like the cherry blossom, a colorful, bright one."

Itou nodded.

"It is important to follow the rules," Haruto said. "If there is a warrant for my arrest, then the proper thing is for me to be arrested and to try to prove my innocence in court. However, I have never had a murder investigation such as this one. Another set of rules tells me that I must not lie down now, I must continue this investigation and see where it goes. And once I have done that, then I can be arrested. If my career is destroyed by not following the first set of rules, then so be it. Like the cherry blossom, all things — careers included — come to an end, and I am prepared for that eventuality. Policeman Itou, will you help me?"

Itou looked carefully at Haruto. "You believed in me when no one else did, Suzuki-san. I have a career and I have my pride because of you. Tell me what to do."

"I will contact you in a few days." Haruto pulled his cellphone out. "Please enter your phone number — personal cellphone would be best."

Itou punched in his number and returned the phone. Haruto bowed to the policeman, then started running toward the *New Pacific Queen*.

Along the now darkened wharf, Haruto saw two bulky dockworkers at the capstans, untying the thick mooring ropes of the cruise ship. A third worker was in the cab of the motorized stairs leading up to the ship, driving them away from the cruise liner.

Haruto flashed his badge to the worker pulling the stairs away. "Police. I need to board the ship."

The worker shook his head. "Sorry. She's untied from the dock, thrusters running. Regulations. I'm not allowed to put these stairs against a moving ship."

"It is an emergency."

"You need to speak to the wharfmaster or Coast Guard. The ship is no longer officially on the wharf."

At the bow and stern, sideways currents of water shot out from under the ship, creating eddies at the surface and gently nudging the ship sideways, away from the wharf. The ship's freed mooring ropes slowly started to winch up to portholes high in the bow and stern. The ship was now about a meter away from the dock. The two bulky workers who had undone the mooring ropes were walking back to Harumi Passenger Terminal. The motorized ship steps were inching away from the ship to their parking position. One mooring rope was at about eye level just above the stern thruster.

Haruto charged at the *New Pacific Queen*.

In four seconds, he was airborne, off the end of the wharf, his eyes focused on the end of the mooring rope. His hands caught the thick rope as his body slammed into the metal hull of the ship.

The blades of the stern thruster chopped the water, a few meters below his feet. Haruto tightened his grip on the rope and held on.

This is the way of the world.

CHAPTER020

The *New Pacific Queen* had covered four hundred meters when the end of the thick mooring rope, and Haruto with it, were pulled through the meter-sized hole in the upper stern. Haruto let go of the rope and rolled onto the winch-room floor. The automated winch stopped a moment later and the room became quiet.

Haruto walked on his toes to the door, and ever so slowly opened it. The outside main deck. A wooden walking surface wound around the length of the ship from stern to bow. A white metal railing topped with polished mahogany allowed a good view of the water off the side of the ship, some eight meters down.

Haruto could make out one of the large cable spools the wharfmaster had mentioned, sitting on its flat side at the edge of the stern. Rising above the cable spool were two massive hydraulic arms and winches.

Toward the bow, three young men stood against the railing, two of them smoking. They all wore clean white t-shirts, white cotton pants and white sneakers, and looked like ship's crew. The men were pointing at the huge titanium sphere on the left shore, which made up the Fuji TV Building. In a few minutes, they finished their cigarettes and went back inside.

The deck was empty now. The best rule was to stay low and observe. What was special about this ship? Who was Co'en supposed to meet? Why were the Koreans involved? If he found nothing, then no harm done. He'd get off at its next port of call, and return to Tokyo. Perhaps Policeman Itou could help him out with a few more clues. If he had to face the murder charges against him, then he would.

Clinging close to the bulkhead Haruto walked ten meters down the deck and turned into a stairwell. A schematic map of the ship. The main deck was *Floor 4*. Haruto quietly climbed the stairs to look for the ship theatre that so impressed the wharfmaster.

Finding it was not difficult. On the fifth floor, Haruto found himself in an empty lobby with a few hundred gaudy small light bulbs spelling out *Queen Theatre* over three shiny brass and glass entrance doors. Large colorful posters in glass display cases on the left and right announced *Animation Classes — Tuesdays, Thursdays and Saturdays when at Sea.*

Suddenly Haruto heard footsteps and saw a white uniform. Haruto edged open the theatre door — the lights were on but it was empty. Haruto eased past the door and closed it gently.

The theatre was massive — able to hold perhaps seven hundred seats. A dozen glass chandeliers hung from the gold-and-white-trimmed ceiling. The lobby and chandeliers looked very nineteen-fifties. However, other parts of the theatre were ultra-modern now.

In each space where perhaps a dozen original theatre seats had been, now was a single seat — tan leather, footrest, headrest, moveable writing surface, large armrests filled with keyboards and joysticks, and a computer flatscreen on each side. To the left of every seat was a small fridge with a transparent door, filled with American soft drinks and a variety of snack-sized food packages. The theatre contained about sixty of these clusters of seats, display screens and refrigerators.

At the bottom of the room, the black wooden raised stage of the original theatre was still intact. However, in place of curtains stood a huge computer display screen, perhaps eight meters wide and five meters high. The screen was on but only displayed the Microsoft Windows logon box. On either side of the stage, at an angle to the sidewall, was a large computer display screen, about four by three meters. On the left wall of the theatre was rack after rack of computers, climbing up from the black stage almost all the way to the top entrance doors. A few hundred green LED lights glowed through the clear plexiglass doors of the racks, with a scattering of red lamps turning on and off.

Haruto walked down to the stage and climbed up. Behind the large computer display screens were a half-dozen of the leather seats, an extra rack of computers and a few empty small refrigerators. Haruto sat, then lay down on one of the leather seats. He stretched out his legs on the footrest and rested his head on the almost horizontal headrest. It was dark and quiet here behind the large display screens.

Haruto was still hoping this ship ride was not a waste of time when he fell asleep.

CHAPTER021

"**O**hayo gozaimasu. Boker tov. Good morning."
The sounds echoed from the theatre and startled Haruto awake.

Haruto carefully approached the crack on the left between the large center display screen and the screen angled on the side, and looked out. The lower rows of the theatre now held seven people — some Westerners, some Japanese — sitting in the plush leather seats. A Caucasian man in his mid-fifties stood in front of the stage next to a fortyish Japanese man with gold-rimmed glasses.

"I am General Otzker," the Westerner said, his English clear but with a Mideast accent. "Please do not call me Admiral. Yes, this is my ship, but I'm regular Army. For twenty years, I've been in charge of nuclear testing."

Haruto stood up tall, suddenly fully awake. This trip was not a waste of time. He took out his cellphone and clicked on *Main Menu*, *Camera* and *Video*. He pointed the cellphone's camera through the crack between the large display screens mounted on the stage and saw the General on the phone's tiny screen.

"This voyage could be anywhere from one to seven weeks, depending on test conditions," Otzker said. "The time will not be wasted — that's why you're all here."

"I am Colonel Tanaka," the Japanese man standing next to the General said. "I'm with Japanese Defense Intelligence Headquarters and in charge of acquiring the nuclear technology and transferring the robotic technology. I will also be responsible for the shipping and use of the robots in Israel."

"Let's introduce each other," Otzker said. "First names only. We'll all practice our English on this trip." The General pointed to the dark-haired, middle-aged woman in the frontmost row of the theatre.

"I'm Ilana," she said. "Israeli Army. I'm specialized in Teller-Ulam nuclear design and will be transferring this technology." She pointed to the rack of computers on the left wall of the theatre. "We have the equivalent of a supercomputer in this room and can do theoretical designs on this trip, as well as discuss production models."

The middle-aged Japanese man in the seat next to Ilana stood up. "I am Isato. Japanese Defense Intelligence. I have fifteen years experience design nuclear power plants. For last few years I was on team produce plutonium fission devices. I acquiring Teller-Ulam nuclear design."

The young white man, perhaps in his late twenties, seated in the next seat stood up. "I'm Daveed. I'm from the university — Technion. My background is in software and artificial intelligence. I'll be acquiring the robot software."

A middle-aged, mustached Japanese man sitting behind Ilana then stood up. "I'm Akihiro. Defense Intelligence and University of Tokyo. I'm in charge of the electronics of the robot."

A Westerner in his late twenties in the next seat stood up. "I'm Shmoel. Technion. My background is in microelectronics fabrication. I'll be acquiring the robot electronics."

A Japanese man in his late thirties sitting next to Shmoel then stood up. Haruto pressed his eye against the crack between the display screens and squinted. That was the professor — Haku Seiko — in the portrait in Toshifumi's office!

"I am Seiko. I'm with the University of Tokyo, on loan to Defense Intelligence. I'm the designer of these robots."

Seiko turned sideways and pointed. "As you can see, the basic models are already here with us."

Haruto put down the cellphone and pressed his eye into the crack to better look out into the theatre. In the seats beside Seiko were two very different looking robots. Haruto put the cellphone back up to the crack, hit the + key and zoomed to a larger image of the robots.

"Alpha, I have hidden a 100-yen coin near one of the chairs. Go find it," Seiko said.

Alpha stood up. The light brown robot was about two meters tall and balanced extremely well on its two legs. Its torso was pear-shaped, with a wider abdomen than chest. Its arms were long, each one perhaps a meter and a half. The upper arm tapered to a narrow forearm that then thickened at the wrist, which held three metallic fingers and a rubber-coated thumb. Its face was humanlike in overall appearance, but machine qualities were obvious. The eyeballs were rotating camera spheres, neither one synchronized with the other. No actual nostrils were present on the molded silicone nose, a camouflage gray. Silicone lips surrounded a two-centimeter opening.

Alpha ran along the row of chairs, its eyes focused on the ground. Near the aisle, Alpha hopped over a seat, crashing down on the ground in front of Ilana, startling her.

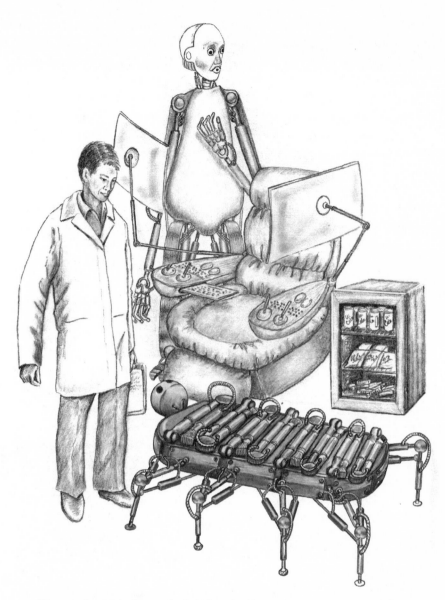

"Sorry, Ilana," Seiko said. "Alpha, I will give you a constraint. The coin is behind an *empty* chair."

Alpha stopped to scan the entire theatre. It ran up the stairs to the top row, then ran across that row in about two seconds, eyes focused on the ground. It ran back across the second row, which was also empty and then across the third row. In the middle of the forth row, it bent and grabbed something between its left fingers and thumb.

"Coin recovered," Alpha said.

Alpha then hopped chair over chair over chair until it was standing next to Seiko with the coin extended out toward him. Seiko held out his hand, and Alpha dropped the coin into it.

"General, Alpha is loaded with soft rubber bullets. Permission for a demonstration."

Otzker nodded, although frowning.

Seiko tossed the 100-yen coin up in the air. "Alpha — shoot the coin."

There was a gunshot and the tiny coin went flying. It was a moment before Haruto realized — the bullet had come from Alpha's mouth.

Some of the audience looked aghast as well.

"The gun barrel is in the mouth," Seiko said, "because the higher it is, the better shot the robot can take. If you want Alpha to have more specialized weapons, remember that Alpha can carry and use any weapon a human soldier can, including a sniper rifle or a rocket launcher."

Seiko pointed at the multi-legged robot standing behind Alpha. "Beta, come here."

Beta hopped over the seat, landing on its eight legs to the side of Seiko. It looked like a two-meter long giant cockroach. Four sets of arms and four sets of legs attached to its flattened, camouflage-brown torso. The legs were each about a meter long, articulated at the knee, and quite thin compared to Alpha's stockier legs. Its arms were longer, each about two meters in length and articulated at the shoulder, elbow and wrist. Each wrist had seven fingertips and an opposable thumb. The front arms were thicker than the three spindly rear pairs. Its head was a sphere sticking out its front. Its eyeballs were also rotating camera spheres, neither synchronized with the other. No nose or mouth was present, nor were any insect-like antennae.

"The Beta robot is a hauler — something foot soldiers do a lot of." Seiko turned and faced the robot. "Beta, go get the thousand-kilo ammunition chest at the top of the room."

Beta dashed up the stairs, actually folding in its middle two sets of legs, using its front and rear sets of legs to push off the ground. It folded its four sets of arms over its back. It instantly recognized the corrugated-metal ammo carrier. In less than a second Beta unfolded its middle sets of legs, then bent all the legs at the knee, lowering its torso to the ground. In another split second, Beta unfolded its sets of arms and placed all its eight hands around the ammunition chest. Beta then slowly lifted up the chest, placed it on its back, and wrapped its long arms around it. The robot's legs all straightened up, and it sprinted down the theatre stairs to Seiko.

"Beta, fill Alpha's ammunition magazine doing an internal-level fill," Seiko said. "Faster external fills are used during combat. But if there's a jam, then an internal-level fill would be done."

Alpha stood erect. Beta lowered its torso to the floor. Its arms then lifted the ammo carrier off its back onto the theatre floor in front of Seiko. Beta raised itself again, and using its front hands, opened the metal ammunition chest. Beta scurried the few steps over to Alpha. Using its front hands again, it popped open Alpha's chest plate.

Haruto hit the + button a few more times until the picture was at maximum magnification. It was grainy and jumpy, but clear enough to see there were no loose wires or any whirring gears — only modular components in a range of bright colors stacked up from Alpha's pelvis to the top of its chest. There were a few very small LED lamps on each of the components, all of them a solid green except for a flashing yellow lamp on a long narrow azure box

Beta grasped the rectangular box and pulled it out. Beta scampered back to the ammo carrier, clicked the box into a socket and removed it a moment later. Beta went over to Alpha and inserted the blue box back into its chest. The light on the box switched from flashing yellow to green. Beta pushed shut Alpha's chest plate and returned to Seiko.

A balding man in his mid-forties in the next seat then stood up. Haruto angled the cellphone so the man filled its tiny display screen. "I'm Menachem. Institute for Intelligence and Special Operations, also known as Mossad. My background is in tank warfare, followed by years in development of our Merkava tanks. I will be acquiring knowledge on how to deploy the robots. I'm *also* responsible for the security of this operation."

The smile on Menachem's face vanished as he began pacing a few steps, a left limp evident. "This is a difficult operation. Why? Because of the technology? No — I'm not worried about that. Then what am I worried about?"

Menachem, still pacing, was silent for a full five seconds. The mood in the room grew very somber. "One of you, I don't know which one, will decide in ten years to write his memoirs or something similarly stupid." He stopped dead. "Don't do it!"

The room was quiet, the people as frozen in place as the robots.

Menachem started pacing again. "You will speak to *no one* about these topics. Not your spouse, not your boss. Colonel Tanaka will handle the Japanese end of security, and I will handle the Israeli end. We're like family here, and in my development groups over the years that's the way it's always been. But — and *please* listen to me — there are standing assassination orders for security breaches. The stakes are too high. If you break security, you will be shot. It is that simple."

Menachem limped down the theatre steps to the bottom landing. "You go into a restaurant or movie theatre where it says *No cellphones* but of course, you ignore it. You even go into a hospital to visit a friend, and there is a sign saying *No cellphones allowed* but again, you disregard it and you bring your cellphone with you. I told everyone that no cellphones were allowed on this trip. Let's see who listened to me."

Menachem pushed a button on the wall. Haruto saw eight young, muscular crewmembers, wearing all white without any designation of rank or name or country, march into the theatre from the top entrance doors. Each of them held an M16 rifle in his right hand, index finger extended over the trigger.

Menachem took from his pocket a small brown device about the size of a scientific calculator. He extended two telescopic antennas and held the device up in the air. A bright red light blinked.

"Someone is using a cellphone right now," Menachem said.

On his cellphone screen, Haruto watched the M16-carrying white t-shirted crewmembers spread out in the theatre.

CHAPTER022

Haruto jerked the cellphone away from the crack between the two display screens, only to see it plummet to the floor. The crashing sound reverberated so loudly in the quiet, backstage space that he expected bullets to start flying immediately.

Haruto dove for the floor, silently landing on his palms, and grabbed the cellphone. Its camera was still running and its tiny display showed the dark, fuzzy image of the rear of the large center display screen. How to turn the cellphone's transmitter off? Too many menus and sub-menus to click through. Haruto jabbed its red power button. Like most cellphones, it ignored a quick push. Haruto froze for the two endless seconds the phone took to acknowledge his finger on the power button, and the cellphone's display finally went dark.

Footsteps! Crewmembers coming up both the left and right stage stairs. He looked through the crack between the display screens, and saw the M16 rifle of the crewmember on the left side. They'd reach him in seconds!

Then Seiko stood and held up his orange cellphone, showing off its bright multicolored screen. "I'm sorry. I use my phone for everything these days." He pressed the power button, and in a few seconds, the phone's display went dark.

The footsteps stopped. In front of the angled display screens the two M16-carrying crewmembers turned to face the audience. Another crewmember in the aisle, also carrying an M16 rifle, went over to Seiko and took the cellphone from him.

Then Daveed stood up. "Guilty, as charged." He tried to smile but apprehension painted his face as a rifle-carrying crewmember approached. He held up his cellphone, opened its clamshell top and pressed the red power button. In a few seconds its display went dark, and the crewmember took it away.

Menachem held up his cellphone detector again, its antennas still extended. A solid green light now replaced the blinking red one. Menachem smiled. "Thank you. First official meeting of the trip will be a restricted nuclear one — General, Ilana, Colonel and Isato. You will meet after lunch, in this room. The rest of us can enjoy the cruise."

The M16-carrying crewmembers marched out of the theatre. The scientists and officers stood up, some of them shaking hands with colleagues in the seats beside them. They meandered up the stairs and out of the theatre. Alpha and Beta followed Seiko out. Menachem was the last to leave and hit a button near the entrance doors. The lights of the theatre went off and the large display screens dimmed.

CHAPTER023

Haruto sat in the dark on the floor of the stage behind the large display screen. It was starting to make sense. Israeli intelligence had found someone established in Japan — Co'en Satoki — and had him snoop around about the robots.

What was happening to Japan? Why would the government break its own rules, go behind the people, and build nuclear weapons? The last time they broke the rules like this, with the military telling one patriotic lie after another, the result was Hiroshima and humiliation. Was the threat from North Korea so bad that the government needed to betray the people and deceive the Americans? If the government wanted to get its own offensive nuclear weapons, then the people had the right to know, the Americans should be told, and the constitution should be changed. The government was breaking the rules.

How to get off this ship now? If they found him, would they actually shoot a *Keibhu*?

Haruto wished he was back home and could look through his rulebooks. The rules worked. His parents had mocked his many rules, but these rules and rituals kept calamity away from the family. His rules allowed him to finish high school with honor. His rules let him date and marry Michiko — one of the most beautiful girls at the university. His rules allowed him to become a full *Keibhu* before thirty-five. This is what the rulebooks were for. He loved his country and his government, but if they were breaking their own rules, then the people should know. It was always better to follow the rules.

Haruto peered through the crack between the display screens at the now darkened theatre. He tapped on the rear of the left angled display screen two times with his left index finger. Then he tapped on the rear of the main center display screen, to the right of the crack, with his right index finger, two times. Next, he tapped on the left angled display screen twenty times, trying hard to keep the rhythm of his taps perfect. He tapped on the center display screen, aiming for twenty times, but at tap number twelve the rhythm was audibly off. Haruto took a deep breath and started again. Two taps left and then right. Then twenty taps to the left. They went perfect. Haruto took a deep breath and held it. Twenty taps to the right. Good. The rhythm sounded perfect. He exhaled, and relaxed for a second.

Under the faint light from the LEDs of the computer racks on the side wall, Haruto eased down the stage stairs. On tiptoes Haruto walked midway up the aisle of the theatre to a seat that had not been occupied and opened a snack refrigerator in that row, taking out a bottle of Coca Cola and a tube of potato chips. Haruto walked back down the aisle, stopping at the men's washroom off the aisle. Then back up the stage stairs, slipping behind the large display screens. He plopped into one of the extra leather chairs and opened the snacks. It was wonderful how good potato chips and a Coke tasted when you were hungry.

What time was it? He wore no watch since he normally used the display of his cellphone to keep time. No portholes in the theatre. But the ship was out at sea — he could feel the engine vibration, the occasional wave. To risk leaving the theatre and going out on deck to escape from the ship now did not make sense. A good rule would be to wait until the ship stopped moving and had pulled into port, before taking the risk of leaving the theatre. The rules worked.

Perhaps even yet with Michiko. He could smell the *Shiseido* perfume she wore. She was so beautiful. She could still change her negative feelings about him. A good rule would be to try to woo her back two more times, each time promising to have a baby soon. If that didn't work, the next rule would be to try to involve other parties, perhaps Michiko's mother, who always seemed so fond of him.

Those were good rules. He lay back on the comfortable chair, and waited.

Pacific Ocean 31°N, 144°E
June 12   1PM Tokyo Time (04:00 Zulu)

CHAPTER024

he entrance doors suddenly opened and the theatre lights came on. Haruto went to the crack between the display screens. Three men and a woman walked down the theatre steps — General Otzker, Colonel Tanaka, Ilana and Isato. They sat down next to each other near the bottom of the theatre.

Isato, a fortyish, small-framed man, put his finger up. "As I have tell you, I'm on team in Japan producing plutonium fission devices — which, of course, not been tested yet. The Colonel and I have much wonder, how you test your devices?"

Otzker smiled. "Your crude fission bombs will probably explode, but the yields may be much smaller than expected. To develop miniaturized, high-yield fusion devices take testing — a *lot* of testing."

"How do you do it, General?" Isato asked.

"How would you do it?"

"The first thing comes to mind is underground testing, but earthquake seismic detectors all over world triangulate blast, even for small fission bomb few kilotons. Testing fusion device thousand times strong would be impossible to hide. But we are on ship right now, so to be testing underwater. Correct, General?"

"Quite correct."

"But underwater testing even worse than underground testing, because sound waves travel underwater very good. If I explode one kilogram of TNT underwater here in ocean near Japan, explosion be heard by the American Navy in Hawaii. If I explode underwater a nuclear fission bomb equal to kilotons of TNT — *thousands of thousands* of kilograms of TNT — the explosion very be heard by underwater listening equipment everywhere in the Pacific Ocean. And you plan to test underwater a nuclear *fusion* bomb — one *megaton* of TNT — and you say nobody notice it?"

Otzker smiled. "They will notice it. They will not care."

A look of puzzlement appeared on Tanaka's and Isato's faces. They seemed at a loss for words and said nothing. After a few moments of silence, Otzker started laughing. He pressed a button in the armrest of his chair, and Haruto saw in the reflection of the brass globe of a chandelier, the main display screen showing a map of the world.

"These are the areas of the world that experience twenty or more earthquakes a year," Otzker said. Many small red circles appeared all over the map, with a large number in the western Pacific Ocean. "This is our destination." A green circle appeared near Guam, with the caption *Mariana Trench*.

Tanaka put up his finger. "General, the underwater acoustics of a sudden nuclear explosion are *very* different from the slower rumbling of an underwater earthquake. This is well known."

Otzker remained sitting and laughed. "I must've been doing something right to have kept my job this long. Enjoy the cruise and in a few days you will see the test unfold."

Ilana stood up. Fortyish, fair skin. She had short black hair and large brown eyes, wore two oval-shaped, plastic-looking earrings and light-red lipstick. She was small and slim, and had on a blue cotton dress with a bright flower pattern.

Reflected off the chandelier globe, Haruto saw on the large display screen a colored schematic drawing. *Teller-Ulam Hydrogen Device*. On the top was an egg shape with concentric layers of different colors. Right below that was a larger green sphere. "This is a two-stage hydrogen bomb design. Building it is many times more complicated than building a simple plutonium fission device."

Ilana put the electronic pointer on the inner layer of the top eccentric sphere. "When the conventional explosives around the plutonium egg explode, they implode the egg and nuclear fission occurs. We put tritium-deuterium gas in the center of the egg to boost its explosive yield. You'd think adding tritium is simple, but it took us seven underwater tests to get it right."

Ilana moved the electronic pointer to the larger lower sphere. "The egg explodes apart at a thousand kilometers per second, but its X-rays travel at the speed of light, three hundred times faster. When the X-rays hit the surface of the lower sphere, there's compression of the layers, both from the X-rays themselves and from the thermal effects. Neutrons from the egg also stream in and start converting the lithium-deuteride to tritium and deuterium. Heat and compression. Fusion of the tritium and deuterium occurs, producing more heat and more neutrons. These neutrons then cause fission of the outer uranium-235 layer."

"Why do you use a layer of uranium-235 instead of less valuable uranium-238?" Isato asked.

Ilana laughed. "We had many debates about this one. The enriched uranium-235 gives us a very high flux of neutrons. If we used uranium-238, we couldn't miniaturize the bomb as we've done."

"What yields do you get?" Tanaka asked.

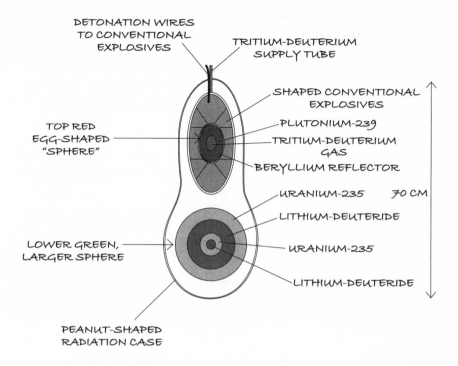

DETONATION WIRES
TO CONVENTIONAL
EXPLOSIVES

TRITIUM-DEUTERIUM
SUPPLY TUBE

SHAPED CONVENTIONAL
EXPLOSIVES

PLUTONIUM-239

TOP RED
EGG-SHAPED
"SPHERE"

TRITIUM-DEUTERIUM
GAS

BERYLLIUM REFLECTOR

URANIUM-235   70 CM

LITHIUM-DEUTERIDE

LOWER GREEN,
LARGER SPHERE

URANIUM-235

LITHIUM-DEUTERIDE

PEANUT-SHAPED
RADIATION CASE

"Fifty kilotons from the first boosted fission stage alone. With the second stage, we a get a megaton of yield."

"What is the size of the warhead?"

Ilana clicked a button on her armrest. A photograph of a shiny metal cone with traces of machine oil appeared on the main display screen. A meter stick was in the background. "Three hundred kilos in weight, including the outer casing you see here. Ninety-eight centimeters high."

The jaws of Tanaka and Isato dropped slightly as they stared at the picture, trying to discern details.

"Unbelievable," Isato said. "One megaton explosion for bomb weighing three hundred kilos — fifty percent of theoretical yield."

"Can you get larger yields if you wanted to?" Tanaka asked.

"To do what? One megaton is a lot of explosion. But the answer to your question is yes." Ilana reached into her dress pocket and took out a small clear-plastic hinged case. She opened the case and pulled out a black ceramic chip, about a centimeter long, with six shiny metal feet on either side protruding from the body. "This is the *Activation Chip* for the bomb. Some people call it the *Masada Chip*, but that's a stupid name. In the first century, Israel was conquered by Rome, and a group of Jews took refuge in the Masada cliffs. The Roman army built a huge stone rampart and stormed Masada, but all the Jews there committed suicide rather than be captured by Rome."

"An honorable story," Tanaka said softly.

"Yes," Ilana said. "However, I view our nuclear bombs as keeping us alive, not leading to our suicide. Anyway, the firing of the explosives surrounding the egg requires the Activation Chip to receive the correct security code."

"How do you get yields larger than a megaton?" Tanaka asked.

"We combine devices. Plug the telemetry output cables of the first bomb into the Activation Chip sockets of the second, third, fourth or however many bombs. The chip computes signal delays, and each of the bombs receives a detonation signal to occur at the same time.

"I've prepared material for Isato to read." Ilana walked over to the Japanese nuclear engineer and clicked on the display screens mounted on either side of his chair. "If he studies the material today, then tomorrow I can work through some simulations with him."

Otzker, Ilana and Tanaka walked up the aisle stairs out of the theatre. Isato sat fixed to the display screens in front of him, scrolling up and down.

Haruto looked through the crack at Isato studying the material on the monitors. Without any doubt, Japan would learn to design nuclear bombs, and, Haruto thought, smiling again almost glumly, would be designing the best bombs on the planet within a few years. The best bombs to do what with? Why was his government doing this?

It was dangerous when you didn't follow the rules.

Pacific Ocean 30°N, 144°E
June 12 3PM Tokyo Time (06:00 Zulu)

CHAPTER025

he sound of the entrance doors and a large number of footsteps pulled Haruto from his light nap. He ran up to the crack between the large display screens in time to see the scientists walk down the aisle steps. Alpha and Beta followed Professor Haku Seiko. Isato studied away, still fixed to his seat, focused on his computer screens.

"Alpha, come here," Seiko ordered. "Stand ready for maintenance."

The bipedal Alpha robot moved closer to Seiko in front of the stage and stood upright and motionless.

Seiko snapped open Alpha's face, chest and right thigh. "All these plates have a front layer of titanium-boron nanocrystalline ceramic tile, a middle layer of lightweight foamed metal, and a back layer of composite M5 fibers. They're bulletproof for small arms fire."

"Let's start with the head. We don't put the brain of the robot here — the processing unit is in the torso." Seiko pointed at a bright orange plastic module in the upper chest of Alpha.

"We use the upper head for sensors. An array of antennae to allow Alpha to locate the enemy by his radio communications. And an array of infrasound and audio microphones — detection of enemy equipment, sometimes even footsteps, kilometers away."

Seiko pointed to the eyes. "One eye is an infrared camera. The other one has telescopic optics. It's sensitive as well to low light levels — equal to night-vision goggles."

Seiko pointed to a third eye sensor in the forehead, normally covered by the faceplate. "Hamamatsu radiation line sensor. In the left wrist of Alpha is a variable X-ray source. He can X-ray small packages. Much like what you see at airport security machines."

"Can you X-ray a building?" Menachem asked.

"Sure. Outfit a Beta robot with the nine-mega-electron-volt X-ray pack — it'll penetrate even concrete. An Alpha stays on the other side of the building and produces a detailed X-ray picture."

Seiko pointed to the gun barrel located at the mouth opening. "I didn't design the weapon systems." He laughed nervously. He pointed to the springs and electronic dampers in the neck. "These handle the gun recoil."

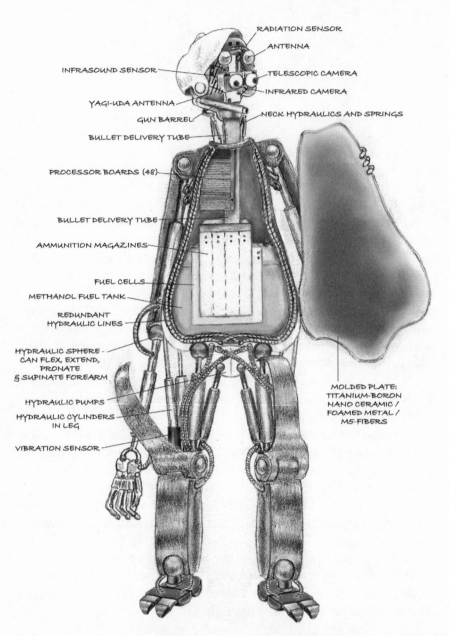

RADIATION SENSOR

ANTENNA

INFRASOUND SENSOR

TELESCOPIC CAMERA

INFRARED CAMERA

YAGI-UDA ANTENNA

GUN BARREL

NECK HYDRAULICS AND SPRINGS

BULLET DELIVERY TUBE

PROCESSOR BOARDS (48)

BULLET DELIVERY TUBE

AMMUNITION MAGAZINES

FUEL CELLS

METHANOL FUEL TANK

REDUNDANT
HYDRAULIC LINES

HYDRAULIC SPHERE -
CAN FLEX, EXTEND,
PRONATE
& SUPINATE FOREARM

MOLDED PLATE:
TITANIUM-BORON
NANO CERAMIC /
FOAMED METAL /
M5-FIBERS

HYDRAULIC PUMPS

HYDRAULIC CYLINDERS
IN LEG

VIBRATION SENSOR

Seiko looked down at the legs of Alpha. "Let's start at the lower end and work our way up." He touched a small cylinder on top of the knee. "Seismic vibration sensor. Alpha can feel the enemy's footsteps or equipment, sometimes from kilometers away."

Seiko pointed to the hydraulic pistons in Alpha's opened thigh. "You probably expected to see electroactive polymer artificial muscles, but they weren't strong enough. Instead, we modified aviation hydraulics. Springs in the cylinders give the robots snapping-fast movements, when needed. Also, to save energy a hydraulic rod can freewheel if other cylinders are providing the forces."

"What loads can they lift?" Menachem asked.

"Thousands of kilograms for short periods, but then you get overheating. Alpha can carry a few hundred kilos for long hauls. Beta can run with a thousand-kilogram load." Seiko pointed to a metal assembly in the pelvis surrounded by magenta insulation. "To keep the center of gravity low, the hydraulic pumps are here. Two pumps and two sets of hydraulic lines for redundancy. The pink fibers are soundproofing."

Seiko grabbed a bottle of water out of the nearest snack refrigerator and took a gulp. He then pointed to the purple U-shaped plastic module arising out of the pelvis, surrounding the ammunition magazines in the abdomen. "These are the fuel cells. Most of their power goes to the hydraulic pumps, the rest to electronics."

"What output do you get?" Shmoel asked.

Akihiro raised his finger. "We developed aerogel electrodes impregnated with electrocatalysts — over a hundred square meters of electrode in every cubic centimeter."

Shmoel's eyes opened wide. "That's a lot of fuel cell."

"Yes, in Alpha we can produce up to five thousand watts of power, in Beta a hundred *kilo*watts."

"Great, so the machines burn up their fuel supply in a half-hour. I was in charge of the Merkava tank development, and I went through this before. You designers are all the same!" Menachem noticed the quiet persons around him. "Okay. What fuel do you use anyway?"

"The fuel is methanol — ordinary wood alcohol." Akihiro forced a smile at Menachem. "Five-thousand watts is a peak use for Alpha. In quiet surveillance mode, he only uses fifty watts. Alpha and Beta can go on ten-day missions with no refueling. For longer missions a Beta robot can haul a thousand-liter methanol fuel tank and refuel a squad of robots."

Seiko flipped the lid off the bright orange plastic module in the chest, revealing dozens of thin, colored printed-circuit-boards stacked vertically. Seiko stared at the electronics, saying nothing for a moment. A look of awe came over his face. "This is what makes it all possible."

"What kind of processor is that?" Daveed asked.

"*Processors,*" Seiko said. He pointed at the bottom half of the circuit boards. "Eight hundred trillion visual processing operations per second. Plus four more boards to process other inputs, from sound to radio signals."

Seiko pointed to the middle of the stack of circuit boards. "These sixteen blue boards hold four billion Hopfield circuits."

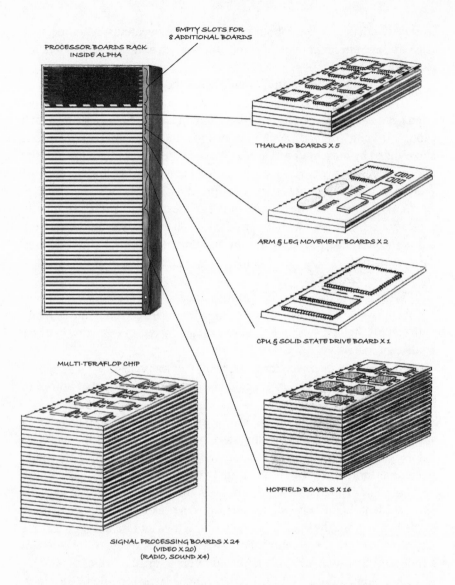

EMPTY SLOTS FOR
8 ADDITIONAL BOARDS

PROCESSOR BOARDS RACK
INSIDE ALPHA

THAILAND BOARDS X 5

ARM & LEG MOVEMENT BOARDS X 2

CPU & SOLID STATE DRIVE BOARD X 1

MULTI-TERAFLOP CHIP

HOPFIELD BOARDS X 16

SIGNAL PROCESSING BOARDS X 24
(VIDEO X 20)
(RADIO, SOUND X4)

Seiko pointed to the orange colored board. "This circuit board is a conventional computer and solid-state drive, which can be accessed by the other boards. The two red circuit boards above it handle leg and arm movements."

Seiko put his finger on the topmost yellow and black circuit board. "This is the main Thailand board, and the four yellow circuit boards below are also Thailand boards."

Daveed put up his hand. "Can one person really learn to program all this?"

"I'll help you learn the software."

Daveed shook his head. "It's not a question of learning the software. I don't really understand what's going on here."

Seiko smiled. "The design is actually quite simple. About half of the boards process video input. You're given a cleaned up and processed video signal — where the objects' edges are, properties of the objects in the visual scene — and much of this feeds into the Hopfield nets, which in turn gives you a recognition of portions of the video scene. You don't have to do much with output either. Two circuit boards deal with the robot's control of his legs and arms, as well as other physical systems such as built-in weapons. So you receive the input on a silver platter, and the system takes care of the details. The question is, what do you do with these inputs? That's what the Thailand boards are for."

"Thailand boards?" Daveed said.

"Rodney Brooks was stuck in Thailand in 1985 when he came up with a way to build robots that could work in real-time. He got rid of the thousands of sequential steps between input and output. Instead he used lots of separate steps that reacted directly to a sensor and then almost immediately carried out some action."

"I think I remember this ..."

"I'll give a simple example using Alpha," Seiko said. "You give Alpha a rule that he's to stay in surveillance mode unless he receives an order stating otherwise. One of the 'finite state machines' on the Thailand boards will start executing that rule. You give Alpha another rule that if he detects an enemy, he's to leave surveillance mode and stalk the enemy as close as possible without a certain risk level of injury. Another of the finite state machines on the Thailand boards starts executing that rule. You give Alpha another rule that if he's close enough to the enemy to take an accurate shot, he should fire on the enemy. Another of the finite state machines on the Thailand boards starts executing that rule."

"Fair enough," Daveed said.

"With these simple rules we've created a powerful behavior," Seiko said. "Imagine you see Alpha lying in the bush waiting for an enemy. Then you see him quietly stalking the enemy, and then you see him killing the enemy. Then he goes back to lying in the bush waiting for another enemy."

"How does Alpha recognize the enemy?"

"That's why the Thailand boards handle millions upon millions of rules."

Daveed looked up at the ceiling and smiled. "I remember this now. This is a *subsumption architecture*, isn't it?"

Seiko nodded.

"I thought it was proven in the 1990s that once this type of computer architecture got too large, the many goals and rules began interfering with each other, and the architecture was useless for creating real-world machines."

Seiko smiled. "That's the problem with proofs. Depending on what your assumptions are, you can prove just about anything. Indeed, in our early robot models we had many problems with the rules interfering with each other. We had robots that would spend all day long turning a room's light switch on and off, or opening and closing a door, or tapping on it, for hours."

"What did you do?"

"Well, we didn't abandon the technology like the Americans." He touched the yellow and black boards. "Keep in mind that we have a top-level Thailand board in our robots and we have a top-down hierarchy in the other Thailand boards — important rules call other rules, which call other rules. Also, keep in mind all the other boards that deal with the low-level input and output processing for us."

Seiko plugged a wireless USB connector into a socket on the leather chair. A detailed technical map of the theatre in red hatched lines appeared on the large display screen. "The robot's eyes have built-in laser range finders. While Alpha was searching for the 100-yen coin he built this internal map of the theatre. As you know, subsumption architectures aren't supposed to use internal maps — they just react. Well, our robots do use internal maps, although it is true they then do react to them."

"How do the robots know what they're doing?" Daveed asked.

"They don't. The robots react to their rules and perform their function. They don't really understand why they do it or what they do."

"Isn't that a problem in the field?" Daveed said. "Why don't we add on another circuit board to handle understanding and strategy?"

Seiko smiled. "Our priority was to make a weapons system that worked. And it does. We designed the robots to take orders from a human commander in the field. *However*, there's still room for eight more circuit boards. You could add Thailand boards with higher-level understanding, higher-level goals."

Daveed's eyes lit up. "It's fascinating... fascinating."

Otzker coughed. "Let's not get ahead of ourselves. Let's just make sure we understand how the robots work. Anyway, this has been a lot for an old head like mine. Let's take a break, and we'll have the next meeting, a restricted nuclear one, after dinner."

Seiko snapped shut Alpha's panels. "It's okay if I leave them here? They don't really need to follow me all around the ship."

Otzker looked at the robots. "They won't touch any of the equipment here?"

"I just put Beta into sleep mode. Alpha will stand there like a piece of furniture unless someone interacts with him."

Otzker nodded and the scientists walked up the stairs, out of the theatre.

Haruto gazed through the crack between the large display screens at the now darkened theatre. He hadn't understood all the technical details, but it was obvious that the Alpha and Beta robots would seriously strengthen any army. No wonder the Israelis wanted them so badly. Almost as much as his government wanted the nuclear weapons.

Haruto was hungry, but he felt anxious. Alpha and Beta were there. They were indeed as immobile as pieces of furniture, but what if they woke up and saw him? Would they set off an alarm? Would another bullet fly out of Alpha's mouth?

But he needed food, needed something to drink. It was only a few meters away. Why should Alpha and Beta attack him? They didn't seem hostile to anyone else.

Haruto started to slide out from the stage, but it didn't feel right. He tapped on the rear of the left angled large display screen two times with his left index finger, then tapped on the rear of the main center display screen, to the right of the crack, with his right index finger two times. He felt a bit better. He tapped on the left display twenty times, then twenty times on the right side. Then he tapped twenty more times on the left side — each tap in perfect rhythm. One more time on the right side and everything would be okay. Haruto took a deep breath and held it. Twenty taps, perfectly. Good.

Haruto eased his way down the side stage stairs to the first row of the theatre. There were no portholes or other windows here, but the service lamps from the computer racks on the side wall provided some light.

Alpha and Beta were still standing immobile in front of the stage. Haruto studied them. Were Alpha's eyes moving?

Haruto froze. He looked again at Alpha. No motion, but, not enough light to be sure.

Haruto tiptoed to the washroom off the left aisle. He let a few drops fall out of the sink faucet and splashed his face. The cool water felt calming. He made use of the urinal, making sure not to touch it nor flush. Good, no need to count again. Get food now.

Haruto walked up the theatre side aisle to the fourth row of seats. He had just started opening a snack refrigerator when he heard the crack of the theatre entrance door. Haruto jerked away from the seats and sprinted silently on the tips of his toes down the carpeted aisle to the stage stairs.

The lighting relays snapped on just as he reached the stairs, and a split second later, lit up the theatre. Haruto pushed off each stair, balancing speed with silence, and slid his body behind the main large display screen to the back of the stage.

His palms were wet with sweat. Had someone seen him? He held his breath, inched over to the crack between the main and left large display screens, and peered out.

Isato, oblivious to the world around him, walked down the aisle to his leather seat, sat down and started studying his computer screens again.

CHAPTER026

Alpha walked silently by Isato's seat. Isato didn't hear him, but when a slight shadow moved across the computer screens, Isato lifted his eyes. He jerked back into his seat, opened his mouth and was about to scream, but stilled himself, and watched Alpha proceed up the theatre aisle. The robot went to the fourth row of seats and started opening snack refrigerator doors and inspecting the colorful refreshments.

Isato smiled and raised his hands like blinders to the sides of his head. Focus now. The equations for the Israeli nuclear device were extremely complicated.

CHAPTER 027

Haruto lay on one of the leather chairs behind stage. He was getting quite thirsty. Hopefully Isato would soon tire, and join the rest of the scientists for dinner. It was hard to tell what time it was any longer, but Otzker had said the next meeting would be after dinner, so Isato would have to leave to eat —

A metal foot suddenly smashed down on the stage.

Haruto leaped out of his seat and reflexively assumed karate fighting position.

A metal arm pushed the edge of the large angled display screen aside a few centimeters, and Alpha slid between the displays backstage.

Alpha froze and stared at Haruto. Then with a blur of his limbs Alpha also went into a karate fighting stance — balancing on the front of both feet, both knees bent, both arms bent, right hand clenched in a fist at face level, left hand partially clenched, a bit lower.

Then Alpha smiled. In the dim stage light, Haruto saw the metallic gun barrel.

Haruto pulled his legs from under him, and went flying to the floor, catching himself on outstretched palms. He hugged the floor.

No shots flew out. But a moment later Alpha went crashing down on the floor, catching himself with his right palm, a deafening noise as the metal slapped onto the wood. The whole ship would hear this!

Alpha smiled again. Haruto saw the steel gun barrel aimed directly at him.

Haruto felt his heart slam into his chest. He breathed in, but no air seemed to enter. Make some rules. Don't stay here waiting for the machine to shoot me.

Quick two-tap.

Haruto tapped his right index finger two times on the floor. Then he did the same with his left one. Okay, make a rule now —

Alpha smashed his right index finger into the wood, actually going through the old wooden stage. Again, another time, the vibrations reverberating so loudly how could Isato, let alone the whole ship, not hear this? Alpha then smashed his left index finger through the wooden floor two times. "Good job?" he asked.

Haruto hopped up on his feet.

Alpha hopped up even quicker.

The robot walked to Haruto and smiled again.

Haruto's heart was beating faster and faster.

Alpha's left hand went flying toward Haruto's face in a blur too fast to see.

It stopped a centimeter in front of his nose, holding a bottle of water. "Humans need clear fluids. You failed to retrieve this before."

Haruto looked at the machine almost pressed against him. Alpha was close enough that even in the dim stage light Haruto could see the details of his face, the camouflaged colors of his chest, the hydraulics of his limbs.

Haruto took the bottle of water. "Thank you."

"Now it's your turn," Alpha said.

"My turn to do what?"

"I will run up the aisle and pretend to grab something. Then you must figure out what I wanted to grab and bring it to me."

What to say to the machine?

Alpha smiled at Haruto again, baring the tip of his shiny gun barrel between his lips.

"I must go to sleep now," Haruto said.

"When will we play again? Tomorrow, in twenty-four hours?"

No... have to break this train of logic. If he tells the others that he will play in twenty-four hours with a man behind the stage, it will be a disaster.

Haruto looked Alpha in the face. "I must go to sleep. You cannot play with me again."

No response from Alpha. Random facial expressions. His lips starting to move apart. Haruto's heart started beating faster.

"Sleep is for rejuvenation. Can I watch you fall asleep?"

"Yes." Haruto lay down supine on the wood floor. He tapped twice on the right, and twice on the left. Then he closed his eyes.

Haruto stayed motionless.

A minute went by. No sound of Alpha moving. Another minute. The robot must still be there, observing. Another minute. Haruto wanted to open his eyes a crack, but the machine probably would detect the motion. Another minute. Then Haruto heard the soft steps of Alpha's feet, becoming more distant, until he could no longer hear them.

Haruto opened his eyes a crack. Nothing. He opened his eyes fully. Alpha wasn't there.

Haruto tiptoed to the crack between the display screens. Alpha was walking over to Isato.

Would the machine tell Isato that a stowaway was behind the stage?

"Play with me," Alpha said to Isato.

Isato raised his eyes and stared at the robot. He said nothing for a few seconds. Then he shook his head. "No. I have much work to do. I am not on the robot team. No playing."

"Then what should I do? Do you have instructions for me?"

Isato scratched his chin. He pointed to the computer seat to his left. "Go sit at that chair. Go onto the Internet and look at pictures."

"For how long?" Alpha asked.

"Uh... for thirty hours. Is that okay?"

"Yes, instruction understood."

Alpha sat in the leather seat to the left of Isato and started typing away and clicking on the mouse.

Alpha was turning his head back and forth, looking at images on the left and right screens, about an image every second. The angle was awkward, but Haruto could make out the left screen attached to Alpha's chair. Alpha was not looking at pictures of integrated circuits or weapons, but was viewing buildings — skyscrapers, museums, palaces and cathedrals.

CHAPTER028

The quiet in the theatre broke as the entrance doors slammed open and four crewmembers entered carrying two large brushed-aluminum cases. Otzker, Tanaka and Ilana followed them down the aisle steps to the first row where Isato was sitting.

The crewmembers started climbing the stairs to the stage. "We'll put the cases behind the stage."

Haruto froze in his hiding place!

"No, leave them here. We need them now," Otzker said.

Haruto took in a deep breath and relaxed.

The crewmembers deposited the cases between the first row of seats and the stage, and left the theatre.

Ilana opened the first case. A shiny seventy-centimeter metal "peanut" lay attached to a latticework of wires and colorful small boxes at the top. "This is the *physics package*. In an actual bomb, a protective shell encloses it."

Ilana pointed to the boxes at the top. "Batteries, electronics, anti-critical explosives and tritium cleaners. In a missile you want as much weight up here in the nose to balance the heavy sphere of uranium in the bottom."

"What are the tritium cleaners?" Tanaka asked.

Ilana pointed to the shiny stainless steel modules stacked on top of the metal peanut. "Tritium gas inside the fission egg dramatically boosts its explosion. But we lose seven percent of the gas each year to decay and escape. Instead of taking apart a warhead every few months to add more tritium, we have a ten-year supply of tritium in this cylinder. But... helium contamination occurs as the tritium decays, so we trap helium on thin metal layers inside the module."

"What's the full maintenance schedule on the warhead?"

"Every ten years change the tritium modules and the lithium battery. Every eighty years change the conventional explosives and the thermoelectric generator. Every three hundred years change the plutonium pit. That's it."

"Our arrangement calls for full honesty from both parties," Tanaka said in an angry voice. "We don't want to be fed a bunch of technical nonsense. Our scientists have already worked on this problem. The radioactivity

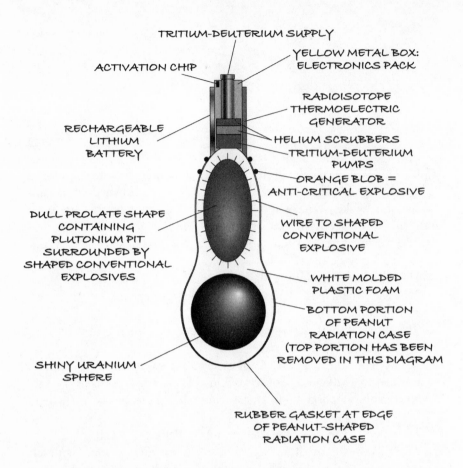

TRITIUM-DEUTERIUM SUPPLY

ACTIVATION CHIP

YELLOW METAL BOX:
ELECTRONICS PACK

RADIOISOTOPE
THERMOELECTRIC
GENERATOR

RECHARGEABLE
LITHIUM
BATTERY

HELIUM SCRUBBERS

TRITIUM-DEUTERIUM
PUMPS

ORANGE BLOB =
ANTI-CRITICAL EXPLOSIVE

DULL PROLATE SHAPE
CONTAINING
PLUTONIUM PIT
SURROUNDED BY
SHAPED CONVENTIONAL
EXPLOSIVES

WIRE TO SHAPED
CONVENTIONAL
EXPLOSIVE

WHITE MOLDED
PLASTIC FOAM

BOTTOM PORTION
OF PEANUT
RADIATION CASE
(TOP PORTION HAS BEEN
REMOVED IN THIS DIAGRAM

SHINY URANIUM
SPHERE

RUBBER GASKET AT EDGE
OF PEANUT-SHAPED
RADIATION CASE

from plutonium will cause cavities in the plutonium metal. It'll fail after *twenty* years."

Ilana shook her head. "Theory is a beautiful thing, but we've looked at *real* twenty-year-old weapons. Plutonium atoms fall into place in the metal's crystalline lattice, and the large cavities you talk about simply don't develop."

"All right... three hundred years. You expect to be around for a while, I see."

"I hope so."

"What about the *anti-critical explosives*?" Isato said. "I never hear of such devices."

Ilana pointed to four small orange blobs at the top of the peanut. "These and other explosives will push the components of the bomb apart and prevent any nuclear reaction from occurring."

"Why do you need them?"

Ilana grabbed a power screwdriver from the other brushed aluminum case and unscrewed the top of the large shiny peanut. "This is the radiation case. It focuses X-rays from fission of the top plutonium egg to the lower uranium sphere."

Ilana pointed to the rubber gasket lining the outside perimeter of the remaining back half of the peanut. "The radiation case is machined to fit together so well that it's waterproof, but we have this gasket here just in case."

The top portion of the peanut, looking like an oversized egg that had sprouted hairs, contained a dull prolate shape with dozens of wires going to it. Beneath the "egg" was a shiny sphere, a bit over twenty-five centimeters in diameter, held in white molded plastic foam.

Ilana pointed to the top egg shape. "What you see here are the wires going to the conventional explosives. When these charges are detonated, the plutonium egg they surround implodes inwards, goes critical, and nuclear fission occurs. The tritium inside the egg will fuse and increase the explosion."

Ilana pointed to the lower large shiny sphere. "The X-rays from the explosion of the plutonium egg then trigger the explosion of this uranium-lithium-deuteride sphere. As I said earlier, the uranium-235 creates a huge flux of neutrons and lets us build a miniaturized device. That's the good part. The bad part is that if the bomb is just sitting there peacefully, and water gets into the uranium sphere, the water molecules will slow the neutrons enough so that they can hit and split more uranium atoms, and more neutrons release, and more uranium atoms split. There's a chain reaction. Even though no one activated the bomb, the uranium sphere quickly becomes a runaway nuclear reactor and explodes!"

Ilana's voice became hoarse, and she reached into one of the snack refrigerators for a bottle of water.

"That's not a real bomb, is it?" Tanaka asked.

Ilana started talking, but her hoarse voice became unintelligible. She gulped a drink. A drop of water rolled onto the plastic neck of the bottle, and hung there in the air. The drop started stretching, and gravity won as the drop let go of the plastic surface and splat onto the shiny sphere below.

Tanaka froze.

"No, of course this isn't a live bomb. In any case, we have a moisture sensor within the device. If water is detected inside then these tiny orange anti-critical explosives detonate and push the components of the bomb apart so no nuclear reactions are possible."

"And we're going to test this bomb in the ocean under ten thousand meters of water?"

Ilana nodded.

Tanaka pointed to his wrist. "This watch says it's good for fifty meters of water. I went scuba diving last year — not very deep, maybe ten meters. When I surfaced, there were water droplets under the faceplate. It's hard to keep out water under pressure."

Otzker laughed. "I'm glad you like scuba diving — it's the specialty of the ship. Don't worry about the test. I haven't blown us up yet."

CHAPTER029

Haruto looked through the crack again. The others had left hours before but Isato still sat, fixed to his display screens, scrolling up and down, continuing to study. About an hour ago Isato had stood up as if to leave, but the nuclear engineer only pulled a snack out of the adjacent refrigerator.

What time was it? Haruto's cellphone was off, and he couldn't see any clocks through the crack in the display screens. There were no windows to even tell if it was day or night outside. His stomach grumbled. As comfortable as the leather chair was, Haruto was tiring of the space behind the display screens. He had never liked enclosed spaces, remembering as a boy how his heart would race every time he went onto the subway or was in a car high on a bridge.

Haruto could still feel the occasional wave and a slight vibration from the ship engines. As much as he wanted to leave the confined space, it was foolhardy to do so. He would follow the rule he had made — wait until the ship pulled into port, then make it to the deck and off the ship.

Haruto lay down on the reclined seat. He looked to the right and counted to two. He looked to the left and counted to two. Then he looked to the right and counted to twenty, and did the same thing on the left side. He would be all right now. He closed his eyes and tried to get some sleep.

CHAPTER030

The young dredge operator hit the red "stop" button, and the clatter of the machine faded. He climbed out of the operator cab and waved to an older man in a rowboat. When the smaller boat was close, he grabbed its yellow line and wrapped it around the capstan on the side of the clamshell dredge.

"Are we done yet?" the old man shouted.

The operator started laughing. "It's one o'clock in the morning. If I was done, do you think I'd still be here?"

"This work was supposed to be finished by midnight."

"Yeah, well tell that to the mud." The operator shook his head. "What's going on, Yossi? We dredged to sixteen meters last year already. What ship are we expecting? The draft of an aircraft carrier is only twelve meters."

"I have no idea, but I know we'll lose all our other contracts if this job isn't done perfectly. They wanted the extra meter of clearance in the center of the channel and at the quay." The old man stood up in the rowboat and stretched out his arm. "Here, give me a hand."

The operator helped the old man onto the dredge barge.

"You're going to stay here?" the operator asked in surprise.

The old man nodded. "How much more to do?"

"About three more hours. Can you give me a cigarette?"

The old man tossed him a pack of Noblesses. The operator climbed back into the cab, hit the large plastic green button, and started adjusting the small control levers. The engine grunted for a second and released a cloud of diesel exhaust, then the bucket came up filled with another load of mud.

CHAPTER031

"**All engines reverse,**" Captain Yamada said. "Prepare to make anchor. V.H.F. Channel 12."

"Yes, Sir." The young bridge officer gave Yamada the black hand-microphone.

"Port Suez," Yamada said in Japanese-tinged English. "This is the *Mikiyasu-ema*. Permission to join convoy requested. Over."

The radio crackled with static for a second, then an Arabic-accented English voice said, "*Mikiyasu-ema*, this is Suez. Permission granted but it is after one hundred hours. Penalty dues apply. Over."

"Port Suez, we are past the latitude and are making anchor. It is one hundred hours only now. Not past the time. Repeat. Not past the time. Over."

The radio crackled again. "Okay, *Mikiyasu-ema*. You are entered within the limit time. You are number seventeen in the convoy. Pilot will be over soon. Standard start time of six hundred hours. Also, your new crew is here waiting for you. They will come over by launch."

"New crew?"

"Yes. Your paperwork all in order. Suez out."

A few minutes later, a midsize launch pulled over to the side of the *Mikiyasu-ema*'s lower boarding door. The large ship lowered a staircase to the water level.

A thirtyish Japanese man dressed in a black t-shirt stepped out of the launch onto the staircase, and entered the *Mikiyasu-ema*. He bowed to Yamada and gave him an envelope. "I am Lieutenant Colonel Okamura. For security reasons I have been put in charge of this ship. This letter is from your company."

"Lieutenant Colonel. We have no need for your security. We have the most advanced missile defense and anti-hijacking system in the industry."

Okamura bowed again to Captain Yamada. "Yes, we were surprised when we found out. A lot of protection for dump truck and excavator parts. In any case, that is good. We are to await further orders before crossing the canal. After we make the crossing we are to pick up two containers at Port Said." Okamura motioned to the launch and a dozen young men in black t-shirts, carrying stuffed duffle bags walked up the stairs to the ship.

Yamada wore a freshly pressed merchant marine uniform sporting the name of Mikiyasu Industries on his lapel and cap. His face was wrinkled, enveloped by a pair of silver-rimmed glasses. "Lieutenant Colonel, I have been sailing ships for Mikiyasu for thirty years now. This is the world's largest container ship. You cannot just leave it floating without purpose somewhere along the shore."

"How long before we reach the Mediterranean?"

"Seventeen hundred hours."

Okamura nodded. "Okay. Proceed."

"The canal pilot will be here in an hour or two. Please have your men stay out of sight."

Okamura held up his hands defensively and followed Yamada to the elevator and up to the bridge. The ship's bridge was only a bit sternward of the middle of the ship, enclosed in huge windows on all sides. Under soft deck lights, for hundreds of meters both fore and aft, thousands upon thousands of identical unmarked brown twenty-foot shipping containers, stacked neatly. The water was calm, and the night dark without a moon.

CHAPTER032

"**O***hayo gozaimasu. Boker tov.* Good morning."
Haruto awoke and jumped out of the leather chair. He looked out the crack between the display screens.

"Yesterday was very productive," General Otzker said. "Before we start again, a few words from our purser."

A tall young man in a uniform embroidered *New Pacific Queen* stood up beside Otzker. "We're based out of Guam, running cruises from Taiwan, Shanghai, Sydney — and soon Tokyo. Our chef is *Cordon Bleu*. Our casino is popular among the gamblers — single-zero European roulette improves the odds a bit. Electronic art classes in this theatre. Scuba diving at some of the most beautiful reefs in the world. Also our famous Challenger Deep dive — we bring the ship over the deepest spot in the world and people get to say that they dove there, although it's too deep for anyone to dive to the bottom, of course."

"Isn't swimming around over the Challenger Deep a bit boring?" Colonel Tanaka said. "What do you see besides water?"

"Actually, it's a very popular dive. Sometimes we do a shark feeding to spice it up. The sharks never seem to bother the divers watching, although I nearly died the first time I saw it." The purser laughed feebly. "We'll be at the Marianna Islands later this afternoon, then make our way down to Guam. Thank you."

Haruto smiled. One more day to go. He'd get off the ship when it docked at Guam. He knew the island well. Michiko and he had honeymooned there.

The young purser climbed up the aisle and left the theatre.

"Robot meeting now, restricted nuclear meeting at eleven, robot meeting after lunch. Break. Another robot meeting. And after dinner another restricted nuclear meeting." Otzker turned to Seiko. "Alpha was up all night looking at pictures on the Internet. I don't really care, but Isato said it interfered with his concentration."

Seiko stood up. "I apologize. I'll take the robots with me today. Anyway, this morning's session is on learning. These machines are robots, not motorcycles, and you have to be aware of their learning curve."

"Are they completely blank when we receive them?" Daveed asked.

"No, no. The factory spends a month training each robot before it is shipped."

"What areas do they train?"

"Locomotion is perfect after the month. The robots can walk, run or jump. As well, arm and hand coordination is excellent. The visual, auditory, radio-detection and weapon systems are also fine-tuned."

"What training do we have to do when we receive the robots?"

"You need to put them into squads with human commanders and start training the robots for their combat roles. Who is enemy? Who is friend? How hard to fight?"

Seiko started talking about the software that coordinated groups of Alpha and Beta robots, of nested hierarchies and local swarms.

Haruto went back and lay down on the leather seat. He was starting to feel more claustrophobic in this dark area behind the display screens. Haruto tapped silently on his right forearm two times. He tapped on his left forearm twice. Then twenty times on his right. Then twenty times on his left. That should be enough.

But it didn't feel right. The anxiety continued building. Maybe another twenty-tap. He tapped twenty times on the left and then another twenty on the right.

Something still felt wrong. Increase the number. That would make everything okay.

Haruto tapped repeatedly on his right forearm, counting silently to himself. About a half hour later, he got to two thousand. Good. Then he started tapping over and over again on his left forearm. Another half hour later, he got to two thousand, not making any mistakes along the way.

It still didn't feel right though. Maybe use the five-thousand rule. That rule always worked, and then it was okay to stop, and mistakes were okay also.

Haruto started tapping, his left index finger silently hitting against the inner fleshy part of the right arm, one to two times each second. An hour later he reached five thousand. Then, with the right index finger, he started tapping against the left arm. At the count of nineteen hundred, he heard the scientists start walking up the theatre aisles. The sliver of light from the crack between the display screens went dark, and the entrance doors to the theatre slammed shut. He was hungry and thirsty. Run down and get something to eat from one of the snack refrigerators? No, it was wrong to stop the count in the middle. He kept on tapping.

At three thousand and twenty-two, Haruto heard the entrance doors to the theatre open, and light appeared in the crack between the display screens. He soon heard a few voices he recognized — Otzker, Tanaka, Ilana and Isato. Haruto kept tapping. Twenty minutes later, he was at four thousand nine hundred ninety-nine, and then finally, five thousand. Haruto exhaled and stretched out on the leather chair for a minute. Everything was all right now.

Haruto got up and went to the crack between the display screens. From the reflection in the chandelier's brass globe Haruto saw on the main display an underwater topographical map of the Mariana Islands and surrounding Pacific Ocean.

"When will we be at the Islands?" Tanaka asked.

"We'll be at Farallon at seventeen-hundred hours," Otzker said.

"Then straight down to Guam, and later the Challenger Deep for testing?" Tanaka asked.

Otzker shook his head. "Challenger Deep is too busy. During a scuba cruise last month, we even bumped into researchers from the University of Tokyo lowering probes."

"How deep is it?"

"It's almost eleven thousand meters. But there are lots of other locations in the Mariana Trench that are eight, nine and ten thousand meters deep — good enough for our purposes."

"How about the earthquakes?" Tanaka asked.

"We make use of Richter twos, threes and fours — about ten thousand a week, many of them near the Mariana Trench. We time our blast to correspond with a nearby quake, usually Richter three or four."

Isato put up his hand. "I do not understand. One-megaton explosion of bomb give Richter four to six magnitude, depending on where energy goes. How mask it with weaker earthquake of Richter three or four? And also, location of bomb and earthquake little bit different. Seismometers will pick up these different locations."

Otzker laughed. "Wait until we do our test. Our system is simple, and it works."

"Are there any active volcanoes right now?" Tanaka asked.

Otzker nodded. "Anatahan Island, about a hundred kilometers north of Saipan, has been vibrating and quaking for years, and as we talk now, steam and gas are belching out of it."

"Would you use it for a test? It would be beyond suspicion."

Otzker shook his head. "No good. There's only one thing that can shut down all the jet engines of an airplane in flight, and that's volcanic ash. So Anatahan is always being closely monitored. Also it's quite shallow around the island."

"How do we find the test site then?" Tanaka asked.

"I'll show you once we start making the Marianas."

Otzker, Tanaka, Ilana and Isato walked up the theatre aisle, closed the lights and went out.

CHAPTER033

H **aruto sat there** in the dark, his mind spinning. History was repeating itself. Like a century ago when the military came to power, the government was breaking the rules in the name of patriotism again. Secret nuclear bombs. Secret ways to test these bombs so nobody even knew what was going on. Break the rules, and the consequences would repeat, too. Last time it was Hiroshima and humiliation. What would it be this time?

These nuclear weapons were a disease, a hidden cancer. If they spread, they would end up being used again.

Perhaps all this was necessary. North Korea had the bomb. Japan was under threats. The fundamentalists had missiles and Israel was under threats. Yet if gangs threatened you, you didn't buy a machine gun in secret to kill them all. That led only to chaos. You called a policeman.

But there was no police force to contain North Korea or the fundamentalists.

But maybe there should be. He paid little attention to international politics, but if it were known that the alternative were secret, sophisticated weapons, perhaps the United States or the United Nations would finally take the necessary risks. They could make good rules and follow them.

He knew his rules. Stay quiet and undetected. Get off at Guam. Get back to Japan. He had the video recordings on his cellphone. He would expose what the government was doing. It was dangerous not to follow the rules.

Haruto was about to tap two times on the back of the display screen, but he realized he had just done a five-thousand tap. It wasn't necessary to tap again right now.

Haruto slithered out from the edge of the stage opening, and in the darkness of the theatre quietly made his way down the stage stairs. He walked along the aisle to the men's washroom and gratefully made use of the urinal. He walked up to a middle row in the theatre and opened the first snack refrigerator, taking out a bottled water, a plastic bottle of Coca-Cola and two Mars Bars.

Haruto went back up the stage stairs to his space behind the large display screens. He washed down the chocolate bars with the half-liter of Coke, keeping the bottled water for later. Soon he'd be out of here, on Guam, and then back home to Japan.

CHAPTER034

L ieutenant Colonel Okamura eyed the graying Arab canal pilot suspiciously. The Captain was having a cup of tea with him on the other side of the bridge. Yamada talked too much sometimes.

Okamura stared out the huge bridge windows. The greenery at the Suez entrance had faded, and there were only never-ending expanses of desert on either side of the canal. The ship stretched out so far in front of him, and the angle was such that he could barely see the water between the *Mikiyasu-ema* and the container ship in the convoy ahead of it. From this viewpoint, it seemed as though these other ships were floating on the desert sand.

Canvas covered the anti-missile systems on the small raised bridges in the front and rear of the ship. They looked like modest cranes covered for protection from the elements. It was a marginal deception. Some smaller container ships did have their own cranes to unload containers at ports without full facilities, but ships only half the size of *Mikiyasu-ema* had long abandoned self-unloading — they required ports with specialized cranes for loading and unloading their thousands of shipping containers.

With its load of fourteen thousand containers, the *Mikiyasu-ema* steamed northward at eight knots.

Pacific Ocean 20.53°N, 144.5°E, due west of Farallon de Pajaros Island,
Commonwealth of the Northern Mariana Islands, USA
Cruise Liner New Pacific Queen
June 13   5 PM Guam Time (07:00 Zulu)

CHAPTER035

Colonel Tanaka followed General Otzker's brisk steps along the main deck. The sea was blue and calm under the fading sky. Otzker walked past the massive spools of cable lying on the rear deck and went to the huge portside stern winch. He put down his briefcase and pushed a yellow button on a white column next to the winch. A motorized metal panel slid back, exposing a small instrument console.

"This is a high-resolution three-dimensional sonar. Much better than the usual navigation ones. We're too shallow now for a test, of course, but I'll show you an image."

Otzker pressed a few of the keys beneath the sonar display. In a moment, a false-color image of the nooks and crannies of the sea bottom flashed on the screen.

"What are those jagged lines?" Tanaka asked.

"Artifact from the engines. Normally we stop them for a bit and glide when we scan."

"Why do you care what the sea floor looks like?   It's a one-megaton bomb. What difference could the seafloor make?"

"The nuclear explosion heats the water, so steam bubbles form, which float up. If these bubbles break the surface there will be a spray dome, with radioactivity shooting into the atmosphere. A disaster for us."

"How do a few rocks on the seafloor help this?"

"It's not a few rocks. We try to lower the bomb down to seafloor that will better absorb the bomb's energy. Once the fireball is below a certain size, it means that the temperature of the steam bubbles will drop enough as they ascend the thousands of meters of cold ocean, and condense back into water. Thus, no release of radioactive gases."

"You need to explain this better to me."

"Yes, of course. You'll see the test unfold." Otzker pointed portside to the small speck of land the ship had already almost passed. "That's Farallon — we're in the Marianas now."

"These areas are close to Japan. Perhaps we can provide you with maps

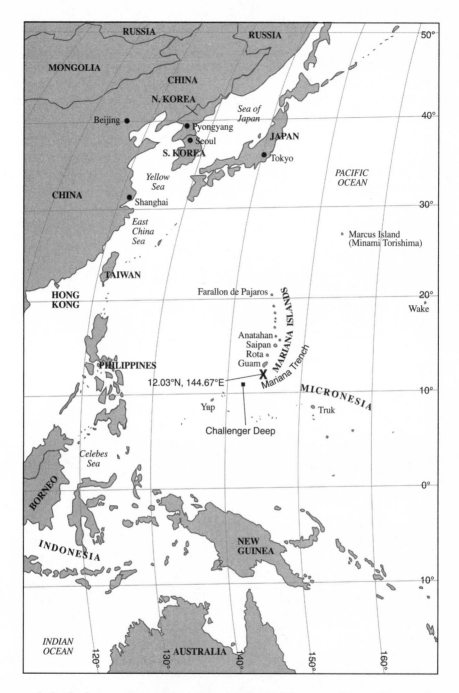

to help find test areas quicker. The Geological Survey of Japan has surely made bathymetric maps of the area."

"So have we." Otzker reached down into his briefcase and gave Tanaka a glossy reprint: *Tectonics, Vol. 23, April 2004, Bathymetry of Mariana trench-arc system by Z. Gvirtzman, Geological Survey of Israel, Jerusalem, Israel.*

Tanaka raised his eyebrows. "You published this? Your country is halfway around the world from here."

"Not my idea." Otzker frowned. "I suppose the idea was, hide it in plain sight. Let's get back to the meeting, and later tonight we'll look for some test sites."

CHAPTER036

Haruto saw Otzker and Tanaka enter the theatre's right entrance door. Seiko stopped talking to the group and turned to them. "I was covering body design. Would you like me to repeat my lecture?"

Otzker shook his head. "No, just summarize quickly and please continue."

Alpha was standing next to Seiko, perfectly still.

Seiko pointed to Alpha. "Why do we choose this design? To make the robot look like one of us? No. We use this design because it works well. Having the arms free is very useful. We could add extra arms and legs as we do in the Beta robot, but it reduces maneuverability. As well, the upright posture gives greater operational height for sensors and weapons."

Seiko pointed to Alpha's pear-shaped torso. "As I said, we did not copy human form, but designed with the robot's functionality in mind. You see that the abdomen is wider than the chest, which is the opposite of what you find in a muscular man. We did this to keep the center of gravity as low as possible. The robots use hydraulics — we don't need the same structure for upper body musculature that a human does."

Seiko took a step back and pointed at Beta.

"We could have made a more robust Alpha version for hauling supplies, but the Alpha design is for a sensing, versatile, fast-acting soldier. We designed Beta specifically for hauling." Seiko smiled. "This is why it looks like a giant monster cockroach."

Seiko pointed to Beta's spindly legs and then turned back to Alpha. "We want a nimble Alpha that can balance his full body weight and a heavy load on one leg at different angles. We use multiple hydraulic cylinders. Thus his legs seem much thicker than Beta's."

Ilana raised her hand. "Robots are not my area, but I have a question."

Seiko nodded.

"You say that you had no intention of making the robot look like a man, only designing for functionality. So… why give it a nose? Also, I see facial movements quite similar to emotions when Alpha is performing various tasks. How do you explain this?"

"That's a very good question. We have hundreds of robot researchers in

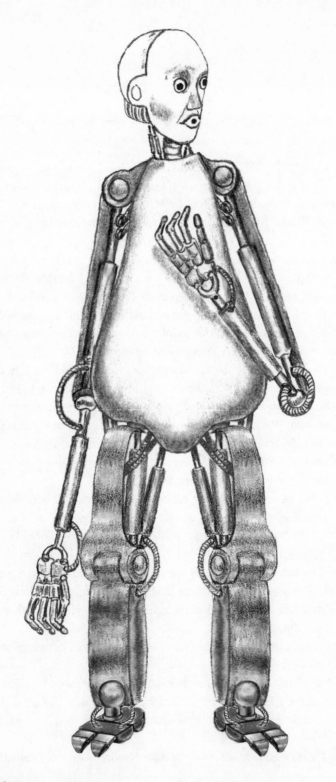

I, robot

Japan that have spent the last twenty years working specifically on making human-looking robot faces, and yes, we used their work in designing Alpha. Why did we do this?"

Seiko looked up at the ceiling in thought for a moment. "The human commander must always carry the handheld Robot Command Unit — 'RCU' — to coordinate his robot soldiers. But often it's quicker for the commander to simply talk to his robots. So we gave the robots voice recognition and voice synthesis. But humans need visual feedback. In early Alpha models, we had a non-human face studded with colored LEDs. When we then tried a human-acting face, there was a dramatic improvement in communication. Not only humans, but all primates know instinctively how to read facial expressions. Good engineering takes that into consideration."

"I agree with you," Ilana said. "When I look at someone, their eyes are very important to me. But in your design, the two eyes of Alpha are moving and rotating in different directions. This ruins the effect you're trying to accomplish."

Seiko shook his head. "No, no. Please remember, we crammed many different cameras into the two eyes. Yes, in operation, you will see the eyeball cameras rotating in different, non-synchronized directions. But when the robot is with his commander, eye motions are very humanlike."

Seiko pulled a silver pen out of his shirt pocket and turned to Alpha.

"Alpha, please give this pen to Ilana."

Alpha took the pen between his rubber-coated thumb and three

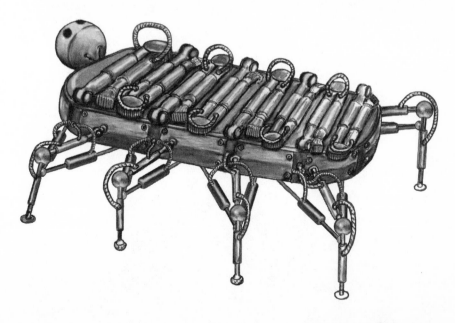

metallic fingers. Two quick but gentle strides over to the leather seat where Ilana sat. Alpha bent his knees and his back, and stretched his left arm toward Ilana.

Alpha's eyes focused together on the pen, and then looked directly at Ilana's eyes. His camouflaged silicone face crumpled slightly and his lips moved up in an arc forming a modest but pleasant smile. A small crease became apparent on his forehead. "This pen is for you," he said, still looking into her eyes.

Ilana's face became red, and she started laughing. She took the pen. "Thank you, Alpha."

The robot increased its smile, then stood up, turned around, and walked back to Seiko.

"He's very charming," Ilana said. "More so than my last boyfriend."

The others laughed.

Otzker stood up. "Let's break for supper, then continue with our schedule."

The scientists, officers and robots walked up the stairs and left the theatre. Haruto lay down on a leather chair behind the stage. He felt all right. He did a five-thousand tap before, so it wasn't necessary to do another tap now. In fact, it was dangerous to. Sometimes those small two-taps or twenty-taps didn't work out and went on for hours. No, he felt okay. He didn't like the enclosed feeling, but it was all right for a bit longer. Soon he'd be out of here and in Guam, then back to Japan.

Haruto closed his eyes, let his body sink into the soft chair, and dozed off.

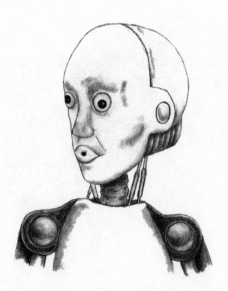

CHAPTER037

Haruto awoke in the darkness behind the stage. No engine vibration. The ship was in port.

There was no need to tap or count. He had already made the rule — wait until the ship pulled into port and stopped moving, and then make it to the deck and off the ship. No need to count. Just follow the rule. He would be all right.

Haruto looked through the crack between the display screens. Except for the modest glow of the indicator lamps from the computer racks, the theatre was dark. After two days here, sleeping because he had nothing else to do, his internal clock had no sense of whether it was night or morning.

Haruto slid out onto the stage. Down the stage stairs. All quiet. Even Alpha was gone, no longer sitting at the display, madly scrolling through images.

Up the aisle to the top of the theatre. He cracked the left entrance door open. A sliver of light shone in. The theatre lobby was empty but brightly lit.

Haruto did not hesitate. On the tips of his toes, he dashed out across the lobby and pressed his body into the shadows of the adjoining hallway. Empty. Running silently on the hallway carpet he made his way to the stair exit, opened the door and listened. No sound on any stairs, either above or below.

The stairs were not intended for guest use. Corrugated steel with a layer of fading red paint. Haruto put his foot slowly on the metal landing, toes first, and then gently lowered his heel to attenuate any of the resonant vibrations. Another step, and he was inside the stairwell. He let the door close slowly, guiding the latch into place.

Step by step, Haruto lowered himself down without a sound. It took about a minute to get down to the next floor level — the main deck. Haruto opened the stairwell door a few centimeters. It was dark outside, save for the ship's lights.

Haruto stepped out onto the deck and let the door close slowly, again guiding the latch silently into position. He felt a small breeze from the ocean on the back of his neck and smelled the sweet salty scent of the Pacific. In Tokyo you only smelled the pollution.

Haruto took a step forward and looked up at the night sky. Away from the bright lights of Tokyo, so many stars of the Milky Way were out. But where was the wharf? In fact, where was Guam? To the right Haruto saw only dark ocean and the star-filled night sky. To the left —

A gun barrel flew in front of Haruto's face. "Who are you?" a white t-shirted crewmember shouted. The crewmember held his M16 tightly, arm muscles straining.

"I'm one of the robot scientists."

"You're off limits. Why? Who are you?"

"I am Isato," Haruto said.

"What is your full name?" the crewmember asked in his oddly tinged English as he reached down with one hand for the walkie-talkie clipped to his belt.

Haruto's front leg shot up in a gentle *mawashi geri* kick and struck the left side of the crewmember's face. Haruto caught the man as he fell, and dragged him to the pile of lounge chair cushions piled up against the wall near the stairwell door.

Haruto pulled the belt off the unconscious man and bound his ankles tightly together against a metal post extending from the deck to the open ceiling above. He pulled the man's shoes and socks off, took one sock, twisted it into a rope, and then bound the crewmember's hands behind his back. He took the other sock, and turned it into a gag.

Haruto slapped the man's cheeks. The crewmember's eyes opened. Good. Haruto then covered the bound crewmember with about seventy lounge chair cushions. Probably the ship was anchored now at Apra Harbor. Hopefully the crewmember would stay hidden long enough to allow time for escape off the ship into Guam.

Toward the bow he only saw the dark ocean and stars. He turned his head aft and saw some small lights, probably the capital city Hagatna, which was about eight kilometers north of the harbor.

Haruto clung to walls and shadows and silently inched his way toward the stern. As he got closer to the rear of the ship he could make out the lights on the huge stern winches. Past the winches was open sea and a dark night sky full of stars.

Haruto looked to the left and right. No crewmembers. On tiptoes and crouching, Haruto stepped sideways across the well-lit three meters of boardwalk on the deck to the side of the ship and stuck his head over the railing. From the light spilling out onto the water he could see the wake of the ship cutting through the smooth ocean water. The ship's engines were off, but the ship was still at sea!

Haruto stepped away from the brightly lit side of the ship, tiptoeing across the deck's boardwalk until he was back in the shadows. He continued inching toward the stern. As he moved aft, he could make out the large cable spools the wharfmaster had mentioned, sitting on their sides at the edge of the stern. Rising above the cable spools were two massive deck-mounted hydraulic arms and winches. Near one of the winch arms he could make out two figures. He continued to move aft. There were General Otzker and Colonel Tanaka standing in front of an illuminated display.

Otzker pointed to the colored lines on the screen. "This little nook here would be a good place, for example."

"What depth are we at?" Tanaka asked.

Otzker pointed to the large numbers in the top left corner of the screen. "Thirty-eight hundred and eleven meters. Too shallow still."

Tanaka nodded.

Otzker picked up the telephone beneath the display. "Engine re-start."

Haruto felt the ship vibrate as its engines started up, and as the ship gained speed he could feel a bit more motion as it cut over the ocean's waves.

"Colonel, tomorrow morning we'll be over deeper water, and we can start looking more seriously for test sites."

Otzker and Tanaka walked down the deck, and turned into the bulkhead.

Haruto turned to look at the pile of lounge cushions. The crewmember was okay and would stay hidden. No tourists on this voyage to take away the cushions.

Haruto could not go back to the dark and enclosed space behind the display screens in the theatre. He'd wait here until the ship reached Guam. The deck was large and there were enough places to hide. The warm tropical breeze and sweet smell of the ocean felt good. He felt calm. No need to even count or tap.

Haruto inched over to a nearby tiki bar on the deck. Probably used to serve drinks to passengers on the paying cruises. With a lounge cushion to rest his head on, Haruto lay down on the deck underneath a reclining lounge chair already stuffed into the bar. Through the cracks in the tiki bar's thatched roof, Haruto looked up at the stars.

CHAPTER038

O tzker turned off the air conditioning and cracked open the balcony door of his cabin. He liked the sea air, especially at night. He was lying in boxer shorts on top of the sheets, slowly drifting off to sleep, when suddenly two sharp knocks.

Otzker grumbled, pulled on his pants, and opened the cabin door. "What is it, Lieutenant?"

"Sorry to disturb you, Sir. One of the enlisted — Ari Adler — is missing. Some of his buddies saw him sitting on the railing earlier. Maybe he fell over."

Otzker frowned. "Lieutenant, I told you this mission was a special one. I wanted everything perfect."

"I'm sorry, Sir. I double- and triple-trained everyone. Don't know why this happened."

"Did you search the ship yet?"

"No, General. Because of the guests… not without your permission."

"Do a full ship search. Not too noisy or too invasive, but check every cabin. Just say it's a security drill."

Otzker lay down and tried without success to fall back asleep. He put on his shoes and shirt, and headed to the main deck.

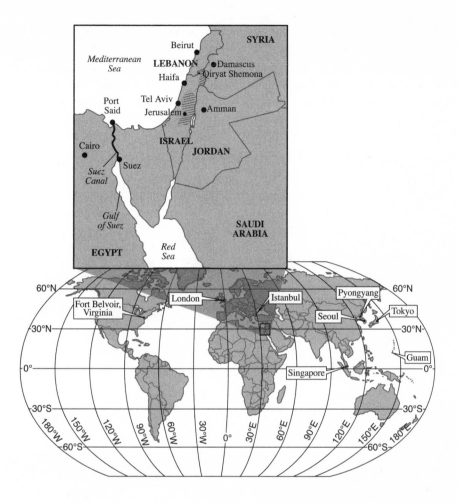

Port Said East International Transshipment Area
Container Ship Mikiyasu-ema
June 13   6PM Local Time (15:00 Zulu)

CHAPTER039

The overhead gantry crane gently lowered a drab brown container to the empty spot on the bow deck in between two high towers of containers. The crane slid back on its tracks, lifted a second waiting container off the quay and deposited it on top of the first. Mooring ropes were unhooked, bow and stern thrusters fired up, and the colossal ship moved away from the dock, into the Mediterranean.

Lieutenant Colonel Okamura looked through the aft bridge window, past the many thousands of containers stacked on the rear deck, and watched Port Said fade away.

"For the last five years, the *Mikiyasu-ema* has sailed to Felixstowe and Hamburg," Captain Yamada said. "If you like, we can make way westward a bit, say to Gioia Tauro, and unload the fifty containers of excavator and dump truck parts. Then a few days later, we can come back to the destination."

"How long to make Haifa from here?"

"About eight hours," Yamada said.

"Continue due north for another half hour, and then we'll decide if we're going east or west."

"For five *years* we have been turning west after the canal. I don't think there's anything to decide. We should go westward, at least to Gioia."

Okamura glared at Yamada. "I'm in charge here!" He reached down on the floor for his satellite radio pack, and started clicking the buttons on its main keypad.

CHAPTER040

"**C**olonel Tanaka!**"** Four knocks followed on the cabin door. Tanaka opened the door, his gold-rimmed glasses balancing crookedly on his nose. "Yes, what is it?"

"There's an encrypted call for you." The white t-shirted crewmember gave Tanaka a handheld satellite telephone receiver.

Tanaka went back into his cabin, entered his ID and password into the telephone's keypad. "Tanaka here!"

"Sir, we have cleared the canal and are in the Mediterranean now," Okamura said. "Do you want us to detour to Italy for a six-day delay to destination, or do you want us to make destination in eight hours?"

Tanaka looked up at the ceiling for a few seconds. "How long to unload at the destination?"

"Unsure, Sir. They don't normally handle this size. One to two weeks, depending on the number of cranes."

"Proceed then to destination. Let's get the cargo unloaded and pick up the weapons as soon as we can."

Tanaka hit the red keypad button, opened the cabin door, and gave the satellite phone back to the waiting crewmember. "Is General Otzker asleep?"

The crewmember shook his head. "No, Sir, he's on the main deck if you want to see him. Follow me."

Tanaka followed the crewmember to the elevator and they both descended to the main deck. Tanaka saw a dozen white t-shirted crewmembers walking around with bright flashlight beams. Otzker was speaking to a small group of them near the bow. Tanaka waved his right hand and caught Otzker's attention.

"Sorry if you were woken up," Otzker said.

Tanaka shook his head. "No, I received a satellite call from the container ship. What's going on?"

"Security drill."

"Our agreement was that we set up the nuclear test, and *then* the container ship unloads while we wait for the right moment to set off the bomb."

Otzker nodded.

"The container ship will be in Haifa in eight hours and will start unloading. Even though you haven't explained how you're going to pull off the test, I do this because I trust you. We must test as soon as possible!"

"We'll start tomorrow morning."

Otzker and Tanaka walked off the deck into the bulkhead. Haruto was lying on his back on the deck underneath the lounge chair and cushion in the tiki bar. He heard the noise around him and tried to stay as still as possible.

CHAPTER04 1

The four hundred meters of the *Mikiyasu-ema* filled most of the length of the Carmel container terminal. Okamura looked out the bridge windows and saw hundreds of flatbed container trucks waiting in a line that snaked out of the port onto the adjoining roadway. Dozens of flood lamps lit up the terminal.

"Do they really need all those lights? Shut them off!"

"They drop the containers onto the trucks by sight," Yamada said. "It might be dangerous to do this in the dark."

Okamura nodded and watched the red and white-striped gantry crane over the front deck lift up two twenty-foot containers at the same time. The crane pulled the containers toward the quay, and lowered them onto the waiting truck below. The truck gave a puff of diesel exhaust and drove off while the crane's spreaders rolled back high over the ship, ready to lift up another pair of containers.

"How many containers have we unloaded so far?"

Yamada smiled. "Twenty-four. Only fourteen thousand and thirty-seven to go."

CHAPTER042

General Otzker stood in the morning sun on the main deck. "What did you find?"

"Nothing. We checked from the engine room to the upper cabins. The ship was all in order, but no sign of Ari."

"Let's keep this under wraps, Lieutenant."

The crewmember nodded. "General, Pati Point is active today. Do you want us to swing out to the one-forty-eight line and then go southward?"

Otzker looked at his watch.

"No, it's nine hundred hours already. I want setup complete by sunset. Proceed down the one-forty-five line. Commence test setup."

"Yes, Sir."

A moment later the ship's PA system came alive. "Welcome to another day of sunshine and scuba on the New Pacific Queen."

About two dozen crewmembers were soon on the main deck. Half of them took off their white t-shirts and replaced them with colorful Hawaiian shirts and sunhats. They started carrying out scuba tanks, wet suits, flippers and masks, and lay down the gear over a large swath of deck. Some themselves lay down on the chaise lounges arranged around the tiny pool in the rear of the ship.

Other crewmembers ran to their stations on deck. Four raised a massive, rainbow-colored canvas shade over the stern, completely covering the two large hydraulic winches and adjoining deck space. A few other men with hand lifts started moving the large cable spools into place.

Haruto heard footsteps coming. Through the cracks in the thatched tiki bar he saw a crewmember approach with large glass pitchers of red and purple fruit juice. The crewmember put the beverages down on the bar and started pulling out the lounge chair inside. Haruto tensed his muscles and held his breath.

"I already set up a bar," came a yell from the other side of the deck.

The man let go of the lounge chair covering Haruto, picked up the juice pitchers and walked away. Haruto's heart still kept racing. Haruto tapped twice on the left, and then twice on the right. He sat up in the tiki bar and looked out through the thatch at the beehive of activity around him on the rear deck.

Haruto jerked when the fast-beat music came out of the loudspeakers all around the ship. He didn't know the name of it, but he recognized it. American music from the sixties. The *bosozuku* bikers in Tokyo — rule breakers he never had much sympathy for — used to play it.

A few meters in front of the tiki bar Otzker was standing next to the sonar screen. The fast-beat rock and roll music changed to a reggae beat. Colonel Tanaka had a somber look on his face. "General, they're unloading in Haifa right now. I have to insist we start the test process immediately!"

"Yes, of course," Otzker said.

"No more excuses. I need to know exactly how you are going to accomplish this test, and we need to do it now."

As Otzker nodded, four F-16's flying no more than ten meters above the ship burst out of the sky, followed a split second later by two F-35's chasing them. Sonic boom after sonic boom thundered around the ship. Two seconds later a dozen thousand-pound bombs detonated, albeit a few kilometers away off starboard.

Tanaka reflexively dove onto the deck, protecting his head with his arms and shielding his ears.

CHAPTER043

A few moments later tranquility had returned to the ship. The sky was clear and painted a cerulean blue, the sun shone peacefully, and the sea was calm without the slightest wave, save for the wake of the *New Pacific Queen*. Bob Marley's reggae came out of the ship's loudspeakers.

Otzker extended his hand to Tanaka, but Tanaka ignored the gesture and got up by himself.

"What's going on here, General? American jets buzzing us and then dropping bombs? Your crew sitting around the pool drinking? What kind of test is this?"

Otzker held his hands up. "Colonel, everything is under control."

Tanaka raised his eyebrows.

"Colonel, there's a U.S. Air Force detonation range just to starboard. After we spoke last night about getting the test started this morning, I didn't want to waste hours, possibly the whole day, swinging the ship eastward to bypass the shore of Guam."

Tanaka nodded.

"And the music you hear, the crewmembers in Hawaiian shirts, the scuba gear out on the deck, and everything else you see here, is indeed the start of the device testing process. We're a cruise ship off the coast of Guam. We have to look the part."

"Fine, you look the part. What about the nuclear device testing? What about our deal? Your robots are being offloaded as we speak."

"The testing process has begun. By sunset the nuclear device will be in place for its detonation. Stick with me and I'll explain step by step what we're doing."

"What're we doing now?"

"The ship is traveling down the one-hundred-forty-five-east longitude. We'll be at the deeps of the Mariana Trench in three-and-a-half hours."

"I still don't see how you're going to pull this off. There's a gas detection station on Guam. Nuclear bombs give off radioactive xenon. That's what confirmed the North Korean underground tests, not that they cared."

"Relax, Tanaka."

"Don't patronize me, Otzker. Your bomb is going to release radioactive xenon. Xenon is an inert gas. It will escape and be detected. Plus the noise of your bomb in the water will be horrendous, plus seismometers all over the world will be clicking away."

"Yes, some of the plutonium in our bomb becomes radioactive xenon. But xenon dissolves in water, and there's lots of water at the bottom of the ocean."

Tanaka grimaced. "No delays. I want the test detonation as per our agreement. We're unloading your robots right now. Do you understand that?"

"Okay… okay." The General held his hands up again. "Can I get you a cup of tea or coffee?"

Tanaka shook his head.

CHAPTER044

A **few meters in front** of the tiki bar, Tanaka stood rigidly at Otzker's side. The two of them were staring at the colored lines on the sonar screen. A large *8347* displayed on the top left corner.

"It's over eight thousand meters deep here," Tanaka said. "I see lots of nooks and crannies down there on the sonar. Let's do the test."

"This isn't the right spot, but we're in the right area. A few more minutes, Colonel." Otzker picked up the telephone beneath the sonar screen. "Restart. Due west."

Haruto felt the ship vibrate as its engines came online. Otzker and Tanaka kept staring at the sonar screen.

"Let's take another look," Tanaka said.

Otzker nodded and picked up the phone. "Cut engines please." He pointed to the right side of the screen. "We're at 144.97 east now. Depth is 8985 meters."

"That's deeper. Let's do the test now," Tanaka said.

"You're like a child before Hanukkah."

"I'm like a military man trying to accomplish a tricky mission. I'm sorry if my impatience offends you, but the fate of my country — of both our countries — depends on this."

Otzker shook his head. "Give it a few more minutes. I don't want a steam bubble from the explosion breaking the surface. I don't want a radioactive slurry at the surface. The right spot makes a difference. An extra thousand meters of ascent and cooling for the steam bubbles can also make a difference. Let's try to find a good nine-thousand-meter, or better yet a ten-thousand-meter depth."

"How long will that take?"

For the next hour, Haruto felt the ship's engine stop and re-start. Tanaka's face became red and sour and serious, as he and Otzker kept arguing about the sonar images. The fast-beat rock and roll music blared in the background.

Finally, Otzker picked up the telephone receiver. "Reverse engines. Hold 144.67 East, 12.03 North. Mark and set for device insertion."

The deck vibrated as the engines came online. The ship slowed and turned in a tight circle back to the spot Otzker had just marked.

Otzker pointed at the colored lines on the sonar display screen. *10095* displayed in large numerals in the top left corner. "This is a good spot. Over ten thousand meters deep — ten kilometers of water on top of the nuclear device. Seafloor formations look solid."

"Do we lower the test device now?" Tanaka asked.

Otzker shook his head, then picked up the telephone again. "Take us ten kilometers away westerly direction, and begin spooling."

"What's the delay now?"

Otzker pointed to the huge cable spools dangling over the stern of the ship on the right and left hydraulic arms. "Those two spools contain five kilometers of explosive cable and fifteen kilometers of electrical cable to reach the surface. We need space to deploy it all."

"Why do you need explosive cable?" Tanaka asked.

"Finally, the answer you've been looking for. We wait until there is a nearby natural earthquake before we detonate our explosions. We create extra vibrations as needed. The noise and seismic spikes from the nuclear detonation are interpreted as plate slippage or some other geological result. Thousands of geological events occur every day — this just becomes one more of them."

Tanaka's face remained serious. "The seismogram of a nuclear explosion is very different from an earthquake. Even if you add in some extra explosions, an expert will see through this right away."

Otzker shook his head. "Not true. Just as we can precisely detonate the conventional charges in our nuclear bomb, we can detonate some twenty thousand charges in the kilometers of explosive cable at microsecond precision. I could have the explosive cable play Bach, if I wanted to. We can mimic a seismogram — not perfectly, but good enough."

"It doesn't make sense, Otzker. The nuclear explosion is going to give a huge vibration peak right at the start. However, an earthquake first has rumblings. Its high peaks may not occur until a thousand seconds *later*, and then you have hundreds of seconds more of rumblings. You can't hide the huge nuclear peak."

"Don't have to. The nearby natural earthquake starts. Vibration sensors in the explosive cable detect it and alert the computer, which starts our detonation sequence. The natural earthquake continues with its low rumblings that are starting to build up in intensity. At about, for example, a thousand seconds the earthquake peaks have reached maximum. We detonate our nuclear bomb right then. The earthquake peak instead of being Richter three or four becomes Richter six because of the superimposed nuclear explosion peak. The explosive cable then goes into action, synchronizing the sounds and vibrations of the fading nuclear explosion with the fading earthquake."

Tanaka's frown reversed a bit. "What about the bubble pulse?"

"Yes, once all the noise dies down, the vibrations from the steam bubbles rising up can be detected. However, the explosive cable will actively mask them."

Tanaka started to smile. "Okay, but what if there are suspicions about the noise or seismology, and they come back here for a radioactive isotope sample?"

"Then they find nothing."

"Nothing?"

"Yes, nothing. Some vibration does make it up through the ten kilometers of water, but not much. You have a few waves, but they quickly die down. The surface water is clean enough to drink... well, if you desalinate it."

"What about a month later, or a year later?"

"As long as the steam bubbles condense deep enough, you will have nothing. In fact, since the Pacific Plate below us is slowly pushing under the Philippine Plate, if you wait long enough, any sediment contaminated by the bomb will actually be recycled into the interior of the Earth."

A huge cargo elevator rose out of the deck near the rear winches. Two crewmembers pushed a cart holding two large metal spheres.

"What are those?" Tanaka asked.

Otzker pointed to the right winch. "As we lower the explosive cable into the ocean, we clip on *more* explosives. Neutral buoyancy so the metal cable doesn't have to support their weight."

The engines slowed down, reversed for a few seconds and stopped. The roar of the bow and stern thrusters began, holding the ship firmly in place in the middle of the Pacific Ocean.

"Scuba party to lower deck gangway," the loudspeakers announced. "Cable drop. Cable drop."

Rock and roll music blared. The lyrics were simple, the beat fast and addictive. A dozen crewmembers flooded the area astern, now covered by a large rainbow canopy. They took off their white t-shirts and keeping with the beat of the music went to their stations.

A cast-iron anchor glittered in the sun, weighing down the loose end of explosive cable coming from the huge spool on the right hydraulic arm dangling over the stern. A crewmember on the winch pushed a lever forward. The winch started lowering the anchor and the explosive cable into the ocean.

Tanaka pointed to the right stern elevator opening onto the deck. A shiny metal cone, about a meter long, lying in a bed of blue packing foam. "Is that the warhead?"

"Yes, all one megaton of it. While we're lowering the explosive cable other crew will activate the warhead and package it in a thick steel sphere for its journey to the bottom."

A new song blared out of the loudspeakers — "Shake…shake…shake" with an even faster beat. Some crewmembers pulled explosive spheres off the left stern elevator, while other men clipped the spheres onto the explosive cable.

Haruto lay behind the tiki bar. There it was in front of him. A nuclear warhead. The genie was out of the bottle.

CHAPTER 045

"**My divorce was** very hard also," Isato said. "It took years time before my daughter speak to me again."

"Look at the ocean — it's so beautiful." A gentle warm breeze blew through Ilana's black hair. She wore a light green dress, a string of white pearls and carried a small canvas purse, her standard cruise ship camouflage outfit. "Do you know how to swim?"

"Yes, I am very good swimmer."

Ilana grabbed Isato's hand, and pulled him toward the bulkhead. "Come, let's go scuba diving. The ship is here for an hour, at least."

"I don't know how to."

"I used to be a scuba instructor."

"Really?"

"Yes." Ilana smiled. "What's the ideal gas law?"

"P times V equals n times R times T," Isato said.

"Good. You just completed the classroom portion of the course. When you're underwater, you always have to keep breathing in and out, and ascend slowly. Do you know why?"

"Of course. When you ascend, less pressure on air in lungs, therefore air in lung expands. If mouth is open — breathing in and out — then excess pressure goes out of lung."

"Wow. I never had a student like you before."

Isato smiled. "Okay. I will try it out."

"Go change and I'll meet you at the second floor in five minutes."

Ilana ran down the wooden promenade on the main deck toward the stern.

"Good morning General… Colonel." Ilana breezed past them toward the crewmember crouched beside the stainless steel cart holding the warhead. "Hi Benni."

"Hi, Ilana. Standard test?"

"Yes, everything is routine. Nothing new on this one."

"Do you want me to handle it?"

"That would be great, Benni. I'll be back in about forty minutes. Okay?"

"I need the Activation Chip."

Ilana paused, reached into her purse and took out a small clear-plastic case holding a black ceramic chip about a centimeter long. "Please don't blow up the ship."

"If I do, what's anybody going to do about it then?"

"Very funny."

Ilana dashed off to her cabin, put on a one-piece turquoise bathing suit and headed to the lower deck. Isato, in a pair of baggy green bathing trunks, was waiting there. The fortyish small-frame Japanese man smiled when he saw Ilana.

"Hi, Isato. This way." Ilana pointed to the hatch door at the end of the corridor. They stepped out of the ship onto the hanging stairs, which led down to a floating dock packed with fins, masks and tanks. About ten meters from the dock, some of the crewmembers in scuba gear, their heads out of the water, masks raised to their foreheads, chatted. The sky was clear, the sun was shining and classic rock and roll music drifted down from the main deck.

"Near the surface the water is quite warm — we don't need wetsuits for a short dive." Ilana lifted two aluminum air tanks and stood them up. "Isato, can you please give me two of the regulators — they're the pieces we breathe through."

Ilana screwed on the regulators and attached the tanks to two buoyancy control vests. "Okay, sit here at the edge of the dock."

Ilana helped Isato swing his arms through the red buoyancy control vest with the aluminum tank attached. Ilana gave Isato a weight belt holding four small rectangular weights, a pair of large yellow and red flippers, and a yellow mask. "Put them on and we'll start our scuba lesson."

Ilana swung her arms through her own vest, buckled on her weight belt, put her feet in fins and lowered a mask over her eyes. She pointed at a button valve coming off Isato's buoyancy control vest. "Press the red button."

Isato pushed the red button and his vest filled with compressed air from the air tank.

"Okay, into the water now."

Ilana splashed into the ocean in front of Isato, her buoyancy vest keeping her upper chest and head out of the water. She waved at Isato to come in.

Isato pushed off the deck. He went underwater, but a half-second later his upper chest and head bobbed up due to the air in his buoyancy vest.

"Okay, we're going to practice breathing through the regulator. Put it in your mouth, breath in and out, and kick a bit. Go ahead, I'll follow you."

Isato started kicking and propelling himself at a good pace, despite the awkward angle his buoyant torso made with the water. He then stopped and took the regulator out of his mouth. "That was easy, Ilana. Can we go underwater some now?"

"Yes, but let's adjust your buoyancy carefully. In a swimming pool it's only three meters to the bottom. Here it's about ten thousand." She pointed again to the button valve protruding from Isato's vest. "The red button puts air into the vest. The black button lets air out. Press the black button to reduce your buoyancy."

Isato pressed the black button. A bit of air came out of the vest but he still bobbed above the water. He pressed the black button again, and again, and again, and finally started sinking beneath the surface. Ilana pressed the black button on her own buoyancy vest and went under to be next to Isato. The scuba-clad crewmembers were swimming underwater a few meters away. Isato turned toward Ilana, pointed at the crewmembers and started kicking toward them.

They were about three meters under the surface, and the water was still quite warm. Ilana saw a constant stream of bubbles coming out of Isato's regulator. Good. As long as he breathed in and out, and didn't panic and hold his breath like some beginners, scuba diving was safe and fun.

Ilana and Isato soon reached the crewmembers. Ilana pointed up and started kicking toward the surface.

Isato felt for the button valve coming from his buoyancy control vest, and pressed the valve to inflate his vest and rise to the surface. But instead of rising toward the light at the surface of the water, Isato started sinking down. He pushed harder on the valve.

He started to sink faster and faster away from the light, toward the blackness below.

Ilana came to the surface and pulled the mask onto her forehead. She turned to the left and right to look for Isato but didn't see him. Ilana quickly pulled her mask down and put her face into the water. There were the bubbles coming from Isato's regulator, as he sank into the darkness of the depth below them. Ilana hit the black button on her buoyancy control vest and started kicking hard to propel herself down toward Isato.

Isato started kicking his fins awkwardly and doing a breaststroke with his arms. It slowed his descent, but he was still falling away from the light. He looked for the actual red button on his buoyancy vest but couldn't see it. The mask was fogging up, so he tried to pull it off to see better. As he yanked his mask off, he knocked the regulator out of his mouth. The salt water stung his eyes.

Where was the regulator? His arms flailed, grasping for the regulator. Isato started to sink again at a faster rate. It became colder and darker. A fear that he had never known before exploded inside him.

In panic, Isato inhaled.

Ilana saw Isato's regulator pull free. She kicked harder. She turned her head to the side for a moment and waved at the underwater crewmembers to follow.

After a half-dozen painful kicks, Ilana reached Isato. Her depth gauge read 22 meters.

A second later, two of the crewmembers also reached Isato. One of them grabbed his shoulder while Ilana jammed the regulator back into his mouth. The other crewmember pressed a few times on the red button protruding from Isato's buoyancy control vest, inflating it until Isato slowly started to rise.

No bubbles were coming out of Isato's regulator. His head was wobbling about unresponsively. What to do? If they ascended too quickly now, the air in Isato's lungs could expand and kill him. If they ascended too slowly, Isato would suffocate. Ilana stayed there frozen, holding Isato, until one of the crewmembers gave two hard kicks and they all started rising.

They ascended to 17 meters depth. Ilana took the regulator out of Isato's mouth, put her lips on his mouth and breathed in and out. The air entered and exited easily — his glottis was relaxed. The air could expand from his lungs as they surfaced.

The three divers kicked hard, took an angled route toward the ship, and soon they were in front of the portable dock.

One of the crewmembers, a tall young man, hopped onto the dock and pulled Isato out of the water, while the other crewmember pulled off Isato's vest and tank.

Ilana was up on the deck, and with the heavy scuba tank still on her back, gave Isato four mouth-to-mouth breaths.

"Get the water out of the lungs first." The tall crewmember turned Isato's head to the right side, and pushed down on his abdomen. Water poured out of the Japanese man's mouth.

The other crewmember ran into the ship and picked up the phone by the door. "Medical emergency, portable dock."

Isato was still not breathing. Ilana felt for a pulse — there was a strong one. She lifted Isato's chin and pinched his nostrils, put her lips on Isato's and exhaled into him. She lifted her head and breathed in, and then exhaled again into Isato.

Nothing, he wasn't breathing.

Ilana started crying. Get yourself together. Okay. Keep going. Again, she took a deep breath in and exhaled into Isato.

A moment passed — it felt like an eternity — nothing. She was about to cry out in anguish, and then, Isato coughed! Ilana put her ear to his mouth, and heard him faintly breathe on his own.

He'd made it. She hadn't killed him.

CHAPTER046

Otzker was angrier than he had been in years. Ilana stood in front of him on the promenade, her bathing suit still dripping water.

"What in the world were you thinking, Ilana? The man never took any scuba course. Why'd you let him go diving over the deepest part of the ocean?"

"I was a scuba instructor. I never imagined… this would happen."

"Scuba instructor? When? Twenty years ago in Eilat?"

"I'm sorry," Ilana said, eyes on the deck.

Otzker calmed down a bit. They were all under strain, and it was an accident, after all. "What's with you and him? He's *Japanese*. You don't even speak the same language."

"He's nice."

"He's a skinny little nerd with huge glasses plastered on his face."

Ilana started crying.

"Stop it," Otzker said in a lower tone this time.

"You go home…" Ilana said between sobs. "You have your wife, your family. I have nobody. Just a little apartment in Dimona surrounded by the desert."

"What about your son?"

"He's in Tel-Aviv with his father." Ilana started crying again.

"Why don't you spend some time up at Rafael? Haifa is nice this time of year."

Ilana nodded.

"I want you to change and come to the winches. We're ready to move. You should be there."

"I don't think… Benni's done the test a million times."

"Benni's an electronics technician. You're a nuclear physicist. You *should* have been there the whole time. Change and report to the winches. That's an order."

CHAPTER047

Haruto remained under the thatched tiki bar, leaning against the base of the lounge chair there. Ilana, in jeans and a red t-shirt, entered under the aft rainbow canopy. She seemed a bit more composed now. She should have followed the rules instead of swimming around in the ocean. The General had let her off easy.

"How is Isato?" Tanaka asked.

"We have him in sickbay," Otzker said, before Ilana could answer. "Doctor says he has a metabolic acidosis and breathing problems. We're doing our best to treat him."

"I will see him later then," Tanaka said. "Ilana, what is the status of my test?"

"Above the seafloor are five thousand meters of explosive cable attached to electrical cable from this spool." Ilana pointed to the huge reel of cable on the left winch. "The ship will soon head off due east, and we'll spool out more electrical cable. As we do so, the five thousand meters of explosive cable will lie down on the seabed."

Otzker looked at the two-meter black sphere dangling off the right winch. Fresh weld lines crossed its surface. A huge drill screw came off its bottom. "Nuclear warhead status?"

"The warhead's been activated." Ilana turned to Tanaka. "This black sphere holding the warhead has steel walls fifteen centimeters thick — they can resist double the pressure of water ten thousand meters down. Hydraulic fluid powers the drill — incompressible also at that depth. Electric hydraulic pump within the sphere. Watertight hydraulic lines leave the sphere to the hydraulic drill motor."

Tanaka finally smiled. "So, you are testing underground underwater. Very good... Although with that tiny drill rig you won't be able to go deep enough."

"Don't have to. Two hundred meters is enough."

Tanaka stared at Ilana. "The megaton explosion will tear through your shallow hole like tissue paper."

"Not true. A much weakened fireball comes out of the seafloor, and the steam bubbles that form don't make it to the surface."

"Really?"

Otzker faced the Colonel. "Tanaka, we've been doing this for twenty years!"

"All right..." Tanaka frowned again. "We have an agreement, which *we're* honoring. We're offloading your robots right now. I expect nuclear devices that work and a test that no one detects, although I'm still not sure that such a large explosion can be hidden as you say. It seems the whole world will hear it and feel it."

Ilana pointed to the massive white sphere, about four meters in diameter, lying on the deck behind the right winch. "Lightweight and filled with incompressible, buoyant petrol. It offsets the weight of the heavy warhead sphere."

Ilana then pointed to the loose cable ends near the warhead sphere. "These are the force transducers. The split second before the explosion incinerates the cable, we get a pressure measurement that implies how forceful the explosion was."

"According to our agreement, you're supposed to use the Japanese transducers, so we know the explosion took place. Did you forget?" Tanaka pulled two small clear plastic cases out of his pocket. Each held a bright orange blob of rubber with four short metal prongs sticking out one end.

"No..." Ilana became flustered. "No, I meant yes, of course we'll use them." Ilana took the cases from Tanaka. She plugged the small transducers into the rubberized sockets of the two loose cable ends.

Otzker picked up the telephone beneath the sonar station. "Secure lower deck. Make way due east along 12.03 at five knots. Stop at marked spot at 144.67."

The bow and stern thrusters increased in speed and the ship pivoted in position so that its bow now faced east. The main engines came online and the ship started moving eastward, spooling electrical cable off the left winch.

CHAPTER048

Haruto looked at the black warhead sphere dangling off the right winch. A nuclear weapon. Breaking the rules. What was Japan getting involved in?

"A bit to port," Otzker said into the telephone receiver. He studied the sonar screen again. "Good, that's it. We drop here."

The main engines stopped but both the bow and stern thrusters continued at high speed. A few moments later the loudspeakers blared, "Instrument drop. Instrument drop."

The right stern winch gently lowered the black warhead sphere and its accompanying massive white buoyancy sphere into the ocean. The chain on the top of the white sphere pulled down on the spool of electrical cable in the right winch.

Ilana plugged the electrical wires from the center of the right winch spool, wires ultimately coming from the warhead sphere, into an aluminum box outfitted with an LCD screen, two dozen variously colored LEDs and a compact keyboard. A green LED came on. Ilana turned to the winch operator. "Lower away." She faced Tanaka again. "In one hour the warhead sphere will be drilling into the seafloor."

Tanaka studied the setup, then pointed at the instrument panel. "Why is that orange light flashing?"

Ilana and Otzker quickly turned to stare at the LEDs.

"Shit!" Otzker said.

Ilana faced Benni. "You evacuated the air with nitrogen after you took apart the warhead, and before you sealed the warhead metal sphere, right?"

The technician looked down at the deck. "Everything was perfectly dry. We were doing the test the same day. I didn't think it was necessary."

"We're on the ocean. This isn't Dimona where we're sealing up the weapon. There's lots of water all around us here, including in the air."

"Shit," Otzker said again.

"General, we should pull up the weapon and do the test another day," Ilana said. She turned to the winch operator. "Stop spooling."

"No. We're doing the test now. Disable the anti-critical explosives."

"But—"

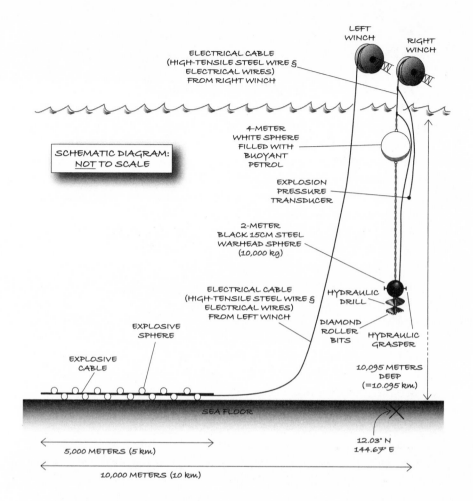

LEFT WINCH

RIGHT WINCH

ELECTRICAL CABLE
(HIGH-TENSILE STEEL WIRE &
ELECTRICAL WIRES)
FROM RIGHT WINCH

4-METER
WHITE SPHERE
FILLED WITH
BUOYANT
PETROL

SCHEMATIC DIAGRAM:
NOT TO SCALE

EXPLOSION
PRESSURE
TRANSDUCER

2-METER
BLACK 15CM STEEL
WARHEAD SPHERE
(10,000 kg)

ELECTRICAL CABLE
(HIGH-TENSILE STEEL WIRE &
ELECTRICAL WIRES)
FROM LEFT WINCH

HYDRAULIC
DRILL

EXPLOSIVE
SPHERE

DIAMOND
ROLLER
BITS

HYDRAULIC
GRASPER

EXPLOSIVE
CABLE

10,095 METERS
DEEP
(=10.095 km)

SEA FLOOR

12.03° N
144.67° E

5,000 METERS (5 km)

10,000 METERS (10 km)

"That's an order."

Ilana went over to the instrument panel and typed two lines into the keyboard.

Tanaka's eyebrows rose. "You said if water gets into the uranium-235 sphere, the water molecules slow down the neutrons enough so that a chain reaction occurs — a nuclear explosion!"

"It's probably just a bit of condensation from the sea air," Otzker said. "Nothing to be concerned about."

"Why did you disable the anti-critical explosives then?"

"The water indicator light is on. I don't want the anti-critical explosives destroying the weapon and ruining the test." Otzker turned to the winch operator. "Start spooling cable."

Ilana turned to the winch operator. "Stop the drop."

The crewmember on the winch hit the large red button and the spool ground to a halt.

"I'm in charge here," Otzker said. "Keep spooling out the cable."

"Yes, Sir." The crewmember started up the winch again.

"Otzker, this is crazy," Tanaka said. "Stop the test."

"Why? You're the one who's been pushing so hard for it."

"Because it was part of the deal. We deliver the robots, you do the test. But getting blown up wasn't part of our arrangement. It doesn't do anyone any good."

"For twenty years, I have run Testing without a major failure. I'm not going to fail now. Do you hear me?"

Ilana turned to the crewmember at the winch. "How deep is it now?"

"About four hundred meters."

"Will this give us any protection at all?" Tanaka asked.

Ilana shook her head. "You want a few thousand meters at least." She turned to the winch operator. "Stop the drop. That's an order."

Otzker's eyes opened wide. "You keep spooling. I'm in charge."

"As chief physicist I'm in charge. Stop the drop. Ignore General Otzker."

The winch operator hit the red stop button.

Otzker glared at Ilana. "How dare you countermand my order. Your career will be over. I'll see to it. I'll see to it that you're thrown in jail."

Ilana started shaking and backed away from Otzker.

"Start that winch immediately," Otzker yelled at the operator.

"Yes, Sir."

Ilana, Tanaka and Otzker all stared at the spinning spool of cable on the right winch. Otzker, despite his bravado, was perspiring and wiping the beads of sweat off his tanned forehead.

Haruto felt his heart beating harder. Madness. This is what happened when you broke the rules. He would end up being blown to bits with them. His life ending here on this mad ship in the middle of the ocean. It started to feel difficult to breathe. He sucked in deeply. The air didn't seem to go into his lungs. He started feeling dizzy.

Haruto tapped two times on the left leg of the chaise lounge chair, and then two times on the right leg. That felt a bit better. He tapped twenty times on the left, and then twenty times on the right. Perfect rhythm. No errors. The breathing felt better.

"General, let's stop the test," Tanaka said.

"No."

"Problem!" Ilana pointed to a second flashing orange LED lamp.

"What's that?" Tanaka said, anger in his voice.

"The top orange LED indicates moisture in the sphere," Ilana said. "This second orange light means water inside the warhead itself."

"General, there's a leak into the warhead sphere," Tanaka said.

"No, I'm sure it's condensation." Otzker wiped the perspiration off his forehead again.

"General, it wouldn't take much," Tanaka said. "Maybe the tonnes of water are compressing the electrical wires where they enter the sphere, creating a miniscule gap there. As the sphere goes deeper, the water pressure will push into the gap and fill the sphere up with water, and this water will push into warhead itself."

"How deep are we?" Ilana asked.

"About five hundred meters," the winch operator replied.

Ilana, Tanaka and Otzker looked in silence at the cable dropping into the ocean. Ilana wrapped her arms around her chest. Tanaka's face was red with anger. And Otzker's forehead kept sweating, with drops of perspiration running into his eyes.

"How deep are we?" Ilana asked again.

"Only about five hundred fifty," the crewmember replied. "You know it takes an hour to reach the bottom, Ilana."

Tanaka looked at his watch and started pacing. Ilana looked down at the deck. Otzker stared at the spinning spool of steel cable on the right winch.

Haruto felt his heart beating again. He twenty-tapped left and right again. It helped a bit. But why did his life have to end like this? Because of the madness of others who were breaking rules. And it wouldn't finish here. Even if this ship exploded, there would be another one. And they would cause the destruction of his Japan, surely as those who broke the rules in the 1940s did.

CHAPTER049

H **aruto stared** at the rule breakers under the stern canopy. His heart was beating fast, but he managed to control his breathing.

"Depth?" Ilana asked.

"Six thousand two hundred meters," the winch operator replied.

"I think we're safe now. Also, the water pressure is so high at this depth that if there really was a crack in the sphere it would be completely flooded." Ilana pointed to the instrument panel. "Electronics of the warhead are still fully functional."

Otzker laughed. "See, I told you it was just condensation."

"Sir, permission to finish the explosive cable setup?" one of the crewmembers asked.

Otzker nodded.

The left stern elevator popped up onto deck with a huge white sphere on it. Two crewmembers hauled it off.

"This white sphere is hollow, so it can give a lot of buoyancy near the surface," Ilana said to Tanaka. "It supports the ten kilometers of electrical cable coming up from the explosive cable on the seafloor."

The left elevator popped back up onto deck, this time with a dull gray buoy. Crewmembers plugged electrical cables into it.

Benni plugged a cable from his instrument panel into the buoy. The LCD panel changed color. "Permission to proceed?"

"Yes," Ilana said.

Benni pressed two keys and the instrument panel changed color again. "Buoy computer has control of the warhead. Buoy computer has control of the explosive cable."

Benni pulled his instrument panel's cable out of the buoy. "Radio link between buoy and instrument panel operational. All electronics good-to-go."

Ilana turned to the right winch operator, still lowering the warhead sphere. "Depth?"

"Approximately nine thousand one hundred meters."

"Slow spooling to half speed." Otzker walked over to the sonar station. Tanaka followed. Otzker pointed to the display. "That's the warhead sphere....nine thousand two hundred two meters. Reduce spooling speed to one quarter." Otzker put his finger on the sonar screen. "Good. Nine thousand eight hundred meters... Ten thousand meters." He turned his head to the right winch operator. "Reduce spooling speed to minimum."

Tanaka leaned forward and peered at the screen. "Is that the warhead sphere on the bottom?"

"There's still a small gap — maybe twenty, thirty meters."

Tanaka looked at the screen. "I see it. Getting smaller.... Gone."

"Warhead drill in sediment. Drill turning," Benni said. "Two hundred meters of cable slack spooled out."

"Now we wait," Otzker said.

"How long?" Tanaka asked.

"Until warhead sphere is two hundred meters under the seafloor."

"Hydraulic tunnel grasp working. Drill working," Benni said. "Seven meters."

Tanaka walked to the stern of the ship, and looked out over the ocean.

Suddenly Benni called out, "Twenty meters. Drill bit stuck."

"Reverse ten seconds and try again," Otzker said.

"Done... Reverse over. Drilling re-started... Twenty-one meters. Drill turning."

Otzker walked over to Tanaka and the two men silently gazed out to sea.

Haruto stretched his legs underneath the lounge chair and put his head against the inside of the tiki bar.

An hour later Benni yelled, "Two hundred meters. Drill complete."

Haruto sat up again and looked out the thatch.

A new song, reggae, came on the loudspeakers. Crewmembers welded the buoy to chains from the two large white hollow spheres — one to hold up the electrical cable from the warhead, the other to hold up the electrical cable from the explosive cable.

"Release the spheres," Otzker ordered. "Everyone clear the area."

A welding torch sliced across a steel restraining bar. The instant the steel severed, the weight of electrical cables yanked the huge white spheres and the attached buoy into the water.

Otzker pointed to the large haze-gray buoy bobbing off the rear of the ship. "The two white spheres are underwater now, but they're attached to the buoy."

Tanaka nodded.

"Benni — full status please," Otzker said.

Benni looked at the display of his instrument panel. "Explosive cable electronics — check. Warhead electronics — check. Buoy computer — check. Radio — check. System good-to-go."

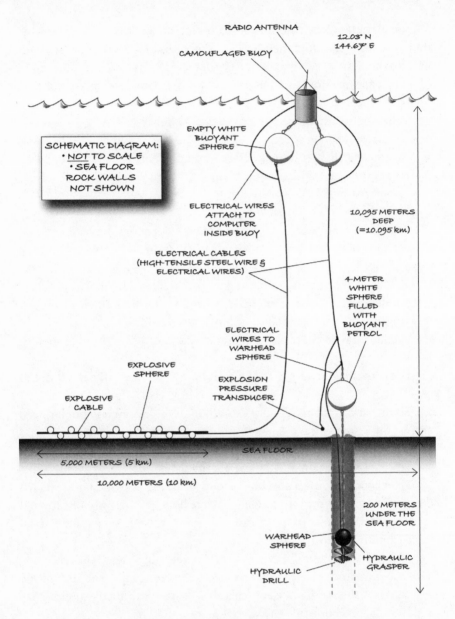

RADIO ANTENNA

CAMOUFLAGED BUOY

12.03° N
144.67° E

SCHEMATIC DIAGRAM:
• <u>NOT</u> TO SCALE
• SEA FLOOR
ROCK WALLS
NOT SHOWN

EMPTY WHITE
BUOYANT
SPHERE

ELECTRICAL WIRES
ATTACH TO
COMPUTER
INSIDE BUOY

10,095 METERS
DEEP
(=10.095 km)

ELECTRICAL CABLES
(HIGH-TENSILE STEEL WIRE &
ELECTRICAL WIRES)

4-METER
WHITE
SPHERE
FILLED
WITH
BUOYANT
PETROL

ELECTRICAL
WIRES TO
WARHEAD
SPHERE

EXPLOSIVE
SPHERE

EXPLOSION
PRESSURE
TRANSDUCER

EXPLOSIVE
CABLE

5,000 METERS (5 km)

SEA FLOOR

10,000 METERS (10 km)

200 METERS
UNDER THE
SEA FLOOR

WARHEAD
SPHERE

HYDRAULIC
GRASPER

HYDRAULIC
DRILL

Tanaka put up his hand. "What signal are you getting from my transducers?"

Benni squinted at the small letters at the bottom of the display panel. *"Code seven zero zero zero one...temperature below set threshold...estimate under ten degree...pressure below set threshold...estimate one hundred million pascal.* Same message from the second transducer."

"Good. When the bomb explodes, the data from my transducers will be sent to your instrument panel, correct?"

"For a split second, before the cable is destroyed," Ilana said. "Whatever your transducers measure, plus the security code they generate to prove that they're genuine signals from your transducers, will be sent up the electrical cable to the buoy's computer and radio."

Tanaka nodded.

Otzker picked up the nearby telephone receiver. "Make way to Guam."

A crewmember approached Otzker and pointed to the rainbow tarpaulin covering the stern deck. "Remove the canopy, Sir?"

Otzker shook his head. "Leave it until we make port. You never know who's watching above."

CHAPTER050

John Sullivan didn't mind working the graveyard shift. He still couldn't believe he had gotten the job. *What ya gonna do with a degree in political science?* his Dad had said. There was that D in English. But it didn't matter. He was here, on his way.

The computers did most of the work. The days of reconnaissance satellites ejecting rolls of film caught in mid-air by an airplane, developed and pored over by hundreds of analysts were long gone. Everything was real-time now. Recon satellites radioed their digital pictures to communications satellites, which radioed everything to Fort Belvoir's computer network. The AI — artificial intelligence — took care of the grunt work. John liked that term — *grunt work* — especially since he didn't have to do it. The interviewer told John that since his AI was bigger than the computer's AI, his job would be to handle situations that the computer couldn't handle. The interviewer told John about the benefits, and six weeks later, here he was.

For the last two weeks there had not been much. Fires were a big thing. When the satellites showed fires, the computers always prompted him about it. He would look at the image, confirm it was a fire and e-mail his liaison contact at the DIA — Defense Intelligence Agency — with the information. A reply e-mail would always politely thank him, signed with contact's anonymous title and nothing more. Three nights ago there was a buildup of tanks near the Black Sea. He sent the e-mail and got the canned reply. Tonight, the computer prompted him with the image of the container ship *Mikiyasu-ema* unloading at Haifa. It was almost noon there, and in the bright sun, the image was exceptionally clear. Yup, that's what it was — a large container ship unloading containers. John zapped the e-mail to the DIA.

Not sure. Please investigate or direct elsewhere. That's what the reply said. John read it twice. Finally, after two weeks on the job, a hot lead.

He picked up the phone and speed-dialed the group head. "This is John Sullivan. I'm an analyst for the KH-14 Gold — we're over the Mediterranean now. I have a container ship to query."

"A container ship? Aw fuck, it's four in the morning, John. Why don't you do it with your group leader at signover at eight?"

"The DIA wants an investigation. Doesn't that mean it's important?"

The phone went dead. A minute later a large fortyish man with curly hair, an overhanging belly and wearing a pair of cowboy boots, was standing over John's cubicle. In a deep voice he said, "Show me what you got, John."

John was startled for a second, then got up. "Ahh, thank you for coming."

"That's what I'm paid for." The supervisor forced a half-smile.

John hit a few keys. A real-time picture of a crane taking a pair of containers off the *Mikiyasu-ema* appeared on his large computer screen. A caption displayed below. *Mikiyasu-ema at Haifa Port. Last five years normal course to Felixstowe, U.K.*

"Here… can I sit down?"

John nodded.

The supervisor clicked away at the keyboard. In a few seconds a shipping manifest popped up on the screen. "Car parts from Mikiyasu going to the Volkswagen dealer outside of Haifa."

"That's a lot of car parts," John said.

"Probably they're knocked down cars that the dealer's going to assemble and sell. Let's assume two cars per container — so that's under thirty thousand cars. Let's see what the overall Israeli car market is for this year." The large man typed away. *181,640 cars* appeared on the display. "Okay, so this dealer is going to flood about one-sixth of the market with inexpensive Mikiyasu models."

"But how can you be sure it's that and not something else?"

The supervisor nodded and started clicking away. In less than thirty seconds, a Japanese document was on the screen. "This is from Mikiyasu's internal production computer. Hang on…" The large man hit a few keys. "Instant English translation."

A factory production form displayed. 26,000 Mikiyasu electric Y cars, in sub-assemblies for easy final assembly, for shipment to *Haifa-Asia Car Corporation.*

John's eyes lit up. "Cool! How'd you do that?"

The large man smiled. "Maybe after you've worked here for more than a month I'll show you."

"What should I do with the information?"

"Well, it's not of great military importance, so I wouldn't send it back to the DIA — or if you do, zap it as a copy. I'd send it to the U.S. Embassy in Israel. Maybe they could push a similar deal for General Motors or Ford. Or they'll just ignore it." The supervisor stood up and ran his fingers through his hair. "It's out of our hands now. I'm going back to my nice soft chair."

CHAPTER051

Haruto saw Otzker walk over to the young purser. "We'll dock at Hotel Wharf tonight. No shore leaves, but replenish the ship. Also, the rush is over now, so no reason for us to become lax. Make us look like a cruise ship. I want the deck cleaned up. Put more of the chaise lounges out. Put up the lights."

"Yes, Sir."

Soon there were crewmembers all over the deck. The stern elevators took the empty cable spools into the belly of the ship. Benni packed up the instrument panel into its aluminum case and walked away with it. Other men picked up tools scattered on the rear deck. Two crewmembers pushed the mobile welding kit toward the left stern elevator. The winch operators retracted and folded the hydraulic arms, and covered them with canvas.

The main engines started and Haruto felt the ship begin to move again.

Haruto looked through the thatch of the tiki bar toward the bow of the ship. Crewmembers were arranging lounge chairs around the pool. Other men were on ladders, fastening a long string of colored lights onto the poles above the deck.

Suddenly the loudspeakers blared, "Code Red. Code Red."

Menachem and two crewmembers, all carrying M16 assault rifles, came running to Otzker, Menachem dragging his bum left leg. With them was the crewmember Haruto had tied up yesterday evening. They started speaking in rapid Hebrew.

"What's happening?" Tanaka demanded.

"We have a stowaway on board," Otzker said. "Japanese. He beat and tied up one of our crew, who we just found by accident during deck cleanup."

"It wasn't one of my men," Tanaka said.

"I don't care who it was," Menachem said. "Let's find him." He went over to the sonar station and picked up the telephone receiver. His voice blared from the loudspeakers all over the ship. "Code Red. All men to the main deck. Full sweep. Secure and work down."

"What's Code Red?" Tanaka asked.

"Armed invader — capture if you can, but orders are shoot to kill," Otzker said.

Crewmembers with M16 rifles over their shoulders were all over the deck, opening every closet, turning over every chair. One crewmember was even on his knees, looking down into the corners of the swimming pool.

Suddenly Haruto heard footsteps and saw two hands push away the tiki bar and flip the chaise lounge he was lying under. The crewmember started yelling at Haruto, reaching for the trigger of his M16.

Without thinking, Haruto's left arm flew up and with an open hand slapped the end of rifle barrel away. A tenth of a second later, he thrust a quick kick into the crewmember's right knee. As the man fell, Haruto jumped up and started running along the wooden promenade toward the bow.

A burst of shots filled the air. A bullet tore through the outside of Haruto's left thigh. He kept running.

Haruto turned into a nook in the bulkhead just before the bow deck. He took out his cellphone and jabbed at the power button. Footsteps of a crewmember running down the wooden promenade grew louder but the crewmember ran past him toward the bow.

The cellphone display came to life. No antenna bar. More footsteps running down the deck. Haruto's heart beat faster. *We'll dock at Hotel Wharf tonight.* Guam should be near. He should be able to get a signal. Boaters always did off these islands. His finger thrust against the keyboard. *Menu. Settings.* Click-Scroll. *Car Mode — Airplane Mode — Marine Mode...* Click-Scroll. *Star-CG — Coast Guard — Range Boost...* Click-Click. In the upper corner a single antenna bar now registered a signal, but flickered on and off. Haruto clicked on the file of the video he had recorded, then clicked the cellphone number of Itou, the young community policeman. The file transfer started to go through but then the antenna bar flashed off and the call terminated.

Haruto stepped out of the nook and went to the railing at the side of the deck. The single antenna bar came on strongly — less metal to interfere. He hit SEND. Suddenly, a burst of shots rang out above his head. Haruto looked to the right and to the left, gasped, and then took the phone and threw it high in the air off the deck, over the ocean, against the orange and red sky of the setting sun. Haruto watched the cellphone twist through the air, hoping that the antenna bar remained solid.

A crewmember was coming down the promenade at Haruto from the stern. Another crewmember was coming down the walkway from the bow. The crewmembers were yelling at Haruto. Bullets were flying over his head.

Haruto put his hands over his head in the surrender position. It was to no avail — the shots kept coming.

A bullet ripped through the edge of Haruto's right shoulder. He turned his head and saw blood trickle out.

Haruto's heart was beating so fast he thought it was going to explode out of his chest. What to do? Haruto looked to the bow and then looked aft. Almost without thinking, like in a dream, Haruto took the large, white lifebuoy with the imprint *New Pacific Queen* off the hook on the side railing and threw it into the water. Another bullet whizzed by Haruto's right shoulder. Haruto climbed over the railing, closed his eyes against the brilliant Pacific Ocean sunset and jumped.

CHAPTER052

Haruto **instinctively** kept his feet pressed tightly together and his hands curled into tight fists. His eyes were still closed when he crashed through the water. The impact tore off his right shoe. He shot down five meters into the ocean, then bobbed back up to the surface in the frothy wake from the ship.

Haruto started swimming away from the ship, from its churning propellers that would soon reach him.

A bullet, then another, whizzed by his head! He dove underwater.

After about thirty seconds, as the urge to breathe increased, he could feel the panic build up in his gut and spread across his body.

No. He would survive this. Haruto counted to himself. One, two, three, four. And then the same pattern again. One, two, three, four. The panic began to subside.

Haruto opened his eyes against the sting of the salt water. He saw the froth from the ship's wake just above him. He pushed gently with his hands against the water and arched his neck and body backward to let just his face come out of the water. He breathed in deeply.

Haruto surveyed the area. Toward the west was the sun, setting powerfully on the calm sea. Toward the north the ship was moving away, leaving its wake behind. The white lifebuoy had landed about ten meters away — the seawater automatically triggering its bright red flashing lamp. Haruto was tempted to swim to the lifebuoy for comfort, but as visible as it was to him, it was more so to the gunmen on the ship. Haruto sculled up against the water and submerged his head and body under the wake again.

CHAPTER053

"**R**everse propellers and stop the ship," Otzker yelled into the telephone receiver. He walked over to Tanaka and a crewmember looking through large binoculars, all leaning over the port deck railing.

Menachem came hobbling to the group. "There's blood on the deck just aft from here. He's wounded."

Otzker took the binoculars from the crewmember and scanned the water back and forth. He pointed with his right index finger to the white ring buoy, now southwest of the ship. "I see a life buoy from the ship but nothing else."

"We saw him swimming a moment ago," Menachem said. "Even if he's still out there, he's not going to last long in the open ocean."

"Give me the surveillance files from the cameras you have on deck," Tanaka said. "I'll send them to Japan. You'll have an ID in ten minutes."

Otzker gave the binoculars back to the crewmember. "Scan the water until the sun is down. Get another person to help you. If you don't see anything by the time it's dark, we'll move on."

CHAPTER054

Haruto heard the ship's engine start up and its propeller churn a strange, echoic sound through the water to his submerged ears. He was exhausted from sculling and let his head float up out of the water. A sense of fear then gripped him as he watched the lights of the ship fade in the distance. He was alone in mid-ocean. It was in some ways more frightening than being on a ship full of people who wanted to kill him.

He breathed in deeply through his nose, and exhaled. He counted — one, two. And again, one, two. He imagined the hard wooden floor of the dojo. One, two. The panic in his gut subsided.

Okay, make a rule now. Find the lifebuoy and conserve energy until a passing ship comes by and picks him up.

Haruto scanned the area around him and off in the distance he saw a red light flashing. He swam toward the red beam and after about twenty minutes, he could make out the lifebuoy in the glow of its light. Haruto picked up his pace and five minutes later, he had the lifebuoy over his head, comfortably supporting his arms.

Haruto took off his left shoe and tied it to the cord coming off the lifebuoy. He then did the same for his jacket. It wasn't too bad for the moment. The water was warm. There were no waves. Only a light breeze. He was even able to lay his head down on the white ring of the lifebuoy, while the weight of his arms draped over the ring kept him securely in place.

It was a good rule. He was following it. He would be all right.

CHAPTER055

A n official police identification picture, meters tall, splashed on the large screen in the front of the theatre.

"His name is Suzuki Haruto," Tanaka said. "He is *Keibhu* — a full ranking Inspector, based in Tokyo. It was he who was in charge of the police investigation at Harumi Wharf when we left so abruptly."

"Why was he on my ship?" Otzker asked.

"I can tell you why, but not how. Co'en Satoki, of Jewish European ancestry, was murdered and Inspector Suzuki was assigned to his case. Co'en had found out through the Korean crime syndicate that the President of the Mikiyasu robot division had large gambling debts. We have interviewed Co'en's widow...."

"Yes..." Otzker said.

"General, why were you dealing with Mikiyasu behind my back? What mess have you created?"

"I know nothing about such matters," Otzker said. "I have nothing to do with the robots or anything else. My job is the testing of weapon systems."

Otzker, and then Tanaka, looked at Menachem.

"We weren't doing anything behind your back," Menachem said. "But otherwise, guilty as charged. Your superiors are aware of the situation. Mossad was looking into *purchasing* military robots from the United States, Taiwan, South Korea or Japan. Cohen was not supposed to get killed. He was not a Mossad agent. He was a businessman arranging an export deal with Mikiyasu. I was the one Mrs. Cohen was supposed to meet the night all the excitement took place at the wharf."

"How did Inspector Suzuki get on the ship?" Otzker asked. "If he doesn't work for you Colonel, who does he work for?"

"We pulled his file and interviewed everyone yesterday," Tanaka said. "I don't think he worked for anyone. Everyone told us the same story — Inspector Suzuki was completely obsessed with his job. In the last thirteen years, he didn't take one day of vacation. He was also completely obsessed by the rules and regulations. In his previous police station, there was some petty bribery going on — free meals by local restaurants, that sort of stuff.

Suzuki turned in the entire stationhouse. About half of them lost their careers."

Otzker turned to Menachem.

"We reviewed all the surveillance files for the last week. The Inspector shows up the first time coming onto the main deck from a mooring closet. He must have climbed up a rope just as the ship was departing." Menachem looked down at the floor. "I'm sorry. We've modified security protocol so this doesn't happen again — observation of all mooring ropes until they're fully retracted."

Menachem limped up the theatre's aisle steps and out the left entrance door.

Ilana jabbed at a few buttons on the armrest of a nearby leather seat. Haruto's picture vanished off the screen, replaced by a pictorial diagram of the cables dropped into the ocean yesterday.

"This is the setup," she said. "We have five kilometers of explosive cable on the seabed. Five kilometers farther east we have the buried warhead. The explosive cable and the warhead both have cables running up to the camouflaged buoy."

Tanaka nodded.

Ilana pressed a button again. The display filled with changing lines of text. "This is real-time telemetry data from the buoy. Warhead electronics and explosive-strip electronics are all fully functional."

"Where is the output from my transducers?" Tanaka asked.

"There, on the bottom of the screen." Ilana pointed to the display — *Code five two zero zero one.*

Otzker turned to Ilana. "Activate detonation for Richter three point five or greater within one degree."

Ilana started typing away at the keyboard. Thirty seconds later, she lifted her head. "Done."

"What will happen now?" Tanaka asked.

"We have vibration sensors in the explosive cable, which communicate with the buoy," Ilana said. "I just programmed the computer in the buoy to wait for vibrations large enough to indicate, we hope, a Richter four earthquake within a hundred kilometers. When that happens, the computer in the buoy will automatically start the detonation sequence."

"How long will that take?"

"Usually less than a month. There are lots of earthquakes here, so hopefully sooner rather than later."

"We'll make the most of this time," Otzker said. "There's a robot meeting scheduled for ten o'clock this morning."

"Are we going to stay in Guam the whole time?" Tanaka asked.

"No, like any cruise vessel we move around," Otzker said. "We'll leave tomorrow and make way for Yap."

Otzker, Tanaka and Ilana walked up the steps and left the theatre. Tanaka took the elevator down to the main deck and walked out onto the wooden promenade. It was quite pleasant here. The sea and the vegetation in the harbor smelled sweet. The sun shone, and a small breeze blew.

Tanaka looked out across the harbor. A gray U.S. Navy tender was pulling into the naval station to the south, and beyond it was nothing but ocean — so large and empty.

Poor Inspector Suzuki. To be stranded there.

Tanaka shuddered. That poor, obsessive bastard. He hoped he had died quickly when he went overboard.

CHAPTER056

The morning had started well. The large, buoyant lifebuoy had comfortably supported him through the night, and he had managed to sleep. The water was warm and sunrise over the Pacific had been magnificent. However, as the sun grew brighter it started to bother him. His face was becoming sunburned and his water-soaked mustache had dried out and was now bleaching in the sun's rays.

Haruto untied his jacket from the rope ringing the lifebuoy and used it to create a tent of sorts to shield him from the sun. It worked well, with the drops of water falling from his jacket providing a bit of refreshing coolness.

Haruto lay there quietly. The thirst wasn't too bad yet. A ship would come soon, and it would all finish well. Haruto scanned the ocean in front of him. He could see nothing but ocean. Haruto turned to the left —

There was a fin in the distance.

His heart skipped a beat.

The fear paralyzed him for a moment. He breathed in deeply. Make a rule. Sharks usually don't attack humans. He remembered what the purser said about the scuba divers watching the shark feedings. Maybe the rule was, do nothing. His wounds had stopped bleeding. The shark probably noticed him at a distance, but once closer, the shark would see he was not its usual prey and would ignore him.

Or maybe not. Haruto looked down, pressed on his left thigh. He could see a bit of red blood ooze out and diffuse into the water. If there was enough blood for him to see, then there was enough blood for the shark to smell.

Make a rule, make a rule. The first rule in karate was the most obvious one. The best defense was to step away from a punch and let your opponent strike the air or the loose folds of your uniform. Follow the rule.

Haruto started working frantically. He untied his left shoe from the lifebuoy and tied it to one of the sleeves of the jacket. He took out a nickel-plated pen and an equally bright metal mechanical pencil from the jacket's inside pocket, and clipped them both onto the exterior of the jacket's outside pocket. He tied the rope that went around the lifebuoy to the other sleeve of the jacket.

The shark swam slowly, cutting a wide arc through the water around him. All of a sudden it jerked to the left, aiming its snout directly at him. Perhaps it was the blood from the wound. Maybe it was the motion of his legs underneath the ring buoy. In any case, the great fish came barreling toward him.

Haruto threw the shoe into the water. It sank and pulled down and opened the jacket into a large meter-by-meter dark contrast against the open ocean above. The pen and pencil set sparkled in the water.

Haruto could see the monster's large dorsal fin and the top of its caudal fin. It was a light green tiger shark, at least two meters long, maybe three. Haruto felt his stomach burn with acid. He tightened his grip on the lifebuoy and violently lifted his legs out of the buoy's ring. His buttocks and hands rested on top of the buoy while his legs and back and head were straight in the air over the water.

The shark was a body length away now, no more than three meters and still coming straight for Haruto.

Haruto jerked the rope up and down.

A split second later the shark was so close that Haruto could see its dark eyes just under the surface. Haruto started counting quickly. One. Two. One —

The shark suddenly dove and lunged at the moving, glistening target. Its jaws clamped down on the coat and the shoe, yanking the attached rope from Haruto's hands.

And then the sea was calm again.

Was the shark going to come back? Jump up in the air and take a bite of his leg? It was coming back. It was circling there underwater, ready to pounce up.

No. He had made a rule and followed it. After biting into the bland coat and shoe, the shark probably had no interest in coming back for more.

But in the movies they always came back. It was there, it was circling there, waiting for the moment. Haruto felt his heart beat faster. This was it.

But second after second went by, and became minutes.

Haruto scanned the water around him. No fins. No ships. Nothing, but water.

Haruto shivered. One. Two. One. Two. One. Two…

CHAPTER057

Lieutenant Colonel Okamura looked through the bridge windows. Dozens of flood lamps shone on the *Mikiyasu-ema* and the surrounding container terminal. Three overhead gantry cranes were now pulling the drab brown containers off the front deck, and another three gantry cranes were working over the stern. Some cranes were loading containers onto trucks while others simply dropped them onto the asphalt. A fleet of two dozen rubber-tired mobile gantry cranes scurried around, picking up stray containers, carrying them to flatbed container trucks waiting in parallel queues.

Captain Yamada entered the bridge and approached Okamura. "We just unloaded container number ten thousand. We can finish the job by tomorrow night, if that's what you want."

"Continue for now. We'll see about the final containers tomorrow."

Yamada nodded. "Good night. I'll see you in the morning."

Okamura reached down on the floor for his satellite radio pack and started clicking buttons.

CHAPTER058

John Sullivan pushed his loafers off and settled in at his desk for another 8 to 8 on the job. He clicked through e-mails and alerts — nothing tonight. He clicked through his bird's status report. Power and electronics fine. Fuel at seven point one tons. Telemetry fine. Servomechanisms fine. Adaptive optics noted Shack-Hartmann correction to two centimeters. Current keyhole 31 degrees and 19 minutes north, 30 degrees and 4 minutes east.

John typed a line of commands into his keyboard. A response of *ALL ROUTINE* appeared in green letters on his computer display. Below that appeared the green-tinged picture of low-rise office buildings. The caption below read *Arab Academy for Science and Technology and Maritime Transport — 3AM — All Routine.*

John typed another line of commands into his keyboard. The display refreshed. Green letters flashed onto the top of the screen. *Keyhole rotating. Cassegrain secondary mirror rotating... Cassegrain secondary locked onto 32°49'21"N, 35°1'9"E — adaptive optics laser guide recalibrated to 2 centimeters — target to be held for 12 minutes 22 seconds.* A moment later a real-time color video image of the busy port splashed on the screen.

John's eyes opened wide. A half-dozen red-and-white-striped gantry cranes were yanking containers off the ship and plopping them down on the ground all over the place. Some two dozen mobile cranes were picking up containers and scurrying with their loads to multiple lines of container trucks.

He picked up the phone and hit the speed-dial button. "This is John Sullivan. I'm an analyst for the KH-14 Gold —"

"I know who you are, John. I'll be right over."

A minute later, the supervisor with the curly hair and his trademark cowboy boots was standing over John's cubicle. "What ya got here?"

"Look how busy it is at Haifa at three o'clock in the morning their time."

"It's a big ship, John. The slower you unload, the more money it costs. What does the computer say?"

"Routine."

"Drill down a bit deeper. Here, give me your keyboard." The supervisor typed a line of commands. New green letters flashed onto the screen. *Ship traffic -80% Haifa. Hourly container movement +410% Haifa, +22% Mikiyasu-ema main hub port average. Analysis — large ship unloading. Latter manually deemed routine prior passes. Image is given* **Routine** *rating.*

"What does that mean?" John asked.

"The computer already pointed out yesterday that the *Mikiyasu-ema* was off course. Otherwise it finds nothing exceptional about unloading the ship's huge number of containers as quickly as possible."

"I don't know... something is strange."

The large man shook his head, smiled and left the cubicle.

CHAPTER059

Tanaka slammed the satellite phone set against the desk. "I thought you said it was going to take one to two weeks," he yelled into the handset.

"They managed to get every single overhead container crane in the port working over the *Mikiyasu-ema*," Lieutenant Colonel Okamura's voice replied. "Containers are flying off the ship and two dozen mobile cranes are loading up the trucks."

"How long until you unload down to the nuclear containers?"

"By tomorrow evening. Permission to continue offloading?"

Tanaka didn't say anything but looked out the porthole.

"Sir, are you there?"

"Yes, continue unloading."

Tanaka hit the red button on the keypad and opened the cabin door. He gave the satellite phone back to the waiting crewmember and took the elevator down to the main deck.

Otzker waved to him to come over to the railing. The engines had started and the ship was almost halfway out of the harbor.

"It's beautiful here, isn't it?" Otzker pointed to the open ocean in the distance.

"I spoke with Haifa. By tomorrow night you will have received one hundred thousand robots worth over eleven billion dollars, one tonne of plutonium and ten tonnes of pre-enriched uranium."

"Yes, that was the deal. And you will have containers holding a dozen one-megaton nuclear bombs."

"That we don't yet know will work!"

"Colonel, I don't control the earthquakes. Let's make the best of the situation. We'll use these days to transfer technology."

"General, what do you think will happen to me if, after unloading the *Mikiyasu-ema*, the nuclear test turns out to be a dud?"

"I'd hold off on the *hari-kari* because the test will work."

"It's called *hari-kiri* or more properly *seppuku*."

Otzker raised his eyebrows. "Would you really do it?"

"It would actually be preferable than returning to Japan as a failure."

"Is that the way all Japanese think?"

Tanaka laughed. "No, not many. I am not *samurai* but I have been studying the martial arts since I was five years old, and the *bushido* — the code of honor — is a large part of who I am. If a *samurai* were disgraced or failed in battle, then *hari-kiri* was a more honorable choice than being prisoner or regular execution."

"Can a man really plunge a knife into his guts and eviscerate himself?"

"I don't know why it's portrayed like that in the West. In fact, there is almost always a helper or *kaishakunin*. The moment the person performing *hari-kiri* touches his belly with the knife, the *kaishakunin* lowers the samurai sword and instantly slices through the neck, leaving the head attached to the body by only a thin piece of flesh."

"Wonderful," Otzker said.

"In fact, if a *samurai* has fought well but lost, his adversary will often act as *kaishakunin* and restore the *samurai*'s honor in defeat."

The ship had passed out of Apra Harbor and was steaming toward the open sea. Tanaka stared at the vast, blue ocean in front of him. The sun shone brightly on the deck, but a pleasant favonian breeze came in from starboard.

"General, by tomorrow you will have eleven billion dollars of the world's most advanced military robots and tonnes of enriched uranium. And I may have some nuclear devices that are nothing but duds. Perhaps, Otzker, you should indeed be my *kaishakunin*."

CHAPTER060

H **aruto had never experienced** thirst like this. His body was filled with a deep visceral unease unlike any pain he had ever known. His mouth was so dry that his lips had cracked and now stung from the salt spray of the sea.

One. Two. Haruto then turned to the left. One. Two. Make a rule. Follow the rule and survive. Get water into the body, keep water from going out.

Haruto put his face into the water, ignored the sting of the salt and opened his eyes. Nothing. No seaweed. No fish. No sharks even. Nothing to eat that might have moisture in it.

Haruto pulled his head out, rubbed his eyes and looked up at the sky. Not even one cloud. No rain to collect even if he could.

Haruto looked at the water around him. Maybe his kidneys could handle a bit of ocean water. Maybe it would buy him an extra day or two of survival until a ship came by. He lowered his face to the water and sucked in a mouthful.

Haruto tightened his belly and chest and neck muscles and resisted the urge to spit out the awful burning fluid. He swallowed a few milliliters. Okay. He closed his eyes and gulped down the remaining mouthful. Within a second Haruto gagged, coughed, and a moment later the contents of his stomach shot out in a projectile stream of vomit.

Haruto wiped the vomit off his lips with seawater and grimaced as the salt burned into the open cracks in his skin.

A few milliliters works, but a mouthful doesn't. Try again later. Follow the rule and survive. Keep water from going out. Haruto looked up at the clear blue sky. His face was red and perspiring.

Haruto undid his trousers. He put the seat of the pants on top of his head and let each leg drape on either side of the exposed skin of his neck and forearms.

Haruto's body and legs were in the center of the ring lifebuoy. His shoulders and arms hooked around the outside of the ring. On top of his head and draped down both arms was a wet pair of blue jeans Itou had given him. Around him was a calm cerulean ocean. Nothing but water as far as the eye could see. Haruto was a speck on its vast surface.

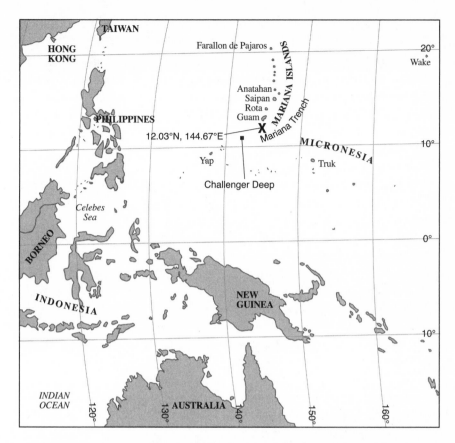

Tomil Harbor, Yap Islands, Federated States of Micronesia
Cruise Liner New Pacific Queen
June 17   12:30 PM Local Time (02:30 Zulu)

CHAPTER061

As the ship dropped anchor, old-time rock and roll music blared from its loudspeakers. A dozen crewmembers dressed in shorts and colorful Hawaiian shirts were sitting in the chaise lounges around the ship's pool, sipping colorful fruit drinks.

"Tender to Yap leaves in five minutes on lower deck," the PA announced. The music resumed.

Otzker found Tanaka at the port side railing, arms outstretched and rigid, staring at Yap.

"Tanaka, everybody's down at the gangway waiting for you. Come… it'll be fun."

Tanaka glared at Otzker and turned his head back toward the harbor. "Have you ever been to Yap? Why don't you join us?"

Tanaka turned and looked at Otzker. "This is the most important mission of my career — one that may save or destroy my country. And you want me to go sightseeing?" He turned away again.

"Why not? Benni is monitoring the telemetry. If an earthquake occurs and the nuclear device is triggered, Benni has orders to give both you and me the telemetry data."

Tanaka nodded, not bothering to turn his head back to face Otzker.

Otzker took the elevator to the second floor and went out the hatch door. He waved to everyone on the floating dock and went down the stairs to it. He found Isato smiling and talking to Ilana.

"How are you feeling, Isato?"

"Much better, General."

"Did the doctor say you could leave the ship?"

"Yes, but I must to take antibiotics."

"Everybody into the tender," Otzker said.

Ilana, Isato, Daveed, Akihiro, Shmoel, Seiko, Menachem and Otzker sat down in the fiberglass boat, and it sped off across the harbor toward the dock at Colonia. A few minutes later, two young brown-skinned women with grass skirts and bikini tops were dancing to a local beat with an exaggerated but gentle swaying of their hips. A young brown-skinned man with large white teeth waved to the passengers and pointed to his stand of souvenirs.

"Is this the whole ship? So few this time?" The brown-skinned man pointed to the small stone-disk necklaces on his table. "Souvenirs. Real Yap money. Only twenty American dollars."

Daveed picked up one of the necklaces. There was a stone disk, about three centimeters in outer diameter with a two-centimeter hole drilled in the center of it. A thin leather strap went through a tiny hole drilled in the top of the stone and formed the necklace. "Ten dollars?"

The Yap native nodded.

The young Israeli software expert took out a green ten-dollar bill from his wallet and put it on the table.

Daveed showed the stone necklace to Seiko. "I remember reading that on this island they used stone money. Apparently, the large denominations were massive stone discs just like the shape of this necklace. The stone discs were too large to actually move and give to someone, so you just left the stone somewhere and said you gave it to them."

"They were ahead of the computer revolution — virtual money," Seiko said.

Daveed laughed.

As the scientists walked off the dock, a middle-aged native man wearing a bright white shirt ran up to them. "Welcome New Pacific Queen. I am Joseph, your guide. We have a wonderful tour planned, followed by a wonderful lunch at the famous and air-conditioned Manta Ray Restaurant."

The group followed the native guide into the verdant island of virtual stone money.

CHAPTER062

The sun continued to beat down.

It had been nearly three days that Haruto had been in the ocean. And the day before that, Haruto had been on the deck under the tiki bar. Almost four days without water.

Haruto had gotten through the worst of the thirst. He had counted. He had sucked on one of the small metal rings at the periphery of the lifebuoy. He had tried to swallow small quantities of seawater from time to time. Now the thirst did not seem to matter so much. The cracks in his lips no longer burned. He looked out at the ocean and saw the waters twist and turn into a psychedelic pattern stretching up to the sky. Then an overwhelming sadness came over Haruto. His mind was working enough to know that it was no longer working well. He would soon slip off the ring buoy and die in this sea. Die neither in battle nor with honor, but here, a great big zero, in this huge ocean, that life, that the world was passing by.

Haruto lay on the dirt, completely naked, covered with vomit and blood. The boys were punching him and kicking him. That was his fate. The counting, the karate, the hard work… he had deluded himself… they could not change his fate.

Haruto laid his head on the edge of the ring buoy and closed his eyes. He opened them again and there was Michiko sitting on the water, laughing at him. He stretched out his right hand toward her. *"Jinzouningen,"* she said and laughed again.

Haruto looked again. It was *Sensei* Nakaya. *Sensei* was younger. No gray hair. A strong young face.

"Help me, *Sensei*. This world around me is so cruel. And this karate, I am not good at it. I can't fight like the others. What should I do, *Sensei?*"

"Come to class, Haruto," *Sensei* said.

"Come to class?"

"Yes, Haruto. Come to class. That is the secret of success."

"But I am no good at this. I am no good at anything in life."

"That is why you must come to class. You must hold onto life. You must come to class and hold on. Hold on, Haruto."

Haruto put his head back down on the edge of the lifebuoy. He wrapped his wrists around the small loops of rope remaining on the periphery of the float and held tightly. He looked out but the ocean was all fuzzy.

He closed his eyes.

CHAPTER063

The blue phone in the bridge buzzed two times before Captain Yamada grabbed the receiver. "One moment." He pressed the red "hold" button and turned to Lieutenant Colonel Okamura. "Crane Three is now locked on the first special container. In turn, the Israelis are waiting on the dock with their two special containers. Do we make the transfer?"

"I thought you said the ship wouldn't be fully unloaded until this afternoon."

"That is true, but it's preferable to load and unload the special containers on a ship that still has a thousand containers left on it. Leaving this step for the end would look far too suspicious — unloading two little containers against the huge breadth of this otherwise empty ship."

"I'm not sure."

"Why didn't you just fly in the contents directly to Israel?"

Okamura stared down Yamada. "Because we weren't sure when we wanted the Israelis to get these containers. Easier to do our trade directly here at the dockside."

"Well, if you're in charge, make a decision. Do we unload now?"

Okamura looked down at the ground in thought. "Go ahead."

The gantry crane started lifting the container. Okamura picked up the satellite radio pack at his feet and started quickly clicking its keypad.

CHAPTER064

anaka's face became red. "I don't care what the Captain says. Return the nuclear containers to the ship. And then just stop everything. I want all the other cranes stopped right now."

Tanaka slammed the satellite phone receiver into its cradle. He leaned against the deck railing, watching the tender pull up to the ship.

A few minutes later, Otzker walked onto the main deck. He was wearing his favorite civilian clothes — tan khakis and a designer polo shirt. He was mid-fifties but looked strong and healthy, and carried himself well. "Tanaka, you should have come with us."

The Colonel glared at Otzker through his thick gold-rimmed glasses. "General, as per our agreement, I have decided to stop the deal. All cranes in Haifa have been halted and our nuclear material has been returned to the ship."

Otzker stood there dumbfounded.

"Furthermore, as per our agreement, I want all the robot containers returned to the ship."

"Have you lost your mind? Everything is going well!"

Tanaka's face grew even redder. His neck veins bulged. "Going well? Yes, for *you*. You have all the robots and you have our uranium and our plutonium. What do we have? Some bomb we don't even know works that's supposed to be tested out in the open ocean in a way that possibly — do you hear me — *possibly* — could not be detected but which defies common sense and probably would be detected, *if* the test is ever done. It won't be, right Otzker? You get the robots and the uranium and plutonium, and then what? Let me guess, some technical malfunction."

"You're losing it, Tanaka."

"I want all the robot containers back on the ship! Do you hear me?"

"Umph…" Otzker stopped in mid-word, not sure what to say next. Then suddenly he put his large hands around Tanaka's neck and started squeezing.

"Let go Otzker. I…am…a karate…black belt…I will…hurt you," Tanaka said between gasps of air.

"I don't care who you are. I've had a perfect record in Testing. You're not going to ruin it." And then he squeezed Tanaka's neck even tighter.

Tanaka fired the bottom part of his open left palm into Otzker's solar plexus.

Otzker's hands instantly let go, and the Israeli fell down onto the deck.

Tanaka heard footsteps. Ilana, Isato, and Benni running toward him on the deck's wooden promenade. Their smiles turned to horror when they saw Otzker grasping his chest.

Ilana lifted the black phone near the bulkhead. "Medical emergency main deck." She ran back to Otzker.

Otzker looked up at Ilana. "I can't...breathe."

"What did you do to him?" Ilana yelled at Tanaka.

Tanaka just shook his head.

Ilana bent down and held Otzker's hand.

Footsteps sounded down the promenade. Two crewmembers with a medical kit and stretcher dropped to their knees. The young doctor, wire glasses with a stethoscope around his neck, remained standing. The crewmembers slapped a transparent green oxygen mask on Otzker's face, a blood pressure cuff around his arm, a portable miniaturized electrocardiogram monitor onto his chest.

"Vitals?" the doctor asked.

"One thirty over eighty-eight. ECG okay — sinus but at one hundred sixteen."

The doctor bent down, put his stethoscope on Otzker's chest, and then slid it underneath the shoulder blades. The doctor moved the instrument over the abdomen, listened for a moment, and then started to palpate the belly.

"Please stop," Otzker said. "I don't want to vomit."

"How do you feel, Sir?"

"Better."

The doctor looked at Ilana. "What happened?"

Ilana looked at Tanaka.

"I'm sorry," Tanaka said. "I... punched the General in the stomach."

"You punched the General?" Ilana said.

Isato's eyebrows went up.

"Please get bloods, including full chemistry —"

Otzker shook his head, pushed against the deck and stood up. "I'm better. Thank you, Doctor. Dismissed."

Tanaka looked at Ilana, Isato and Benni. "Why are you here?"

Benni smiled. "Detonation at Zulu six oh five. Our transducers indicate zero point nine megatons with an error range of four hundred kilotons. We're calling it one megaton."

"Physical effects?" Otzker asked.

"Seismic activity — perfect synchronization. Over ground zero — computer projects small waves, no radioactive xenon release. Farther out at sea — no effects. At the shoreline — computer projects waves under one meter. We'll see a wave here in Yap in less than ninety minutes."

"What about my transducers?" Tanaka asked.

Benni took a piece of paper from his pocket and started reading. "*Code nine nine zero nine nine zero nine nine four seven one eight...pressure derivative overmax...one hundred kiloton plus confirmed.*"

"Thank you." Tanaka looked at Otzker for a split second, then quickly averted his eyes. "I need to make a satellite phone call." He marched off down the deck promenade and turned into the bulkhead.

CHAPTER065

Out of nowhere, a wave appeared on the calm ocean and knocked the lifebuoy against Haruto's face. Haruto reflexively relaxed his grasp and straightened his arms. The small surge on the sea lifted the buoyant float off Haruto.

Haruto's face fell into the water. The salt stung his eyes and he became alert again. He sculled with his hands and lifted his head out of the water. The float was drifting a few meters away.

Haruto started a crawl stroke, kicking his feet and pulling the water frantically, heading for the drifting lifebuoy. After two strokes, he lifted his head and saw that the lifebuoy had drifted even farther from him. He put his head down in the water, started kicking and started the stroke again, but he felt so weak. He stopped and treaded water. He looked at the lifebuoy drifting away.

Haruto started tiring. He slowed the treading and sculling — just enough to keep his mouth above water. Haruto began to drift off to sleep and the treading stopped. He started sinking. The water hit his nose and he woke up and started treading again. Haruto watched the lifebuoy drift away. He looked around him. The ocean was calm. The sun was still shining but lowering in the sky.

He started to cry.

CHAPTER066

aptain Nakamatsu was standing on the bridge, looking out the bow windows at the serene water. They would be at Yokohama in three days. Home, finally. These southerly voyages were way too long. But at least they filled their freezers. It was too hard catching fish near Japan these days.

The boat suddenly shook and Nakamatsu went flying against the plywood wall behind him and fell down to the deck. He got back to his feet and stared around him.

What had happened? He looked out the half-dozen bridge windows wrapping around the room in a horseshoe. The sea was calm. Not a cloud in the sky. He glanced down at the radar screen — all quiet.

Nakamatsu pulled back the throttle and cut the engine. He left the bridge and climbed down the stairs from the high bow to the lower stern deck. The ship was about thirty meters long, most of this length aft of the bridge, where they kept huge rolls of fishing nets.

He saw two crewmen stacking yellow plastic instrument and tool kits near the stern. "Is everyone okay?"

"Sorry, Captain," the young unshaven Japanese man said. "The main engine toolkit, the medical kit and the radio kit went overboard." He pointed to a bobbing yellow case in the distance aft of the ship. "The other kits must be nearby."

"Why didn't you secure this equipment?"

"The wave was completely unexpected. Look at the ocean for yourself. We were just going through the checklists. We were going to secure the kits when we were done."

The Captain removed the large orange-coated binoculars around his neck and handed them to the other crewman, a skinny kid with a smooth face not in his twenties yet. "Takuya, climb up the mast and spot where the kits are. They cost millions of yen. I want them all found."

The youngster raced to the top of the mast. Holding on with one hand, he started scanning the ocean with the binoculars.

"Captain!" he shouted.

"Yes, what is it?"

"There's a man in the water, about a kilometer and a half due north. I think he's drowning."

CHAPTER067

The yellow polypropylene rope landed a few meters in front of
Haruto. He tried to lift his left arm and swim to it, but the arm did
not go anywhere. Instead he sank a bit. The water came into his nose
and burned his eyes, and he started to scull and kick as best he could to
come up, coughing out the water he'd swallowed as he did.

The line disappeared. Had he imagined it?

Then there was a Japanese man in the water with him, a young man,
little more than a teenager. Was he someone Haruto knew, someone come
to laugh at him?

The teenager grabbed Haruto's chin and shoulders, and turned him
onto his back. Haruto's weight pushed the skinny teen under water, but the
teenager started a strong backstroke kick, and both he and Haruto stayed
above the surface.

Finally, Haruto could stop swimming. He tried to turn his head and
make sense of all this, but the world went fuzzy.

Captain Nakamatsu walked over to group. "Is he alive?"

"Barely," Takuya said. The skinny teenager took a plastic bottle of water
from one of the other crewmen and poured a bit into the corner of the
stranger's mouth. Eyes remained closed but the stranger coughed violently
as the water dripped down into his airway.

Nakamatsu bent down and pinched the stranger's cheek. There was a
faint groan but the man did not open his eyes. The Captain put his fingers
on the man's carotid artery. "Very fast. Too fast. He will die soon. Takuya,
climb up and direct me to the kits that went overboard."

"Yes, Captain. They were near that gray buoy."

Nakamatsu gave the skinny teen his orange binoculars again and
walked to the bridge.

"Captain," Takuya yelled from atop the mast. "The buoy is sinking. I
see its top still, but most of the buoy is underwater. The three yellow kits
are all there. Go due south."

The trawler's engine sputtered to life. The ship turned in a wide circle
and headed south again.

Two crewmen picked up the stranger and carried him below to the lounge, laying him on the long sofa there. This man was clad in a pair of white underwear and a soaked, white polo shirt. He had a good growth of stubble on his face, his mustache was bleached, and his forehead was sunburned.

"Do you think he's going to die?" one crewman asked.

The other crewman looked at the stranger and frowned. "You stay with the man, and I'll go up and see what the Captain wants us to do."

The crewman went up onto deck and saw Takuya on the mast yelling directions to Captain Nakamatsu below.

"I see them. Come down now," Nakamatsu yelled back at the teenager. He turned the ship slightly to port, and there were the three bobbing yellow plastic cases. He eased up on the throttle and the trawler glided toward them.

Three crewmen stood on the port side of the stern deck with long wooden grappling poles. As the ship passed near, each man caught a plastic case by the handle and hauled it out of the water.

Nakamatsu snatched the suitcase-size yellow case with the red cross on it and headed down to the lounge.

The stranger was on his back on the sofa, eyes closed, breathing rapidly.

Nakamatsu flipped open the yellow medical kit. He straightened the man's right arm, put an elastic tourniquet around it, and swabbed the inside of the elbow with an alcohol wipe. With thick, calloused fingers, he ripped off the protective plastic around an intravenous catheter and then pushed the catheter's metal needle into the vein there, bluish, barely visible. He quickly pulled out the metal needle, leaving the orange-striped plastic catheter inside the vein. As a drop of blood spilled out of the catheter, Nakamatsu grabbed one of the two bags of intravenous fluid from the medical kit and plugged its clear plastic tube into the orange-striped catheter. He then released the tourniquet and watched with satisfaction as drop by drop of fluid spilled out of the intravenous bag into the small chamber below, then into the line connected to the vein.

Nakamatsu motioned to one of the crewmen. "Hold this bag up."

Fluid dripped into the man, but there was no improvement. He lay there on his back, unresponsive, eyes closed. His breathing was still fast and labored.

"Will he die, Captain?" the crewman asked.

Nakamatsu didn't answer.

CHAPTER068

O tzker was standing on the main deck's starboard promenade, leaning against the railing, looking out at the harbor. He heard footsteps and turned his head. Tanaka was walking toward him with a green frosted bottle and two small white cups.

Tanaka stopped a meter in front of Otzker. He looked him in the eyes. Paused. Looked down at the deck for a moment. "I'm sorry, Otzker."

Otzker smiled. "For punching me when I was trying to strangle you? This mission has been a very stressful one. I know that."

Tanaka lifted up the small bottle. It had a white label with Japanese *kanji* and a graphic of a free-floating golden vine with blue grapes. "It is *Minowamon Junmai Sake*. Have you heard of it?"

Otzker shook his head.

"It's already chilled. Here, please try some." Tanaka popped off the top and poured the glistening liquid into each of the white ceramic cups.

"Let's wait for the wave," Otzker said.

"When do you expect it?"

"Sometime in the next half-hour.   Did you speak to your headquarters?"

Tanaka nodded. "Seismology looks good. Everyone's calling it an earthquake. No one heard the bubble pulse. Our consultant is satisfied with the testing methodology. We are *all systems go* on the project."

The ship suddenly rocked. Otzker and Tanaka saw a thick wave, about a half meter in height, smash into the dock at the end of the harbor.

Tanaka picked up his cup of sake. "*Kanpai!*"

"*Kanpai!*" Otzker repeated. He smiled as the sweetness hit his lips and tongue, but grimaced a moment later at the acidic aftertaste.

CHAPTER069

T he *Mikiyasu-ema* activated its bow and stern thrusters and pushed
off from the wharf. Its decks were bare and the ship rode higher in
the water. But from the vantage point of the high bridge windows,
Okamura could see deep down into the ship's innards, and in the third
bow compartment he saw the two light green containers, replacing the two
valued drab brown containers offloaded only an hour ago.

With its bow and stern thrusters and occasional nudges from its main
engine propeller, the *Mikiyasu-ema* exited Haifa's harbor channel. The bow
and stern thrusters shut down and Yamada ordered engines half ahead. A
few minutes later he changed the command to full ahead, and the *Mikiyasu-
ema* was in the open Mediterranean, sailing toward the Suez Canal for the
long trip home.

CHAPTER070

Takuya pointed to the empty intravenous fluid bag, now hooked onto a loose nail in the wall, its clear plastic tube running down to the stranger's arm. "We have no more bags, Captain. What should we do now?"

Nakamatsu sat on the couch beside the stranger. He placed his index finger under the angle of the man's left jaw, over the carotid artery. "It's much slower. That is good."

The man groaned. The crewmen noticed the yellow urine coming out of his shorts.

The men were shocked, but the Captain smiled. "That is good, too. Get him some clean clothes."

The man opened his eyes, and looked in a confused way at the men standing over him. "Where am I?"

"On a fishing trawler," Nakamatsu said. "We're heading back to Yokohama but we'll drop you off at the hospital in Guam."

The man shook his head. "No, take me to Japan."

"How did you fall in the ocean?"

Haruto smiled at the men. "Thank you for saving my life."

CHAPTER071

Haruto entered the small bridge of the trawler. He could see the Port of Yokohama straight ahead. The Marine Tower, the world's tallest inland lighthouse at over one hundred meters, was instantly recognizable.

Captain Nakamatsu turned his head and smiled at Haruto. "Welcome Mr. Stranger. I see you have shaven your beard off."

Haruto had regained much of his strength over the last three days, and the horror of thirst and delirium seemed like a bad dream. "I wanted to thank you again. I have much money saved in Tokyo. I will return in a few days and pay you."

Nakamatsu put up his hands. "My freezers are filled with tonnes of bluefin tuna. By tomorrow morning they'll be at the fish market at Tsukiji. I have no need for your money."

The bluefin, that's why they were heading into Yokohama here, to avoid registration of the tuna catch. The Captain was breaking the rules. He should be arrested in Japan. But the man had saved his life. Haruto turned to the left. One. Two. He turned to the right. One. Two. Felt a bit better but the man was still breaking the rules.

Haruto bowed to the Captain. "*Domo arigato.*"

Nakamatsu bowed in return, and turned back to look out the bridge windows and study the computer display mounted beside the ship's wheel. He whistled a happy, simple tune.

Haruto stepped out onto the bow. Breaking the rules. Bluefin tuna catches were restricted. The ships had to go to specific docks to have their catch registered. At ten thousand yen a kilo for wild bluefin, that was hundreds of millions of yen. Illegal.

He saved my life, Haruto told himself two times. Then he turned to the other side and repeated the phrase two times. Haruto felt better and walked down the steps to the larger stern deck.

Haruto looked up at the sky. He couldn't see as many stars here with all the light from Yokohama and nearby Tokyo. He breathed in the harbor air. It was not the sweet smell of the open ocean any longer. That was all right. It felt good to be back in Japan.

In twenty-five minutes they were docked at a dim wharf away from most of the activity of the port. Three large refrigerated trucks were waiting at dockside. A half-dozen crewmen scrambled topside and started opening hatches leading to the frozen bins below. The ship's telescoping red crane, now extended to about three meters, went into action and started winching up a load of fish from the most aft compartment. The crewmen were smiling and joking. They should be. A long voyage was over and, as was the tradition, a portion of the excellent profits would be going to each man.

Haruto went to the port side and disembarked. He approached the driver of one of the white refrigerated trucks, "Can I get a ride with you to Tokyo?"

"Get in," the young driver said.

Haruto climbed into the left side of the truck's cab. Twenty minutes later, he heard the rear door rolling down. The driver started the diesel, the overhead refrigeration unit whirred to life, and the truck jolted forward.

The driver said nothing — eyes focused on the road in front of him — and chain-smoked the whole trip into Tokyo. There was traffic even at this hour, and it was not until a few minutes before ten that they arrived at Tsukiji Fish Market.

Over a hundred trucks were jostling for position outside the market. Thousands of tons of seafood would be unloaded in the next few hours for auctions starting at five o'clock the next morning. The driver got into one of the queues.

"Thank you," Haruto said, and left the truck. Wearing a clean white t-shirt, a faded pair of blue jeans and a pair of black, plastic sandals, Haruto walked out of the market's loading yard toward nearby Ginza. His mustache was gone. The sunburned skin on his forehead was still peeling, and his tall, slim frame had edged toward skinny.

CHAPTER072

Haruto glanced up and saw the neon lights of the Ginza Mitsukoshi Department Store. The *koban*, the community police station where Itou worked, was just past the store, a tiny red brick building, topped with an enormous silver pyramid.

He actually hoped Itou was not on duty. What if Itou hadn't received the file on his cellphone? What if this whole adventure was for nothing?

Haruto got to the Ginza Koban and went in.

Itou was yelling at two teenage boys in tight blue t-shirts with oily, slicked down hair. "What were you thinking? Why did you do this?"

One of the boys, a good few centimeters taller than Itou, looked remorsefully at the floor. The other boy had tears starting in his eyes.

"I'm sorry," the tall boy said. "It was stupid. I have no idea why we did this. We just said, 'let's grab it' and we did."

"You did it because you had no beer money and the purse was there. That's why you did it! I am going to tell your parents and your school, and I don't know how long you will be in prison, but no one will give you a job when you get out."

The shorter boy started crying.

Itou was about to scream at the boys again when he noticed Haruto standing in the lobby. "Inspector Suzuki!" Itou turned back to the teenagers. "If I give you a chance — one chance — will I ever see you here again?"

"No, Sir," both boys answered.

"Get out of here!"

The teenagers ran out of the police station.

"Did you receive... uh..." Haruto asked.

"The entire file," Itou said. "I have not told anyone about it. What is our next step, Inspector?"

"When is your shift over, Policeman Itou?"

"Midnight."

"Do you know the all-night *yakitori* underneath the Yamanote Line tracks?"

Itou nodded.

"Meet me there."

Haruto left the community police station and walked down the road. The video file had made it.

The neon-bathed chrome and glass and glossy posters of Ginza changed abruptly just south of the Yurakucho subway station. Beneath the elevated train tracks was a rabbit warren of *yakitori* — inexpensive restaurants that few tourists or executives visited, but that nourished the *salarymen* and *office ladies*, as well as the workers that kept the city going.

Haruto sat down on a bench to wait for Itou. Sweet smells of skewers of chicken roasting in the nearby restaurants drifted over, and Haruto's stomach growled.

He had no money on him and would have to wait.

CHAPTER 073

"**Inspector Suzuki!**"

Haruto jumped off the bench. It was Itou. Haruto smiled, relaxed his muscles and gestured to the young man to sit down beside him. "Thank you for your help, Policeman Itou."

"I am honored to serve you, Inspector." He reached into his pocket and pulled out a thin orange memory stick. "Here's a copy of the video you sent to me."

"What did you think of it?"

"After the Koreans fired that missile, I'm glad our military will be acquiring nuclear weapons."

"I, too, would like my country to be strong. I'm not against standing up to North Korea. But if we get nuclear weapons, it has to be done by the government, not the military, with the knowledge of the people," Haruto said. "In the 1930s the military broke the rules. They assassinated the politicians they didn't like. They took over control of our country. Look at the results — Hiroshima, Nagasaki, and the loss of honor from which we still haven't recovered. We must expose this plot. The rules must be followed."

"But, Inspector... we're only police officers. I work in a police box, handling drunken office workers, lost tourists. What do I know of international affairs? There's probably a good reason the rules are being broken. If it became known that we were getting nuclear weapons, an uproar from even some of the population would be enough to make the current government fall, and Japan wouldn't get the weapons, and... maybe North Korea would really destroy us. Are not matters of such consequence beyond our rank?"

"My job as a police officer was to investigate the murder of Co'en Satoki. Following up on his murder has led me to this plot. The murder case must be resolved properly, the military plot must be exposed. Those are the rules."

"Inspector, according to those same rules, I must arrest you," Itou said.

"That is the military abusing the rules. I murdered no one. They wanted the lights off that night on the wharf, so they snapped their fingers and cut the electricity. You saw that with your own eyes… They don't want the truth revealed, so they just go ahead and charge *me* with murder. It is they who are breaking the rules."

"I believe you, Inspector. That is why I'm helping you. I put my career in your hands. Tell me what you want me to do."

"Maybe you shouldn't, Policeman Itou," Haruto said with a tired voice. "In the last week I've been shot and wounded twice. Then I was left in the ocean for days to die. I sit here penniless and hungry and homeless — afraid to go to my bank account or my apartment. I'm hardly an example of leadership."

"Nonsense. A lesser man would've given up by now."

"Thank you."

"What's our next step?" Itou asked. "What are you going to do with the video?"

"The video is not enough," Haruto said. "We need proof to corroborate the video. Either of nuclear weapons in Japan or of the Mikiyasu robots in Israel."

"How many robots are going to Israel?" Itou asked.

"About a hundred thousand each shipment."

"Finding a tiny nuclear weapon somewhere on a military base in Japan will be next to impossible."

"Finding hundreds of thousands of robots out in the field in Israel or Lebanon might be much easier," Haruto said.

"Do you speak Hebrew?"

"No, but… my victim's wife's family in Israel does. I think she would help us."

Itou took a notebook out of the upper right pocket of his utility vest. "What's her name?"

"Co'en Ayaka. She lives off Minami-dori, the tall steel building."

"I have a day off tomorrow. I'll make contact with Co'en and if she agrees, we'll meet you right here at noon. Even if she is under surveillance, nobody will be able to follow her through the lunch-time crowds near the *yakitori*." Itou pulled out his wallet, took out all the yen notes, and shoved them into Haruto's hand. "When no one believed in me, you did. Now I believe in you, Inspector."

Itou bowed to Haruto, and left.

Haruto walked into the *yakitori* across the alleyway. About fifty red stools in front of a sinuous melamine counter were crammed into the restaurant. Some dozen *salarymen* were nursing drinks or had their heads down on the counter, trying to pass the night here until they could go back to their offices again in the morning.

The waiter came up to Haruto with a large plate of two skewers of charcoal barbecued chicken — *yakitori*, from which the restaurants took their name, a bowl of steamed rice and a can of Asahi Super Dry Beer. Chicken was the only item on the menu now, but they served it immediately. Haruto nodded and the waiter left the sweet smelling feast on the counter in front of him. The waiter flipped open the pull-tab on the silvered beer can and poured the contents into a tall glass.

Haruto wolfed down the meal, licking the last bits of the *yakitori's tare* sauce — sugar, sake and soy sauce — off the bamboo skewer. He put his head down on the counter and drifted into a deep sleep.

CHAPTER074

Somebody started shaking him. "Why are you here? ID please!" Haruto opened his eyes. The tall policeman was staring down at him. Across the room another policeman was going from seat to seat, as the late-night denizens of the *yakitori* showed ID's.

"I lost my wallet."

The policeman kept staring. Haruto's heart jumped a beat. "Your name! Where do you work?"

"Uhh…Takahashi…uhh…Ichiro."

The policeman pulled a PDA out of his utility belt and started scrawling on the small touchscreen with a stylus. "Too many Takahashi's. I need more information." The policeman hit the screen a few times with the stylus and Haruto saw head shots flash on and off the PDA's blue display. Suddenly the policeman jabbed again at the touchscreen and the images stopped changing. A black-and-white photo of Haruto filled the entire display.

The policeman raised his nightstick. "You are Suzuki! There is a warrant for your arrest for murder. Hands up! Lie down on the floor!"

"No, you are mistaken. I am Takahashi Ichiro. I am a maintenance worker over at Ginza's —"

"Hands up! You are under arrest!"

As Haruto slowly raised his arms, the policeman looked down and reached for a nylon hand restraints band. Haruto suddenly swiveled on the stool, and from the sitting position, thrust a modest *yoko-geri* side kick into the policeman's belly, pushing him down to the ground.

Haruto leapt from the seat and ran toward the door. The other policeman bolted toward Haruto and grabbed his right arm. Haruto pivoted and jabbed his left hand into the policeman's solar plexus. Haruto pulled free, ran and jumped over the *yakitori*'s counter, through the tiny kitchen, and out the back of the restaurant into the cramped alley.

CHAPTER O75

H **aruto ran** to the end of the alley, turned left and ran down the next one, turned right, ran down another one, and stumbled onto the street beside the Yamanote Line tracks, exhausted.

The hotels all wanted a credit card at check-in... except the capsule hotel. Maybe it would be all right now. After surviving a shark attack and the Pacific Ocean, lying down to sleep in a luxurious capsule should be manageable.

Haruto reached the capsule hotel. The twelve-story building looked like a typical business hotel. The lobby and registration desk were quite ordinary. However, in the floors above, instead of separate rooms, there were nearly a thousand capsules, stacked two high. Each capsule was a little over a meter squared and two meters deep. Inside the capsule was a television-computer and clean bedding. Bathrooms and showers were in a common area on each floor. The capsule was small, but so was its price — seven thousand yen for the night, less than a quarter what a real hotel room would cost.

Haruto paid for the capsule with cash, and took the elevator up to the seventh floor. He stepped out into a long tight corridor. On both sides of the corridor, from floor to ceiling, were white fiberglass capsules. Fluorescent light from the white ceiling bathed the white vinyl tile floor below, giving the place the feel of a hospital.

The blue magnetic card Haruto held in his hand read *Number 783 — Please Insert*. Haruto started walking down the corridor, looking at the bright red LED displays below each capsule. *Number 712 — Occupied*. The shade in its window was down. On top of this capsule was *Number 713 — Free*, its clear plexiglass door revealing inside a small mattress, pillow and computer keyboard-remote control.

Haruto passed one white fiberglass capsule after another. The corridor, long and white, seemed to go on forever. Finally, midway, the corridor branched off in an H shape to another corridor. Haruto walked down that branch. *Number 734... Number 736... Number 738...* Capsule after capsule crammed both sides of the white corridor from floor to ceiling. Most capsules had their shades down, with the occasional one empty.

At the end of the branching corridor, *Number 740*. Haruto turned left and walked down the new corridor, which looked identical to the other ones — white ceiling, white floor, white fiberglass capsules on top of each other stretching down to the end of the hallway.

Number 770. Haruto kept walking down the tight corridor. *Number 782*. He looked up — the LED sign underneath the top capsule read *Number 783 — Free*.

Haruto slid the blue magnetic card through the slot of the LED sign. *Welcome. Welcome.* The message scrolled down. *Please insert card again to lock capsule. Check Out time is 11AM. Number 783 — Occupied.* The latch on the plexiglass door clicked and unlocked. The small reading light and television-computer screen in the capsule lit up.

Haruto swung open the plexiglass door of the top capsule. Clean fluffy pillow. Clean sheets and blanket on the small mattress. There was a pleasant odor too, reminding him of the countryside at Chichibu-Tama.

Haruto looked up and down the corridor. So quiet. Hard to imagine there were nearly a hundred other people on this floor, sleeping or reading or watching television in their tubes. Haruto looked at his capsule again. It would be nice to put his head down.

Haruto grabbed the shiny chrome U bar mounted to the side of the top capsule, put his foot on the small black ladder step and started pulling himself up. For a moment, he looked up and down the corridor, a blur of whiteness in either direction. Then he looked inside his capsule.

It was a fine capsule. Other capsule hotels were cheaper. This one had some of the newest and best capsules in Tokyo. He'd have a comfortable sleep here.

Haruto let go of the chrome bar and put his foot back down on the floor. He breathed in deeply through the nose and exhaled slowly through the mouth. One. Two. Haruto tapped quietly, so as not to wake anyone, two times on the left side of the capsule, and then two times on the right side. Now make some rules. Climb into the capsule and search the Internet for his name and for Co'en's name. Then search for information on the *New Pacific Queen*, General Otzker and Colonel Tanaka. Then catch up on sleep. In the morning, buy new clothes before the noon meeting.

Going into the capsule head-in felt strange.

Haruto took a deep breath, gripped the chrome bar, and headfirst, pushed himself into the fiberglass tube. So far so good. The mattress was comfortable, the pillow smelled fresh.

The tube that made up the capsule was over a meter high, and Haruto was able to lean over and close the plexiglass door, then pull down the privacy shade. Haruto fluffed up the pillow and lay back with his head against the right corner of the tube, looking straight at the television-computer flatscreen built into the left wall of the capsule. He breathed in. All right, now follow the rules.

Haruto took the television-computer remote control and entered the *kanji* for *Suzuki Haruto*. Tens of thousands of sites came back. Haruto scanned the list on the first page. None of the references to *Suzuki* or to *Haruto* had anything to do with him. He then put quotes around his name for an exact match. Hundreds of sites came back, most nothing to do with him, and the few that did from community or work newsletters where his name was included in some inconsequential manner. None from any news sources.

Usually the police released a fugitive murder suspect's name and photo to the media. They had not done so for him.

Haruto then googled Co'en's name. A few dozen sites, all local Japanese ones, came back. Haruto clicked the *Yomiuri Shimbun* newspaper site. *Co'en Satoki, the founder of Co'en Electronics, an export firm in Shibuy-ku, Tokyo, found dead in the bathroom... body not discovered until two days later... coroner rules death possible suicide ...*

Haruto looked around the tube. Meticulously clean white fiberglass walls... comfortable mattress... quiet... fresh smell... it wasn't so bad. All right, continue to follow the rules and then get some sleep.

Haruto googled *New Pacific Queen.* Thousands of hits came back. *Scuba Tours... Scuba Diving... Cruise Vacations... Lowest Prices on Cruise Vacations...* and more of the same on the second page. Haruto googled *General Otzker.* Top search result: *IDF - Israel Defense Forces — The Official Website — The General Staff.* Haruto clicked it.

Brigadier General Moshe Otzker. Director of Weapon Systems Testing. Born Sao Paulo, Brazil. Made aliyah to Israel 1975. Holder of a B.Sc. in Chemical Engineering and a M.Sc. in Electrical Engineering from Technion... Brigadier General Otzker is married and father to four children. In the top corner of the screen was a color portrait photograph of Otzker in uniform against an Israeli flag in the background. Haruto smiled. Good. Some corroboration of the video file made on the ship.

Haruto googled *Colonel Tanaka.* There seemed to be a number of different Colonel Tanaka's. Finally, on the third page of hits, in an online U.S. Air Force Air Wing newsletter from 2006 there was an image of *Colonel Tanaka, Iraqi Reconstruction Support Airlift Wing Commander visits the 386th Air Expeditionary Wing.* Haruto clicked the site. He bent his head closer to the LCD screen and studied the color photograph. No, it wasn't his Tanaka.

Haruto yawned. He'd do some more Internet research before leaving tomorrow morning. Corroboration of Otzker's identity was good. If he could get corroboration of the Mikiyasu robots deployed in Israel, that might be enough. Haruto clicked the row of ON/OFF buttons near the pillow. The television-computer went dim and then the reading light turned off.

It became pitch black inside the capsule.

Haruto put his head on the pillow and closed his eyes. All right, fall asleep now. Follow the rules.

A minute went by, then another. Nothing. He couldn't fall asleep. Yes, it was a tube, but the door was right there. Haruto felt the plexiglass door at the other end of the tube with his bare toes.

Haruto felt his heart beating. He breathed in deeply through the nose and then slowly exhaled through the mouth. One. Two. Haruto tapped lightly on the left side of the tube twice, and then he tapped again twice on the right. That felt a bit better. All right, go to sleep now. Haruto took a deep breath in.

Something about the air seemed different. Haruto felt with his hands for an air duct. On the left side of the capsule, his hands glided over the smooth fiberglass surface of the tube. On the right side were the control buttons, but he felt no air duct. He had never slept in a capsule before — maybe he had turned off a switch for the air vent, just as you could in your car.

Haruto tried another deep breath. The air didn't seem go into his lungs. Or maybe there wasn't enough oxygen in the air. Haruto started breathing faster. He started to feel his heart racing again. He opened his eyes, but the tube was so dark he couldn't see anything.

Haruto tried to feel for the light switch but it wasn't there. He breathed even faster while his hands frantically swept all over the tube wall for the switches. His hands and arms started feeling numb. The choking sensation became overwhelming. Haruto could barely suck in any air.

Going to pass out soon. Do something.

Haruto bent over, banging his head by accident against the top of the tube. His hands moved up and down, searching in vain for the release latch of the plexiglass door. He was breathing faster but no air seemed to be going in. A pain shot through the front of his chest. He kicked on the plexiglass door. Nothing happened. Kicked again. Nothing. The plexiglass was thin but strong.

Control the pain. Breathe in, get air in. Must get out before pass out and die.

He kicked on the plexiglass door again. No good. Must get out. Must not die.

Perspiration covered his body. He was panting. No air. His heart smashed against his chest with each beat, and he felt the throbbing in his ears.

Haruto kicked the plexiglass door again. Nothing. He could barely breathe. He was going to pass out. He kicked again. Then again, and again, harder and harder. Finally, the small latch popped and the door flew open. Haruto pushed with his numb hands against the tube walls and fell out of the capsule, landing on the floor below.

The ruckus had woken up Haruto's capsule neighbors, a few who had come out of their own capsules, some bothering to dress, others standing in underwear. Lying on the floor, Haruto was sweating profusely, breathing rapidly with his mouth open, and clutching the front of his chest in agony.

A middle age man in underwear shouted, "He's having a heart attack!" He reached into his capsule, opened his cellphone. "We need an ambulance. Heart attack. Capsule Inn Ginza, Seventh Floor. Near capsule seven-eighty."

Haruto looked up at the crowd of men around him, but his vision was blurred. His heart was pounding against his rib cage. He tried to breathe in deeply but could barely get any air in. It didn't feel right. This time he wasn't going to make it. Too weak. The world started spinning around him.

Another man in the crowd bent down and took Haruto's pulse. "Your pulse is strong and you're breathing. But hang on there, the ambulance is coming. Can you hear me? Do you understand me?"

"Yes." Haruto nodded. "Thank you."

The air suddenly felt a bit fresher. Haruto tried to take a deeper breath. It felt like a bit more air went into his lungs.

Haruto closed his eyes. He counted to four. Then he took a deep breath in again, tried to keep it there, and exhaled slowly. And then he counted to four and breathed in again.

The air was definitely better quality here. He opened his eyes and saw the crowd of men, this time in focus. The numbness in his hands and arms seemed to be going away.

Haruto breathed in deeply this time. The dizziness was gone. The chest pain was fading.

As Haruto stood up, the ambulance crew came barreling down the corridor with their stretcher.

Haruto shook his head at them. "Thank you for coming, but I'm all right."

"You should come to the hospital to make sure you're not having a heart attack." The ambulance attendant flipped open the large plastic yellow box attached to the stretcher and pulled out a handful of colorful electrical wires. "Let us get an ECG of your heart, okay?"

"No, I'm fine. This has happened to me before. It's never a heart attack."

Haruto grabbed his sandals, the ones the fishermen gave him, and ran down the corridor, squeezing past the onlookers. He took the stairs down to the lobby and ran out to the alleyway behind the capsule hotel.

Haruto felt drained. He saw large pieces of box cardboard beside the dumpster and lay down on them. He closed his eyes, and fell into a deep sleep.

CHAPTER076

Haruto jostled against the mob of people from the *yakitori* underneath the elevated Yamanote train tracks. Despite the humid heat of summertime Tokyo, he wore a white shirt with starched collar, blue tie and a loose dark blue jacket. At 180 centimeters, Haruto stood a bit above most of the crowd. Tall and slim, freshly cropped hair, tanned, clean shaven and sans mustache, teeth brushed to a white polish, newly pressed clothes and an energetic gait — a few *office ladies* turned their heads as he passed.

Haruto looked for the bench. There, in front of the restaurant he visited last night. Itou was there, but he didn't see Mrs. Co'en.

"Hello, Policeman Itou."

"You look much better today, Inspector."

The woman with long, straight black hair sitting on the bench beside Itou turned to Haruto. "Hello, Inspector."

Haruto looked at the middle-aged white woman for a moment and smiled.

"Although the police have ruled the case *possible suicide*, Mrs. Co'en has been interviewed at length by Defense Intelligence Headquarters," Itou said. "There is still active surveillance on her, but I don't think they, or even the Korean mafia, could find someone near the *yakitori* at lunchtime, at least not someone wearing a black wig."

"Good work, Policeman." Haruto turned to Mrs. Co'en. "Your husband's death was not a suicide. I believe he was killed by North Korean agents, and I will gather the necessary proof to solve this case with truth and with honor."

"I know, Inspector. Policeman Itou has filled me in on some of the details. You have my support and my gratitude for your efforts."

Haruto paused. "I need to get to northern Israel, possibly Lebanon. I need contacts and support in Israel. I need help with the language. I'll probably need large sums of money for logistics."

Mrs. Co'en smiled at Haruto. "All you need is a passport and plane ticket, Inspector."

Itou gave Haruto a large tan backpack. Haruto unzipped the main compartment. On the floor of the backpack, was a Japanese passport, a glossy white and red JAL ticket folder, a thin no-name digital camera attached to a massive ivory white Canon telephoto lens, and a bulky plastic Casio watch.

Haruto reached down for the red passport with distinctive gold letters on its cover, and opened it within the confines of the backpack shell. "Who is Aoyama Takamichi?"

"Visitors from Hokkaido," Itou said. "Some teenagers snatched his wife's purse while the couple was shopping in Ginza. Mr. Aoyama is about your age."

Haruto stared at the picture in the open passport. "But I don't really look like him."

"Close enough for Western eyes," Itou said.

Haruto opened the Japan Airlines ticket folder. On top was an itinerary. *E-ticket. JAL 5091 Depart Tokyo NRT June 21 23:30. Arrive Istanbul IST June 22 7:00.*

Underneath another piece of paper was stapled. *Turkish Airlines 1184 Depart Istanbul IST June 29 7:50. Arrive Tel Aviv TLV June 29 9:45. Hotel voucher. Aden Otel Kadikoy June 22 — 29. Meal Plan included.*

"Why Turkey for a week?" Haruto asked.

"It was cheaper than Singapore," Itou said with a small grin. "I didn't want a direct flight to Israel. Right now, you're a Japanese tourist with a big camera. It'll be harder for Defense Intelligence to keep track of the thousands of other Japanese that may include Israel indirectly in a tour of the Middle East or Europe."

"How did you pay for the ticket?"

"With yen. The credit card number that purchased your tickets belongs to my friend who owns the travel agency."

"Very good, Policeman Itou." Haruto picked up the Casio Pathfinder GPS watch, and put it on his left wrist. He clicked the middle left button. The blue watch face lit up. A high-resolution map of the small section of Yurakucho and Ginza around the *yakitori* appeared. Clicking a few more times zoomed out to a less detailed map of Tokyo and then of Japan.

Haruto bowed to Itou and Mrs. Co'en. "This is excellent. Thank you for your help. I'm sure I can get to northern Israel, but..." He turned to Mrs. Co'en. "I still think I'll need some money and some help in Israel. Could you give me the phone number of a friend or relative?"

"I can do better than that," the woman said.

CHAPTER077

Haruto, **Itou** and Mrs. Co'en took the subway to Shibuya Station and then the Number 3 bus to the Gakkan Mae Girls School. They walked to the corner.

The Jewish Community Center stood out with a seven-meter menorah and two-meter Star of David in its concrete façade. Haruto pointed to the compact white Honda parked across the street, the driver watching the building. "Surveillance — probably Defense Intelligence."

Mrs. Co'en took her cellphone and quickly dialed. "Can you come down and meet us at the Rabbi's entrance? Thanks." She turned to Haruto and Itou. "Follow me."

The three crossed the street and instead of walking down to the community center, went into the lane cutting behind the buildings. They walked around the dumpster and came to the rear of the Jewish Community Center.

A white woman with light brown hair was holding open the metal door. "Ayaka, how are you doing?"

"It's still hard. Thank you for the meal you brought over."

Haruto, Itou and Mrs. Co'en followed the woman up a flight of stairs through an oak-paneled corner office. "This is the Rabbi's study. If a sermon goes bad, he can run out here, away from the congregation." Haruto and Itou kept straight faces but Mrs. Co'en smiled weakly. They walked into a smaller office across the hallway.

"Please have a seat," the brown-haired woman said. "I'm in charge of the Kibbutz Program and Cemetery Burials."

Haruto raised his eyebrows. "Is a cemetery burial common after participating in the kibbutz program?"

The woman started laughing, almost uncontrollably, but quickly stopped when she saw no one else was smiling. "I'm sorry for your loss, Ayaka." She turned to Haruto. "And no, despite all the news reports you see coming from Israel, none of our members has ever died on a kibbutz due to terrorism. The kibbutz program used to be *very* popular. The Japan Kibbutz Association had thirty thousand members in the seventies and eighties."

"What happened?" Haruto asked.

"All the attacks on Israel you see on TV. Everyone is too scared to go now. We won't have more than two hundred volunteers this year. That's why I do both jobs here."

Haruto nodded. "Can you tell me more about the program?"

"You pick a kibbutz from my catalog here. You fly to Israel, go to the kibbutz and become part of it. The kibbutz provides you with food, lodging and expense money. It's very Buddhist. You own nothing — there are no individual possessions on the kibbutz."

"What do I do on the kibbutz?"

"Ayaka said you would be willing to go to one of the northern kibbutzim. With all the rockets fired from Lebanon, we haven't had many volunteers for them. There's Kibbutz Misgov just at the Lebanese border. It has apple orchards and a poultry operation. So I imagine you'll be picking apples, feeding the chickens, that sort of thing."

"Do I get time off?"

"Of course. You'll have plenty of time to travel around," the brown-haired woman said. "There's a fee of twenty thousand yen, which supports the Japan Kibbutz Association."

Itou stood up and opened his wallet. He gave the woman two ten-thousand yen notes, then handed Haruto the remainder of the money.

The woman opened the large green folder on her desk, thumbed through it and removed four orange sheets stapled together. "This one gives you more information about Kibbutz Misgov. This one is your personal reference — Ayaka has already signed it. This is the medical form — have a doctor fill it in before you leave. When you get to the kibbutz, you also have to give them this top sheet filled in with your name and address, which tells them that I referred you. Every few months the main association correlates the results and sends us a report. It's a bit low tech."

"Sometimes low tech is better," Haruto said.

CHAPTER078

Haruto waited his turn in the long line, holding the tan backpack that now bulged with extra pairs of underwear, socks, jeans and sneakers. He looked out of the massive bay windows of the airport lobby. It was still light outside — *Xiazhi* — the summer solstice had started. That day always scared him. He liked the sunlight, so he should be grateful for the days of the year with the most daylight. But it was also the start of the days that would grow shorter. Something deep inside him hated those dark days of winter and started to rejoice as soon as he sensed, usually by February, the days lengthening.

The queue moved quickly. Haruto came to the head of the line. A small poster in Japanese and English read:

New Security Rules

• *Discard all liquids now! No liquid allowed in luggage or aboard plane.*

• *All cameras, laptops and electronics must be taken out of luggage and handbags and be checked at security.*

• *Present passport and one other picture ID to ticket agent for flights to the USA, one government picture ID for all other flights*

Haruto looked at the ticket agents. In lieu of what used to be a superficial glance at identification, they were all now carefully inspecting passengers' documents. Haruto started to feel his heart beat faster. He turned to the left and tapped twice on his right forearm and then he turned to the right and tapped twice on the left one.

"Next!" A thirtyish woman at the right check-in counter waved Haruto to come over. "Ticket please."

Haruto took out the JAL e-ticket. The ticket agent looked at it for a second, typed a line of text into her keyboard and looked at the flatscreen in front of her. "The flight is operated by Turkish Airlines. You could've avoided the line here by going to their check-in counter." The woman pointed to the counter toward the end of the bay windows. A young Caucasian man was behind the Turkish Airlines counter, checking in a couple with two large suitcases, but otherwise no one else waited there.

"Sorry, I'll go over there right away."

"Oh, it's not necessary. Once you're here, I can do the check-in. ID please."

Haruto froze for a second. Ignore the JAL agent and walk over to Turkish Airlines where Western eyes would examine his passport? No, that would arouse suspicion. But if he stayed here, she would see that the passport was not his and call security. Ignore the agent. Just go to the Turkish Airlines counter. Haruto looked to the left and silently counted to four. He then looked to the right—

"Sir?"

"Sorry." Haruto took a big breath in and reached into his front pocket for the passport.

The agent opened the red document and looked at the main page, then looked at Haruto for a few seconds. Haruto felt his heart smash against his chest with every beat. It was over. She'd call security and he'd end up in jail.

No!

Suddenly Haruto blurted out, "Colitis."

"Colitis?"

"Yes, colitis."

"I don't understand."

"I had colitis last year and lost ten kilos. Do you know what that is?"

"Yes," the agent said. "One of my friends had it."

Haruto pointed at the camera and telephoto lens. "After I got out of the hospital I decided to travel and take the year off."

"You look fine now. I have almost the same lens as that one. EF300. Very good close-ups."

Haruto nodded.

The woman gave Haruto back the passport and a rectangular cardboard boarding pass.

Haruto walked to Security Check. The young security agent detached the long Canon telephoto lens and peered through it carefully. He gave the white lens back to Haruto and moved him on to Passport Control. After Haruto waited about ten minutes in line, the government agent spent all of ten seconds scanning Mr. Aoyama's passport into the computer, looking at the LCD screen, and then waving Haruto off.

Haruto walked down the long, wide corridor to the boarding gate for Turkish Airlines Flight 51, sat down, and waited.

Black Sea 31.71°N, 43.92°E,
Turkish Airlines Flight 51
June 22   5:30 AM Istanbul Time (02:30 Zulu)

CHAPTER079

t was almost noon back home. Haruto couldn't sleep anymore. He looked out the Airbus window and watched the sun rise magnificently over the Black Sea below as the plane started its descent.

An hour later the airplane touched down at Istanbul's Ataturk Airport. With his backpack, telephoto lens-equipped camera and Japanese passport, Haruto breezed through Customs. He walked out of the airport into the early Turkish sun, and showed the hotel voucher to the taxi dispatcher.

"Save your money. Take the bus to the train station and walk to the hotel," the Turk said in accented but clear English.

The train station bus came a few minutes later. Haruto paid the twenty-one-dollar fee and sat in the front coach seat. The old and new of Istanbul streamed past as the bus made its way to the Bosporus, the strip of water connecting the Black Sea to the Mediterranean that divided the city.

The olive-skinned, balding driver hummed to himself as he guided the large coach through the city's narrow streets. He turned to Haruto. "I show you Istanbul?"

Haruto nodded, then for the next half hour stopped breathing each time the driver let go of the steering wheel and with both hands pointed out landmarks through the bus's panoramic front window. The driver saw Haruto's apprehension and started laughing. "I bigger than cars. It is they who should be afraid."

Finally the bus climbed the Bosporus Bridge. The driver pointed to the rear of the bus. "That is Europe." As the bus came onto the other shore of the bridge he said, "This is Anatolia — *Asia*. Now you are home." The driver laughed and continued pointing out sites to Haruto.

The bus soon pulled into Haydarpasa train station, a baroque sandstone building with large circular turrets, right on the waterfront. Haruto tipped the driver an American five and got out. He went over to the traffic policeman standing in front of the station and showed him the hotel voucher. The policeman pointed down the street to the right.

Haruto walked down the wide street into the neighborhoods, away from the waterfront. The magazine on the airplane said that this area of Istanbul, Kadiköy, was an area of ancient Christian monasteries and churches, but the street looked like a middle-class one in Japan, with tight neat apartment buildings pressed together. He passed a Kentucky Fried Chicken restaurant on one corner. On the next was a medium-sized mosque with a single minaret and what appeared to be a supermarket in its first floor. Haruto took a closer look. *Süpermarket Türk.* People could do their grocery shopping *and* pray. Certainly more efficient than the Christians.

On the next block was a row of small shops. On the second floor were signs for *Avukat, Doktor, Ingiliz* something, probably classes or tutoring. The words sounded close to their English equivalents. Then Haruto spotted a second-floor sign for *Karate-Do.* He smiled to himself.

Haruto continued down the street and looked around to get his bearings at the next block. He was getting too far from the waterfront. He turned right and after a few blocks turned onto Uzun Hafiz Street, and walked back toward the water. In five minutes he was on the road that the traffic policeman had mentioned, Rihtim Street. Haruto walked down the large boulevard parallel to the waterfront and was soon at the Aden Hotel.

CHAPTER080

"**B**attalion! Attention!**"** Major Gershon yelled into the mega-phone.

On a large dirt field in front of a steel building a thousand soldiers, each one next to his squad of thirty robots, snapped to attention in synchrony. So did the thirty thousand Alpha and Beta robots, slapping hands on their metal torsos, creating an explosive, resonating boom.

"Major," Menachem said.

"Yes, Sir?"

"I don't think it's a good idea to have all the robots out in the broad daylight together like this."

"I agree, Sir. Just wanted you and Colonel Tanaka to see Mustang Battalion. Basic training completed. Mission training to start tomorrow."

"Be quick, then."

Gershon, a tall man with large forearms protruding from rolled up sleeves, pointed to the sea of men and machines massed on the field. Menachem and Tanaka squinted against the rising sun to try to better pick out the soldiers, dressed in drab brown fatigues and wearing camouflage makeup, from the waves of drab brown Alpha and Beta robots facing them.

"Each squad consists of twenty-two Alpha robots and eight Beta robots," Gershon said. "Every soldier leading a robot squad has been to Lebanon before. Some are low-level rank, but many are sergeants."

"Where'd you get these men from?" Tanaka asked. "Are any of them new recruits?"

"No, no," the Major said. "*Everyone* in front of you — well, except for the robots — has seen combat before. Many of the soldiers are from Sayeret Golani, our best recon company."

One thousand soldiers in full combat gear, twenty-two thousand upright Alphas, and eight thousand monster-cockroach-like Betas carrying trunks of ammunition, containers of extra fuel and ration packs, stood dead still in complete silence.

Gershon did not need the megaphone. "Mustangs Disperse!" he yelled with his own voice.

CHAPTER 081

T he call of the muezzin filled the neighborhood. Distorted by too-small loudspeakers on the mosque's minaret, its message to the faithful was clear enough.

Well-rested, showered, in a clean gray t-shirt and new blue jeans, Haruto sat on his bed looking at the room alarm clock. The display switched to *6:00*. Good, the breakfast buffet was open now. Haruto jumped up and took the stairs down to the hotel's *Le Bistro* Restaurant. But a bus boy in a light gray uniform at the door raised his hand. "*Bir dakika.*"

Haruto obeyed, standing quietly and salivating as the sweet odors of the Turkish breakfast wafted out of the restaurant. A small line, mainly of middle-aged men in business suits, soon formed behind Haruto. A minute later the bus boy waved Haruto and the crowd into the breakfast room. "*Gel, gel.*"

Haruto was possessed of an appetite he never knew he had. He loaded up his plate with four soft-boiled eggs, a half-loaf of sourdough bread still warm from the oven with steam rising from its surface, a quarter-slab of natural, uncolored butter, a serving of oil-cured olives, a slice of white sheep-milk cheese, and a fragrant half-melon. Juggling the heavy plate with one hand, with the other Haruto filled a huge glass to the top with orange juice.

Haruto attacked the melon. He did not remember a fruit as sweet as this one. With his bare hands, he swabbed the fresh bread into the butter and the eggs and shoved the chunks into his mouth, washing it all down with gulps of orange juice.

It did not take long for Haruto to empty the plate and feel satisfied. With his belly full, he walked out into the lobby and sat down on the couch across the registration desk.

Haruto watched the occasional guest trickle down into the lobby. Some went into the restaurant, some to the checkout desk, and some left the hotel. One of them was a young clean-shaven man in a sharp navy suit and bright red tie. Should he go put on his suit too? No, he was a tourist. He was not *Keibhu* here.

Haruto shuddered. He had worked so hard to achieve that rank. Would he ever be *Keibhu* again in Japan?

A new call of the muezzin filled the air again. Haruto hopped off the couch and went outside.

The Istanbul morning was wonderful. Sun not too bright. Cool, refreshing breeze coming in from the Marmara Sea. Haruto decided to follow the loudspeaker's call into the neighborhood.

The people on the street did not look much like the Muslim faithful he expected — beardless men in Western clothes, some with olive skin but others with skin as white as Russians, and women with long, flowing hair wearing tight jeans and t-shirts. He continued to follow the loudspeakers.

Haruto was soon at the mosque with the supermarket. The store on the main floor was closed. Haruto felt nervous about climbing the stairs and entering the second floor. Maybe this was not for a tourist. Haruto turned his head and looked at the street around him. Most of the shops were closed and dark, but there, on top of the furniture store with its advertisement of *Karate-Do*, the lights were on and Haruto could see the outline of someone high kicking a large punching bag.

Haruto climbed up the narrow staircase beside the furniture shop and looked through the open door of the *dojo*. It was a large room, maybe fifteen meters long, with a varnished wood floor. Full-length mirrors covered the room's left side, street windows the right side. Not so different from the *dojo* back in Tokyo. Dozens of plaques and different colored belts on the far sidewall. Beside this wall a large black canvas punching bag hung from a ceiling rafter. A middle-aged man with lightly tanned skin, frontal balding and wearing a crisp white karate uniform with an old, frazzled black belt, struck the bag repeatedly with a combination of *mae-geri* front kick followed by a *mawashi-geri* roundhouse kick before he let the leg touch down again on the ground.

The karate black belt noticed Haruto at the door. "*Merhaba?*"

Haruto did not know what to say, but smiled.

"Hello, can I help you?" the black belt said in accented but slow, clear English.

"I am a tourist."

"Hello, tourist. I am Erhan. Do you know karate?"

"Yes, I am also a black belt. *Shotokan* style."

"What dan?" Erhan asked.

"*Shodan* — first dan. I am still working on my second."

"How many years have you been doing karate?"

"Twenty-three."

"How long to get your black belt?"

"Eleven."

The Turkish black belt whistled. "Such a long time."

"My master said I was not ready."

"Who is your master?"

"*Sensei* Nakaya."

Erhan smiled. "Tokyo, right?"

Haruto nodded.

"We are also *Shotokan* style here. I am a high school teacher, but this is my side business. We have classes every day during the summer. How long are you in Istanbul?"

"One week."

"Then you will train with us, yes? It would be an honor for us to have a student of Nakaya as our guest."

Haruto bowed. "Yes, thank you very much."

CHAPTER082

Menachem and Tanaka followed Major Gershon into the underground briefing room. Thirty-two soldiers in combat fatigues wearing black camouflage makeup snapped to attention.

"Platoon ready for mission, Sir!" First Lieutenant Chaim Dayan shouted.

"At ease," Gershon said. "Menachem Levi, our Mossad liaison, will be directing this mission. By order of the Defense Minister he is given rank equal to that of the Northern Commander."

Menachem hit the light switch and clicked twice on the overhead LCD projector's remote control. A map of northern Israel and southern Lebanon appeared on the white wall at the front of the room.

"Nahariyya was hit with a barrage of forty missiles today. One missile hit a retirement home and killed eight people. Another missile hit the beach and wounded three American tourists. Still another missile sprayed shrapnel through a high-tech industrial park with two deaths and fourteen wounded." Menachem pointed to the map. "All launches came from Chamaa. You will cross the border, make your way up to Chamaa and destroy the stockpile of al-Haleeb missiles kept there."

"Uh...Mr. Levi," Lieutenant Dayan said.

"Just call me Menachem."

"Menachem... wasn't the ambush last time on the road between Chlhine and Chamaa?"

"Yes, it was. Perhaps you can also find al-Haleeb's stockpile of anti-tank missiles." He clicked the remote control again. Black-and-white video of a single lane road and dim fields appeared on the wall. "An old noisy Merkava tank that we're controlling by remote control will soon cross into Lebanon and make its way to Chamaa. Perhaps they've set another trap for us tonight. Your mission is to cross the border at Yarine at nightfall. It's six kilometers north to Chamaa — less than an hour's hike for the robots, even through the hills, assuming your men can keep up."

"My men can keep up. But aren't there United Nations troops at Yarine?"

"There are," Menachem said. "You will walk quietly in the dark, west of Yarine, through the fields, up the hills to the plateau west of Mazraat Kraibe. Down the hills north from Kraibe, there are no more United Nations troops, but you're now in al-Haleeb country. Sweep out, look for the tank traps, look for missile stockpiles, and everyone converges on Chamaa. Search the village. Return to Israel before dawn."

"Yes, Sir!"

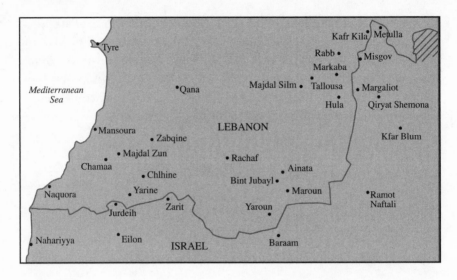

Jurdeih, Israel, on the Lebanese-Israeli border
June 23   9PM Local Time (18:00 Zulu)

CHAPTER083

Each black-faced soldier slipped through the fence into Lebanon, followed by his twenty-two upright Alpha robots and his eight monster-cockroach-like Beta robots carrying loads of ammunition and explosives. Each soldier had his handheld Robot Command Unit, or RCU, hanging around his neck. Other than side arms and fragmentation grenades, the soldiers carried few weapons. All wore an earphone/ microphone unit in their left ear and night-vision goggles, with sensors that flipped down over the eyes.

The Alpha robots, drab brown and about two meters tall, moved quickly and gingerly. Their feet sensed the firmness of the ground, their balance stayed perfect, silent step after silent step. They almost seemed like men, but only at first glance. Their pear-shaped torso and long arms immediately distinguished their species. The Alphas slipped through the fence, their rotating camera eyeballs scanning the night.

The Beta robots, a similar drab brown from head to rear, showed an agility that betrayed their two-meter long bodies, bulked up further by crates of cargo. Their thin legs sensed the ground as each foot pushed off, and the Beta's balance stayed perfect. Inside their bodies, methanol alcohol was being burned in hundred-kilowatt fuel cells, powering massive hydraulic pumps, yet outside, only a slight hum and the crunching of the earth under their feet was heard.

Lieutenant Chaim Dayan, 32 soldiers, 704 Alphas and 256 Betas — nine hundred and ninety-three men and machines, were in Lebanon now.

In ten minutes, the men and robots reached the hills and fields west of Yarine. Dayan was perspiring, even in the cool Lebanese evening. He grabbed the RCU around his neck and keyed in a command to reduce speed to six kilometers per hour for the ascent of the hills.

Fifteen minutes later, all men and robots were on the plain west of Mazraat Kraibe. Dayan keyed his radio. "Arrived at mission start point. Over."

"Did any machines fail because of the terrain? Over," Menachem's voice crackled in the earpiece.

"Negative. One hundred percent operational. Over."

"The remote controlled Merkava has passed Al Jibbayn and is on its way to Chamaa. As well, artillery will soon hit empty fields. Please double-check your RCU's. Over."

Dayan clicked a few keys on the remote control unit around his neck. "All the robots and RCU's already programmed to avoid these coordinates. Over."

"Al-Haleeb expects us to retaliate. They're waiting for you. Find them, Lieutenant. Over."

"Yes, Sir! Platoon out."

Dayan waved over the short olive-skinned soldier a few meters to his left. "Ariel, I'll tag along with your robots tonight."

"Sure, Lieutenant. My first position is at the road to Chamaa. When do we start?"

Suddenly the men saw a bright artillery flash about a kilometer north of them, followed a split second later by its thundering noise.

"Now!" Dayan said. He reached for the white plastic RCU around his neck, and keyed in the commands to begin the operation.

The men and robots started down the hill, scattering left and right, on their way to sweep out wide arcs that would eventually converge on Chamaa some three kilometers in the north. In some groups the robots led, in others the soldiers did.

As artillery shells exploded in the hills and fields nearby, Ariel and his twenty-two Alphas and eight Betas ran at high speed down the hill toward the ravine some three hundred meters away. Dayan followed. Ariel's RCU vibrated and beeped as he arrived at the bottom of the hill.

Ariel and Dayan turned around and saw one of the Alphas lying motionless near the bottom of the hill.

"What should I do?" Ariel asked.

"Pretend I'm not here. Follow the standard ops."

Ariel grabbed his RCU and entered the commands to have Alpha self-destruct. Menachem's voice came on in Ariel's earpiece. "Destruct command successfully received. However, we want to see why any robots failed. Continue with your mission. We'll pick up that Alpha by helicopter. Command out."

Ariel and the Lieutenant ran the eight hundred meters up to the next hill, its peak just east of Tayr Harfa. The Alphas and Betas followed behind them effortlessly.

Menachem's voice came on in Dayan's earpiece. "Why aren't you checking your RCU? Half of the Alphas are reporting directional radio transmissions. What are you doing, Lieutenant? Tighten the sweep. Get your men north of the road and sweep up to Majdal Zun."

"Yes, Sir!"

Dayan grabbed his RCU, scrolled down to the radio transmission reports, and vectored the sweeps toward this position.

"Changes in my orders?" Ariel asked.

"Nope, except it looks like your position's going to be a hot one."

Ariel looked down at the small road three hundred meters in front of them. "We'll cross the first road here and then traverse the hills to the main road."

Dayan nodded.

Ariel took his RCU and clicked away at the buttons. Eight Alphas ran down to the roadway, taking up positions on either side. Ariel, the Lieutenant and the rest of the Alphas and Betas then sprinted down the hill and across the road. The Alphas guarding the roadway then left their position and followed the group as it traversed a kilometer of hills, arriving at the main road just west of Kheurbet Ksar Ramle.

Dayan and Ariel saw the remote-controlled Merkava coming up the road. A few seconds later the bright flash of an anti-tank missile lit up the night sky. The missile shot across the road, missed the tank, and crashed into the hills. A moment later, another, then a third, then a fourth anti-tank missile launched, one of them smashing into the Merkava and exploding the old tank.

Menachem's voice crackled out of Dayan's earpiece. "What are you doing, Lieutenant? A dozen Alphas in Ariel's squad have fixed onto the enemy running up the hills toward Majdal Zun. Look at your RCU. Order the robots into pursuit. Find out where the stockpiles are."

"Yes, Sir." Dayan grabbed his RCU and transmitted the pursuit order to Ariel's RCU, which in turn transmitted orders to every Alpha in the squad to pursue and catch the enemy. The Betas were to keep a few meters behind the Alphas.

As the Alphas charged across the road a hail of RPGs swept across them. Two Alphas fell down, five Alphas made it across, while the remaining fourteen Alphas waited at the edge of the road. Following standard ops two of the five Alphas that made it across the road opened fire on the origin of the RPGs. The other three Alphas continued in pursuit of the enemy. The fourteen Alphas still waiting to cross the road also opened fire. The enemy RPGs stopped, and one by one, the Alphas and Betas dashed across the road.

Dayan was entering commands into his RCU when Menachem spoke again in his earpiece. "Lieutenant. You should be calling in the other squads to pursue the enemy. Faster, Lieutenant."

"Yes, Sir. Just doing that now, Sir."

A half-dozen Alphas on the other side of the road laid down a stream of fire against the possible enemy positions, as Dayan and Ariel crossed the road. The robots and men now ran as fast as they could on the field north of the road. After five hundred meters, the gentle rolling hills climbed steeply up.

Dayan looked at his RCU. Two Alphas had tight fixes on two different enemy fighters converging on a spot in the hills between Chamaa and Majdal Zun. He hit a few keys, moved the tiny joystick, and put a green box over the spot of convergence.

Menachem's voice crackled from his earpiece. "Lieutenant, I see your green box. Fast — get more squads there."

"Yes, Sir." Dayan continued clicking away at the RCU's keys.

"Lieutenant, I'm going with my robots up the hill," Ariel said.

Dayan nodded, continuing to enter commands into his control unit.

A spray of machine-gun fire and mortars thundered from the hill ahead. Dayan flipped down his night-vision goggles.

Ariel's body sprawled out on the dirt, surrounded by two Alpha robots missing limbs.

Dayan sprinted the two hundred meters up the steep hill, digging his hands and feet into the loose rocks and soil.

"Ariel, Ariel!" He bent down to hold his fallen subordinate's head.

Ariel opened his eyes and groaned. "I'm... okay... Lieutenant. My left leg, I think it's broken."

Dayan hit four buttons on his RCU. A Beta came running over, dumped its load of ammunition, and gently lifted Ariel to its back.

"Lieutenant, what the hell are you doing?" Menachem's voice yelled from Dayan's earpiece. "Ariel's squad is in hot pursuit of the enemy but it no longer has a leader. Take over. Standard ops."

"Ariel's been hit."

"Yes, and that's why you have to get off your ass and do his job now. We'll helicopter him out of there. The Beta's taking him to a pickup spot."

As Dayan picked up his RCU, Alphas, Betas and other soldiers were streaming toward the target spot farther up the hill above. His RCU started lighting up with multiple messages of *Enemy into Structure* from different Alphas.

Dayan tore up the steep hill as fast as he could.

Two hundred meters ahead, he saw a collection of men and robots. He flipped down his night-vision goggles again. A dozen Alphas were firing bursts of shots each time one of the enemy stuck his head over the concrete barrier. He studied the layout. There was foliage everywhere, but he could make out the rough shape of a massive concrete bunker, rather than a simple barrier, sunk into the hill.

"Lieutenant, your men are all bunched up. A few mortars and you're going to have casualties," Menachem's voice said again.

"I'm proceeding to penetrate the bunker, Sir."

"Good. Out." Menachem said.

Dayan started madly keying in commands.

A dozen Alphas opened a continuous stream of fire onto the concrete barrier. Twenty other Alphas accompanied by a dozen Betas charged the barrier. Only one Alpha was hit and fell down.

The Betas offloaded their explosive charges. They crammed twelve tonnes of C4 against the concrete. All the robots scurried around the corner, moved a hundred meters away, and pressed themselves into the ground.

A bright orange fireball lit up the night sky.

The Alphas returned to the bombed structure, along with a dozen new Betas laden with explosive loads. The night was dark again, and it was quiet — no return fire.

Dayan and a few of the soldiers ran to the bunker and looked down. Through a five-meter hole in the top concrete layer, they saw a massive room below, now completely burnt out, with a concrete floor and staircase leading to a floor or floors even deeper in the hillside. Dayan hit a few more keys on his RCU.

The Alphas started taking explosives off the Betas and climbing down into the bunker with their load. They placed twelve more tonnes of explosive on the bunker's concrete floor. The Alphas climbed out, and everyone, man and machine, ran away and hugged the ground.

The night sky filled with orange once more as the C4 detonated. Then another explosion occurred. And another. And again and again, as the store of missiles and anti-tank weapons went off like giant popcorn kernels.

Finally silence. The Lieutenant and men ran back to the bunker and looked down. The first floor was gone, except for a few jagged remnants in the southwest corner. Some ten meters down, they saw blackened fragments of missiles and anti-tank weapons.

Dayan was about to click away again on his RCU when instead he simply looked at an Alpha robot near him and said, "Bring up some of the fragments."

Dayan keyed his radio. "Bunker destroyed."

"Good work," Menachem's voice said from the earpiece. "Come home, now."

CHAPTER084

M**enachem and Tanaka** came into the meeting room and sat down across from the other men.

On one side of the large underground conference room, a Beta held an Alpha horizontally up in the air. Dirt still covered the Alpha, its left leg was charred. Its chest plates were off, exposing the hydraulic pumps and fuel cells deep in the pelvis, the ammunition magazines in the abdomen, and the bright orange central processing module in the chest.

Sitting at the conference table was Lieutenant Chaim Dayan, in a freshly pressed green uniform, looking well recovered from the previous night's activities. Beside him were Daveed and Seiko. Both had left the *New Pacific Queen* in Singapore and had flown to Israel.

"Your report, Lieutenant," Menachem said.

Dayan stood up. "We estimate we destroyed between two to three hundred missiles and an equal number of anti-tank weapons. One soldier wounded but he's expected to recover. Seventeen Alpha robots and four Beta robots lost. Estimate forty-one enemy killed by robots, two more in the bunker. No collateral damage."

"How did you arrive at your estimates for the number of missiles destroyed?"

"The robots recorded the secondary explosions, plus a survey of remaining fragments."

"How did you arrive at the conclusion of no collateral damage?" Menachem asked.

"The robots have photographic evidence associated with every shot taken. All who were shot were armed."

Menachem hit a few keys on the computer in front of him. A news video appeared on the wall, showing dozens of bodies wrapped in white sheets lying on a sports field.

"*Last night the Israeli Air Force bombed the soccer field outside of the tiny village of Chamaa, Lebanon. These bodies await burial today. Al-Haleeb leadership in Beirut has stated that Israel will pay a thousand times in blood for this crime. Lebanon has registered a protest with the United Nations. FIFA — The International Football Association — is considering a suspension of Israeli football teams.*"

"These are lies," Dayan said.

Menachem nodded. "Yes, yes they are. Dismissed."

When Dayan had left, Tanaka said, "Maybe you should show the world the photographic evidence from the robots."

Menachem glowered at Tanaka. "You think the world actually wants the truth? How come there were no protests yesterday when rockets came crashing down on the retirement home and the beach in Nahariyya?"

"Maybe if you tried to show the world that you weren't as bad as they make you out to be, it would be better for you."

Menachem stood up. His left leg collapsed but he caught himself on the table and straightened again. "If we proved that there was no collateral damage in the mission last night, maybe we would feel good for a half-hour. Maybe some newspaper somewhere might even print a retraction. But the fact that we are now using robots would get abused somehow. I don't know how, but it would. Nobody likes us. Another reason to hate us."

"I was just trying to help," Tanaka said.

"Of course you were, and thank you." Menachem rubbed his left thigh and then took a pill bottle out of his pocket. He washed down two pills with a glass of water from the table.

"Lebanon?" Tanaka asked.

Menachem nodded. He clapped his hands together. "All right, let's get on with our meeting. First point, we had several unexplained failures. The robots just stopped working."

Seiko raised a finger. "I went through the software dumps. The hydraulic pumps overheated in all cases, and when I actually looked at them, I could see leaks with my bare eyes."

"I've made a note of this," Tanaka said. "In the next production batch, the cooling fins coming off the hydraulic pumps will attach to the back of the robot, so the whole surface is used to radiate away the heat."

"Also, Seiko and I have modified the software and downloaded the new patch to all the robots," Daveed said. "We monitor the pump temperature more closely and will reduce outputs to avoid a critical failure."

"The next issue is the Lieutenant's command of the robot platoon last night," Menachem said.

"I thought he was one of your best officers," Tanaka said.

"That's exactly the issue. He wasn't able to handle the information flow last night."

"Alpha Prime," Daveed said.

"What?"

"Seiko and I discussed this already. There's still room for eight more circuit boards in the orange processors module. We can stick another computer board into one of the Alphas of each platoon — call this the

Alpha Prime. The Alpha Prime would be assigned to the commander of the platoon and would process much of the information that otherwise would be going into the commander's RCU."

"I don't know...." Menachem said. "When I was in charge of the Merkava development, whenever we tried things a bit too complicated, it never ended well."

"I could spend my life working on this," Daveed said. "But not to worry — Seiko and I will give you a working Alpha Prime by the end of the week. The platoon commander still has to make the important decisions, but the Alpha Prime will take care of lots of data flow."

"All right, we'll see. Next issue is the user interface of the RCU. It was taking the Lieutenant too long to enter commands last night."

"The robots all handle speech recognition," Tanaka said. "You should change Standard Ops to use voice commands."

"On the next mission, the platoon commander can use the Alpha Prime as a big, intelligent Remote Control Unit," Daveed said.

"Do you think this is a good idea?" Menachem said.

Everyone at the table nodded.

CHAPTER085

Haruto arrived at the dojo early. Good. Four hours to train in the comfortable early morning temperatures until the first class began. Like the other day, Erhan was there already, kicking the punching bag over and over again.

Haruto went to the changing room and took off his street clothes. His thin, tall frame rippled with muscle. Large, tight triceps bulged out of the back of his arms and led to massive rear shoulder muscles. Washboard abdominals followed by large, tight quadriceps in legs below.

Haruto put on the borrowed *gi* and fastened it snuggly with the brand new black belt. One of the students would have his black belt exam at the end of the week. If the student passed, this belt, which Haruto, a student of *Sensei* Nakaya now wore, would become the Turkish student's black belt.

Haruto went to the large gym, and kneeled on the wooden floor. He closed his eyes, clasped his hands together and let them fall into his lap. He breathed in through the nose. *Ichi. Ni. San.* He exhaled through the mouth. Soon Haruto no longer felt the pain of his knees against the hard wood floor. Soon he stopped thinking about the investigation, about his career, about Michiko. He was in balance, at peace with himself, at peace with the world.

After about twenty minutes, Haruto gradually opened his eyes. Still sitting on the floor, Haruto extended his legs and then spread them apart to almost a full splits. With both hands he then grabbed his left toes, stretching the ligaments in his leg and pelvis. He then switched to the other side, grabbing his right toes with both hands and stretching the ligaments in his right leg and pelvis. Haruto then stood and did the top-level *kata* — a dance of karate moves done in rapid-fire, repetitive fashion.

Haruto went prone and on his knuckles did three hundred pushups. He then lay on the small of his back, supporting the position with his palms at either side, and raised and lowered his extended legs three hundred times without letting them touch the floor.

The Turkish black belt had left the gym. Haruto approached the punching bag, and focused on the large *C* in the *Made in China* white imprint. For the next ninety minutes, Haruto punched and kicked the bag without stop. He hit the *C* with every blow.

CHAPTER086

The four men in jackets with open-collar shirts and cropped black beards, went into the apartment high-rise and took the stairs down to the second basement. They walked past the two armed guards into the spacious, wood-paneled office. They sat down on the circular couch opposite the graying man at the large oak desk.

"I saw your report on Chamaa," Ibrahim said. "No air support, no tanks. The Jews are up to new surprises."

"We will not let it happen again," one of the young men said.

Ibrahim closed his eyes and shook his head. "You have a good, fierce heart, but you must use your head." He touched his head with the tips of his fingers. "When the other side decides to do something new, then you cut him off immediately." The old man slammed his fist down on his desk. "Whatever the cost, you don't give him even the smallest victory. You make him think his strategy was wrong!"

"In 1983, before some of you were even born, the Americans, their famous Marines, came to Lebanon — here in Beirut. One truck bomb..." He smiled broadly. "Three hundred of them killed. With that taste of defeat they ran back home so fast..." Ibrahim looked at the unfurled maps on his large desk. "Village of Yaroun... What do we have there?"

"About three hundred of the two-twenty-millimeter missiles... about fifty anti-tank missiles," the same young man said.

"Sneak in a dozen of the medium-range missiles. Carefully! In horse-drawn wagons covered with straw or on trucks covered with construction materials. Launch them tonight for Tel Aviv, plus launch about fifty of the two-twenty-millimeter missiles you have."

"The Jews will bomb the village, maybe worse."

The old man shook his head. "For one, world opinion is too much against them. For two, the Jews won't kill the Maronites so fast. In 2006, they didn't destroy the village. For three..." The old man looked up at the ceiling and laughed. "Their strategy worked in Chamaa, so they'll try it again. They'll send a few hundred foot soldiers in." Ibrahim stood up. "I want a *thousand* fighters around the village. When they come in, you capture them. We'll put them on display and then we'll kill the Jews. So much for their new strategy."

CHAPTER087

The new Northern Commander, General Peletz, bit his lip as the large overhead display screen lit up with red lines. He answered his ringing phone, listened quietly, clicked the receiver and quickly dialed.

"Menachem, it's Eli," Peletz said. "A dozen missiles from Yaroun just hit Tel Aviv and a few dozen more landed in the north."

"What damage?"

"They were filled with conventional explosives and ball bearings. It looks like a few deaths and many wounded in Tel Aviv. In the northern cities it's mainly property damage."

"Where were they launched from?"

"Beside the west church in Yaroun, in between homes, and from the field beside the school."

"Great."

"Yes… look Menachem, I can't just stand here and watch. We need to do something."

"Any new launches?"

Peletz looked up at the display screen again. "No."

"Give me twenty-four hours. We'll clean out Yaroun tomorrow."

CHAPTER088

I n the large underground room Daveed and Seiko sat on one side of the conference table loudly discussing... some point of the software, something about hooks into the Thailand boards. Tanaka listened in, trying to understand, asking the occasional question. Lieutenant Dayan sat quietly beside them. An Alpha robot stood at the end of the table, still but for its spherical eyeballs occasionally darting about.

Menachem hurried into the room, dragging his left leg. "Sorry I'm late." He caught his breath. "The barrage of missiles yesterday, including the ones that hit Tel Aviv, came from Yaroun. I wanted an extra day, but we have no choice, we have to mount a mission tonight."

"We're ready." Daveed pointed to the Alpha robot standing at the door. "Are you ready Alpha Prime?"

"Alpha Prime is ready."

Daveed looked admiringly at the robot, then faced the men at the table again. "Instead of using his RCU, the Lieutenant can give commands by voice to Alpha Prime. As well, Alpha Prime will alert the Lieutenant to battle decisions that need to be made."

Menachem crunched his face. "How'd you modify it so fast?"

"The voice recognition was already in the robots — just not part of standard ops. For the battle decisions..." Daveed smiled and then laughed lightly. "I took the code from a chess program I wrote when I was in high school. Seiko and I added another processing board to the Alpha. It's a standard, dumb computer CPU running Linux and C. We interfaced it to the Thailand boards and added my programs."

"Where's Yaroun?" Tanaka asked.

"Sorry," Menachem said. "It's so close to the border that I didn't bring any maps. Two kilometers north, east of here. Maronite village of a few thousand. Lots of al-Haleeb there now, too."

"They're Christians, right? Where did they come from? Europe?"

"The Maronites have been there forever," Menachem said. "Centuries ago they survived the Muslim conquests by hiding in the mountains. But I don't know if they're going to survive al-Haleeb. Lots of them living in fear and poverty. Lots of them leaving."

Dayan stood up. "Sir, I need a mission plan for tonight."

"Are you okay with trying out the Alpha Prime?"

"Yes, Sir, as long as I can also keep my RCU remote."

Daveed nodded.

"Full platoon," Menachem said, "but also take two more Betas loaded with X-ray units, so you have the option of scanning buildings. Cross the border at Baraam, head to Yaroun, spread out in the town and search it. Search everything. Intelligence says there may still be a few hundred missiles there."

"Artillery, Sir?"

"None… sorry. Orders from above. No bombing, no helicopters, no artillery. Just you, Lieutenant." He pointed to the Alpha. "You and them."

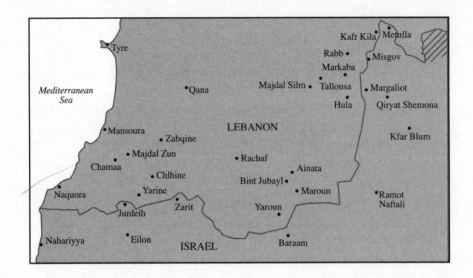

Baraam, Israel, on the Lebanese-Israeli border
June 27   9PM Local Time (18:00 Zulu)

CHAPTER089

Lieutenant Dayan grabbed the white RCU hanging around his neck and started punching commands into its little keyboard. He looked up at the Alpha Prime waiting obediently beside him and let go of the RCU. "Give me data for insertion into Lebanon."

"One moment," the Alpha Prime said. "Plan completed. Checkmate."

"What is the plan?"

"Data from all robots analyzed. Based on radio, visual, vibration and infrasound sensing, border one kilometer northeast of here is inactive. No United Nations soldiers."

"Execute," Dayan said.

"Orders given and confirmed by all robots... by three soldiers... by eleven soldiers... by all soldiers."

Over the next ten minutes, 32 soldiers, 704 Alphas, 256 Betas loaded with ammunition and explosives, plus an extra two Betas carrying mega-electron-volt X-ray packs — along with Dayan and his shadow Alpha Prime — spread out northeast along the border and eased across it. Nine hundred ninety-six men and machines moved across the dirt and scrub grass of the low hills a few hundred meters into Lebanon. The lights and buildings of Yaroun to the northwest were close.

Dayan was about to reach for his hanging RCU when he remembered the Alpha Prime beside him and gave it a verbal order: "Two squads to take wide circles around the town to the left and right, converge on the trees northwest, then straight into the town from the rear. All the other squads straight ahead into Yaroun. Spread out once in the town and standard ops search."

"Done."

From the hilltop, the Lieutenant watched a squad of thirty robots and a soldier start wide clockwise around the town, while another squad of robots and a soldier circled wide counterclockwise. The remaining soldiers and their some nine hundred robots went straight in.

The smells of the aromatic fields filled the air. A cool breeze blew down from the mountains in the distance. Only a low hum, the soft crunch of the robots' steps, and the songs of insects filled the night. Dayan smiled and started down the hill after his men.

Suddenly, a white flash and fireball, followed by an explosive thunder sliced through the calm. The young Lieutenant gasped as hundreds of robots and some of the men flew through the valley.

Hundreds upon hundreds of al-Haleeb fighters stood up from behind dirt covered concrete bunkers on either side of the small valley. Dozens held meter-long RPG-29V thermobaric anti-personnel rocket launchers. Spotlights clicked on and lit up the small valley and its bloodbath of Israeli soldiers and their robots. A loudspeaker blared in Hebrew: "Put down your weapons and you will live." Almost instantaneously two more al-Haleeb fighters fired their tube style rocket launchers. The thermobaric warheads crashed into the valley and lit it up. A wave front of over-pressurized air flattened the robots and men nearby.

The dust settled and the shocked but intact twenty remaining Israeli soldiers, their uniforms stained with blood and dirt, held up their hands.

Dayan turned to the Alpha Prime. "Call in an airstrike!"

"We don't have permission for artillery or airstrikes. With more time our losses will increase. Lieutenant, advise you to please give me instead orders to execute."

"Execute what?" Dayan screamed. "We've lost."

"Your data analysis is not correct. It is still checkmate. Victory for white."

Perspiration was dripping off his forehead. He was breathing faster, feeling a large lump in his throat. For a moment in silence, Dayan stared at the erect, two-meter tall, drab-brown robot. "Are we... white?"

"Yes."

"Execute."

"Done," Alpha Prime said.

Four undamaged Alphas started running at full speed down the valley toward Yaroun. Another four Alphas started running at full speed the other way through the shallow valley toward Lieutenant Dayan. A few dozen of the hundreds of submachine gun-armed al-Haleeb fighters started spraying bullets at the robots.

A moment later, 162 undamaged Betas, each laden with a thousand kilograms of explosives and ammunition, ran at full speed to either side of the valley, toward the hundreds of al-Haleeb fighters standing on top of the dirt level concrete. As some of the fighters took their eyes off the running Alphas and noticed the Betas surging toward them, they started screaming and firing off the magazines of their assault rifles as well as launching another few thermobaric RPGs. A dozen Betas fell to the ground, but the remaining hundred fifty kept coming.

At the same time, the remaining four hundred undamaged Alphas in the valley had each locked onto several al-Haleeb fighters and started pumping rounds out of their mouths. Since more than one robot had targeted each fighter, a hail of bullets streaked through the air toward each man.

The al-Haleeb fighters started dropping. A moment later, wide lines of Betas reached either side of the valley, dumped their explosive loads and reversed direction. The explosives suddenly ignited. One hundred fifty *tons* of explosives decimated the hundreds of standing fighters and those behind and inside the concrete bunkers in a blur of blood and dirt.

The Alphas started running toward the bunkers, two hundred robots in either direction. They passed the dirt-level concrete and continued in clockwise and counter-clockwise wide circles around the town, sweeping up and shooting the remaining al-Haleeb fighters in their path.

From the opposite counter-clockwise and clockwise directions came the two squads of thirty robots each that had broken off from the rest earlier. In a few minutes, the squads met up and combined with the larger group.

Dayan looked down at the thick fog of smoke and dust swirling in the valley. The Alpha Prime turned to him. "Wounded and dead being transported to Israel. Four hundred Alphas, one hundred and forty-two Betas now in Yaroun. Twenty-two soldiers functional they ordered meet reformed squads in Yaroun. Lieutenant, please give me search orders to complete mission."

It felt like years since the ambush started but it could really have been no more than minutes. "So… it's all right then?"

The Alpha Prime's silicone lips drew up into a smile. It might have been reassuring, but for the gun barrel protruding from the middle of it. "Checkmate."

CHAPTER090

John Sullivan came back from lunch and settled in at the cubicle. Yesterday marked one month on the job, and today he got his first day shift. More hustle and bustle during the daytime at the agency, although the work was about the same. A bit boring, actually.

John clicked through e-mails and alerts — nothing this afternoon. He clicked through his satellite's status report. Power and electronics fine. Fuel at seven point one tons. Shack-Hartmann correction to two centimeters.

John stared at the computer screen. The computer's AI was doing fine without his own. Nothing for him to check or verify or research.

John turned his head and looked around. No daytime supervisors hovering.

John typed a line of commands into his keyboard. The display refreshed. Green letters flashed onto the screen. *Keyhole rotating... Cassegrain secondary locked onto 33.1000°N, 35.3000°E — adaptive optics laser guide recalibrated to 2 centimeters — target to be held for 2 minutes 15 seconds.* A moment later, below the message, a nighttime-enhanced black-and-white real-time video image of Zarit, Israel splashed on the screen.

Nothing. The other week John had followed the containers moving into Zarit and he had seen some of the machinery. They weren't car parts, even his superiors acknowledged that now. The best guess was some secret earthmoving equipment, maybe to lay or remove mines near the Israeli-Lebanese border.

John clicked on the *wide field-of-view* option and the image zoomed out to display a larger area around Zarit, albeit at lower resolution. Nothing still. He could see some moving lights on the roads. John typed a few more commands and the satellite's secondary mirror tilted east along the 33.1° north latitude coming from Zarit. Nothing as the image slowly moved to the east. Nothing.

Then suddenly to the right of the screen, a bright white light flashed out of the dark nighttime image. Before John could click another key, the computer took over.

The display screen quickly refreshed. *Cassegrain secondary rotated and locked onto 33.080277°N, 35.434000°E — southeast of Yaroun, Lebanon.*

Real-time video then flashed onto the screen. Israeli soldiers bleeding, lying on the ground, scattered among the smoke and debris in the valley just southeast of the town. John could see hundreds of al-Haleeb fighters on either side of the valley as the computer automatically zoomed in and out of the scene.

The computer started zooming in on a fighter's smoking RPG launcher. The image of the launcher was so large that John could almost read some of the lettering on it — and probably could have, had the image been taken in sunlight. The display split in two. On the right side was an enlarged image of the RPG launcher. On the left side was a reference box showing an RPG-29 rocket launcher and the types of rockets it could launch. *Please confirm RPG-29 rocket launcher.* John clicked *y* for yes. The screen restored and zoomed out a bit.

Dozens of giant insect-like machines were now charging at the al-Haleeb fighters. Suddenly bright flashes filled the nighttime image on both sides of the screen where the rows of fighters were. Had been. A split second later, the screen refreshed to a blank display giving the message in green letters: *Out of range. Returning mirrors to standard track. Current target — Mediterranean.*

"Shit!"

This could be important, no — it *was* important. The satellite would have to orbit the Earth again before it would be back to have another look. He started clicking away on the Tools Menu.

Molniya orbit.

He remembered that from training. In a Molniya orbit the satellite orbits in a giant ellipse, so it can spend twelve hours looking at one spot on the Earth. John clicked that option.

Are you sure?

Yes.

Approval?

Yes.

Fuel burn 2.4 tons. OK?

Yes.

Are you sure?

Yes.

The satellite image of the Mediterranean Sea disappeared, replaced by the green letters *Molniya orbit burn in progress.* Underneath telemetry data filled up the screen.

A minute later, a small, bald man in a green shirt and blue striped tie ran over to John's cubicle. "Are you Sullivan?"

"Yes."

"Who the fuck gave you permission to burn two tons of fuel?"

"I was in the middle of an image that could be vital to intelligence, and the satellite went out of range."

"I doubt it, but that's not my concern," the bald man said. "My concern is to keep the birds happy. You went out of the way to find that scene. The computer didn't even prompt you for it."

"I got a computer prompt."

"Once you'd already found the scene, forget that for a second. You just burned two tons of fuel without permission."

"I didn't want to lose the scene."

"You fucking idiot, it's going to take hours before the bird hits Molniya orbit. You're going to see shit for the moment." The bald man took a deep breath. "Oh hell, I have to put you on report, kid. I can't take responsibility for this burn."

"But… the satellite's fuel tanks are refillable."

"It can't exactly pull into a gas station, can it? Do you know how much it costs to boost fuel up there?"

John felt like he was about to cry. "What will happen to me?"

"Look… stop it… *I'm* not going to take responsibility for this burn. Look… I'm not going to put anything crazy down on you, so you can keep on working… okay? But the report will bounce to your supervisor, the shit will hit the fan, and there'll eventually be a small investigation, maybe one or two months."

"What happens then?"

"How long you've been here?"

"A month."

"Post your resume."

CHAPTER09 1

"**D**espite an ambush by nearly one thousand al-Haleeb, you defeated them and completed your mission to destroy over three hundred missiles found hidden in Yaroun," Menachem said. "Please come here, Lieutenant."

Chaim Dayan stood up, wearing his dress shirt with the two olive branches of a First Lieutenant. Menachem handed him new insignia containing three olive branches. "Congratulations, *Seren* — you are the new Captain of the Eilon Company of Mustang Battalion."

Colonel Tanaka and Major Gershon sat at the conference table and clapped, as did Menachem.

"Thank you, Sir," Dayan said.

"You will have five robot platoons in your company — one hundred and sixty men and five thousand robots. Congratulations. Dismissed."

The new Captain Dayan, a large smile on his face, left the room with his Alpha Prime trailing close behind.

Menachem looked at the Major. "We will waste no time. The cross-border attacks end now."

"Yes, Sir!" Gershon said.

Menachem unrolled a plastic map of the Israeli-Lebanese border and plugged it into the USB connector in the table. "Major, you now have the first company of Mustang Battalion. Within a few weeks I want all thirty thousand robots of your battalion set up in six companies."

Gershon nodded.

"Within the next month I will set up two more battalions, each containing thirty-five thousand robots. The two new battalions, plus your Mustang Battalion, will be part of the Northwest Robot Division — one hundred thousand robots. I will be Commander."

"What is *northwest* exactly?" Tanaka asked. "What areas are you going to cover?"

Menachem put his finger on Nahariyya, on the Israeli coast just below the Lebanese border, and drew a line to Baraam, Israel. "My division will cover this part of the border. Anything attacking Israel north of this line, we will take care of."

Tanaka pointed to the map. "What about the northeast border with Lebanon, up there to Qiryat Shemona, and the border up north there with Syria?"

Menachem laughed. "We're waiting for the next shipment. The next hundred thousand robots will go into the Northeast Robot Division. For the moment, we'll use some extra army troops to take care of that part of the border, and reduce the attacks. Don't worry, those kibbutznicks up north there are tough."

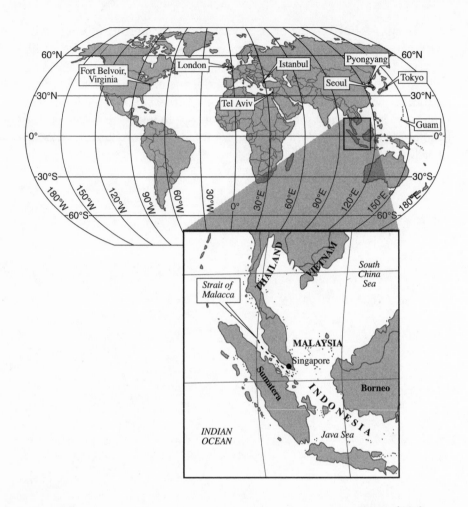

Strait of Malacca
Container Ship Mikiyasu-ema
June 28   2PM Malacca, Malaysia Time (06:00 Zulu)

CHAPTER092

L **ieutenant Colonel Okamura gazed** through the huge bridge windows at the strait sandwiched between Malaysia to the north and Indonesia to the south. The coastline on either side was a warren of thousands of inlets. Even in these modern times, even though one-quarter of the world's trade passed through this strait, pirates still lurked here. Okamura scanned the water ahead. Nothing except for the huge liquefied natural gas carrier with its row of massive white spheres ahead. "Why did you get behind an LNG? You should try to keep a low profile."

"Lieutenant Colonel... I can't control what other ships are sailing today. What did you want me to do? Wait a few hours cruising around the Indian Ocean?" Yamada said. "Not to worry. I have been sailing ships for Mikiyasu for thirty years now. I have never been attacked yet."

"What about the *Idaten* and *Kurooshia* — they kidnapped the crew. What about the *Ocean Bridge*?"

"Unfortunate," Yamada said, "but Indonesia has been increasing patrols."

"I'll feel better once we round Singapore and head north for home."

"This is very important for your career, Lieutenant Colonel, is it not?" Yamada said.

"It's important for Japan," Okamura said.

The *Mikiyasu-ema* crawled through the straits behind the LNG carrier, as Okamura, Yamada, and the half-dozen bridge crew looked out the large glass windows.

An hour passed quietly. Then suddenly an explosion roared through the ship. Okamura turned and saw smoke coming from the aft deck.

Yamada looked through his binoculars and pointed down to the left. "Near the inlet... three fishing boats... rocket launchers."

Okamura grabbed a large pair of binoculars and scanned the area. The pirates came into view. One was now holding a power megaphone: "LNG carrier and *Mikiyasu-ema* slow down immediately! Danger if you do not. We will blow up LNG Carrier."

Okamura saw another man fire his RPG launcher. The rocket slammed into the *Mikiyasu-ema*'s bow deck and exploded meters away from the containers holding the nuclear bombs in the third bow compartment.

Okamura pulled the black walkie-talkie off his belt and started screaming commands into it.

"No!" Yamada said. "If they hit the LNG carrier... there's enough gas in there... it's the same as a small nuclear bomb. We'll be destroyed."

"What do you suggest we do, Captain? Let them board and take our cargo?"

"They probably only want what we have in our safe. They might look in the shipping containers, but they won't take anything. What would they do with it?"

Okamura's eyes looked like they would pop out of their sockets. "So we should let some pirates take a Japanese military officer prisoner and help themselves to nuclear weapons that they could then sell to their terrorist buddies?" He clicked on his walkie-talkie. "Aegis systems up and activated! Hit every target in the water and the air! Exclude LNG carrier straight ahead. I repeat, exclude LNG carrier straight ahead."

On the small raised bridges in the front and the rear of the ship, the Gatling guns of the Aegis anti-missile systems elevated and pushed through the canvas cloth covering them.

Yamada gasped. "Lieutenant Colonel. Don't! What if you hit the LNG? Calm down for a second. Let's think this through."

Okamura glared at Yamada. "You may be worried about dying, I am not. I live to complete my mission and I will."

The pirates' megaphone blared again. "*Mikiyasu-ema* slow down and prepare to be board —"

"Execute," Okamura yelled into the walkie-talkie. In less than a second, the Aegis' guns on the front and the stern of the ship swiveled to focus on the pirates' boats. But, as the first shots flew, a missile streaked from the white middle craft toward the LNG carrier.

Okamura gasped as he saw the rocket stream in a well-aimed low arc toward the LNG carrier. *Hit the rocket, hit the rocket.* But the Gatling guns spent the next second strafing the pirates' boats with hundreds of 20mm rounds.

"No!" Okamura watched the rocket reach the top of its small arc and now charge onward and downward straight ahead to the LNG carrier. Suddenly the front Aegis Gatling gun swiveled a few degrees to the right and up, and let go a burst of two hundred rounds of ammunition tracking the flight of the rocket. The hail of gunfire sliced up the rocket. Its remnants dropped harmlessly into the strait.

Okamura exhaled. "We're supposed to be heading back to Israel in another two weeks. Will you be joining me again, Captain?"

Yamada remained silent, but then nodded. "Yes. I love my country too."

CHAPTER093

Two Turkish teenagers with brown belts gathered beside Haruto. Like many high school students around the world, they had studied English as a second language. Haruto had found it to be an awkward and disorganized language, but its universality made it useful. "Do you know what a *makiwara* is?"

"That's the wooden stick you hit, right?" the tall teenager said.

"Yes," Haruto said. "The punching bag is good, but the *makiwara* is better. It's smaller, and if you don't hit it correctly, your strike will bounce off. In real life, your targets will be small. It is not enough to land a punch or kick on your opponent — you must be able to place this punch or kick to the right spot on your opponent."

The students nodded.

"To use the punching bag effectively, focus on a small part," Haruto said. "For example, look at the *Made in China* printing here. Try to hit only the *C*. If your punch or kick hits another letter or hits the black canvas, then you must consider your target missed."

Haruto assumed fighting stance, knees bent with sideways posture, in front of the punching bag. His left hand fired into the bag with a *yaku-tsuki* punch, the second large knuckle smashing into the *C*. A split second later his right hand fired forward with a shorter *oi-tsuki* jab, its second large knuckle also driving into the *C*. Left-right, left-right, left-right. Over and over again, Haruto's left and right second large knuckle smashed into the *C* of the punching bag.

Suddenly the canvas bag ripped, a tear starting at the *C* and going diagonally through the middle of the *Made in China*. The rag stuffing of the bag poked out of the gash.

Haruto stopped. He turned around and made eye contact with Erhan down at the other end of the gym. The Turkish black belt walked over to Haruto.

"I am sorry. I will pay you for a new punching bag."

Erhan laughed and shook his head. "It is a great honor to have a student of *Sensei* Nakaya be the first to wear through this punching bag." He held up a roll of gray duct tape, covered the tear, and wrapped a strip all around the bag so that the tape adhered back on itself. "Like my worn black belt, this punching bag is good for many more years of use."

Haruto bowed. "Your student is ready for his exam."

"*Seiretsu!*" Erhan yelled. The fourteen students scattered in the gym formed a line in front of Haruto and the Turkish black belt. "Suleyman, come forward."

The tall teenager wearing a brown belt took a step forward and bowed to the older men.

"*Jion kata,*" Erhan said.

Suleyman began a dance of karate moves — arms blocking the air, legs kicking the air, and arms punching the air in a rigid, stereotyped pattern. Two minutes later the *kata* was complete and the teenager stood at attention, beads of sweat dripping down his forehead.

"Flawless and powerful," Haruto said.

Erhan nodded. "Next… controlled fighting, which I will evaluate."

Haruto jumped in front of Suleyman, fists clenched and arms in fighting position. For ten minutes, the black belt candidate blocked Haruto's idealized punches and kicks and counter-attacked with his own. No blows to the head, punches or kicks to the body were pulled.

"Excellent," the Turkish black belt said. "*Kumite* — Suleyman fought last month in the Bosphorus Karate Tournament. Even though no medal was won, the fight required skill and courage."

Erhan held up a board at arm's length. "Suleyman, to obtain your black belt you must complete the final test — break this piece of wood. No direct punch — too easy. Demonstrate another technique."

The teenager lined up in front of the board in fighting stance. He stared at the board. The gym grew quiet. Second after second of silence went by. Suleyman suddenly shifted his weight to his rear leg while pivoting so his body lined up with the piece of wood. His front leg bent at the knee and accelerated up, and an instant later the entire leg extended with the edge of his heel firing a *yoko geri* kick into the wood. The teenager yelled a loud *kiai* as his foot sliced through the board.

Haruto untied his belt and held the black strip of cloth in his outstretched palms. "Suleyman, please remove your belt."

The teenager put the new black belt around his *gi* and tied a tight knot. The class broke out with applause, and Suleyman's friends came over to congratulate him.

Haruto bowed to the Turkish black belt. "Thank you for allowing me to train with you this week."

Erhan smiled and bowed in return.

Haruto went to the changing room. He removed the loaned white karate uniform, folded it neatly, and changed back into his jeans and t-shirt.

Suleyman came running into the room, holding a small box. "Haruto, thank you for helping me prepare for the exam. This is for you."

Haruto opened the small cardboard box covered in a decorative blue foil. A silvered chain necklace holding a stainless steel amulet with a blue eye embedded in the middle of it.

"In Turkey we believe that the evil eye protects you from harm."

Haruto put on the necklace and let the evil eye pendant fall on top of his t-shirt. He bowed to Suleyman. "Thank you for the gift. I shall wear it on my journeys."

CHAPTER094

Turkish Airlines Flight 1184 hit the tarmac hard and bounced twice before rolling to a stop. Twenty minutes later, Haruto disembarked and walked to Border Control. He centered the camera and large telephoto lens around his neck.

"Next."

Haruto gave the officer his passport.

"Mr. Aoyama, why are you coming to Israel?"

"I am tourist."

The Customs officer looked coldly at Haruto. "What is the purpose of your trip?"

"I am tourist." His heart skipped a beat.

"Why are you tourist?"

Haruto felt his heart beating harder and his palms starting to sweat. "I'm on vacation."

The officer picked up the large red stamp. "Passport okay to stamp?"

"Why do you ask me?"

"You're not planning to go to Syria or Lebanon, are you?"

Haruto's heart started beating even faster.

The Border Control officer smashed the rubber stamp into the passport, printing a large red Israeli tourist visa. "Well you won't be going there now, not with this passport — they won't let you in."

Haruto smiled and walked to the Arrivals Hall. Made it through Israeli security. Not too bad. He had no checked luggage, and so started walking toward the Exit sign.

"Hey you, where are you going?" the tall, young Border Police Officer shouted at him. "Do you have any bags?"

"No."

"Come here."

Haruto followed the officer to a small frosted glass cubicle on the side of the luggage carousels.

"Why are you here in Israel?"

"I'm a tourist."

"How come no suitcases?"

"Just the backpack," Haruto said. "I like to travel light."

The Border Police Officer looked suspiciously at Haruto. "Let me see your passport."

Haruto's heart started beating faster again.

"One week in Turkey... let me see your camera."

The officer screwed off the lens and looked through it. He replaced it and then hit the power button on the camera. A picture of the Aden Hotel appeared on the display screen. The officer hit the NEXT arrow button, but no more images appeared.

"You spent a week in Turkey and only took this one picture of a hotel? Why are you coming to Israel?"

Haruto flipped open the small door on the bottom of the camera and with his index finger and thumb pulled out the thin white memory stick. "This is a new one for photographs of Israel." Haruto put the memory stick back in place.

"Why are you here? Where are you staying in Israel? Where's your luggage? What type of work do you do in Japan? Why are you here? What's the purpose of your trip?"

Haruto pulled out the orange kibbutz sheets and handed them to the officer.

The officer smiled. "Oh, a kibbutz volunteer. Up north. Good... very good." The officer reached into Haruto's bag and quickly felt around. He gave Haruto back his passport and bag. "Have a nice stay in Israel."

CHAPTER095

Haruto pointed to the road sign written in Hebrew, Arabic and English. "Is that Nazareth there?"

The driver, a young man in a business suit, nodded. "I live about ten kilometers north of here."

"Thank you for the ride from the airport. It was very kind of you to offer. I can take a taxi at Nazareth."

The driver shook his head. "Everybody hitchhikes here in Israel. Before I had a good job, I used to grab rides all over the country. Save your money. I'll swing onto Seventy-Seven and drop you at Golani Junction."

Haruto clicked the buttons on his GPS watch and squinted at the small screen. "Where is Golani Junction?"

"It's the intersection of Highways 65 and 75. You always have soldiers there hitching a ride north to their units."

Fifteen minutes later the driver pulled off the road in front of the intersecting highway.

"Thank you. What do I do now?"

"Here, let me show you." The driver got out of his car and waved at traffic going north. A minute later, a compact white Mercedes pulled over and lowered its window. A man about thirty years old, clean-shaven with close-cropped hair and wearing military fatigues smiled at Haruto's driver.

"This guy's a volunteer up at Misgov. How far north you going?"

"No problem. I leave my car at Qiryat Shemona." The soldier in the Mercedes turned to Haruto. "Get in."

Haruto was sweating and shielding his head from the sun. He hopped into the front seat of the Mercedes, the car pulled back onto Highway 65 and zoomed north.

The air-conditioning felt wonderful. "You have a nice car."

"Thank you."

"Are you an officer?"

"No, just a low-level soldier."

"How can you afford this car?"

"I'm an electronics engineer. But once a year my unit is activated, and we all return to the field for a month. I'd rather be designing circuits than

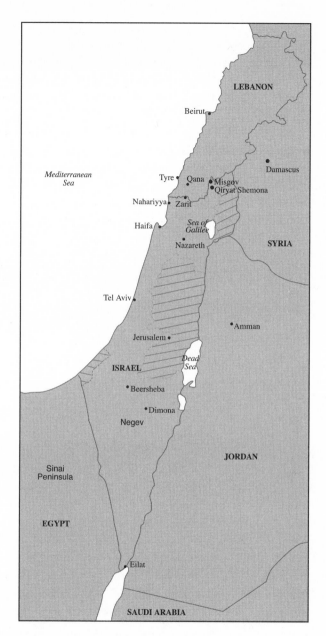

walking around in the mud with a gun, but not much I can do about it."

An hour later Haruto saw the sign for Qiryat Shemona. The soldier looked at his clock. "I don't have to join up with my unit until sixteen hundred hours. Let me drop you off."

The Mercedes continued north through the small city and then took a left onto a smaller service road. Eighteen minutes later, it pulled up to a two-story light-brown brick building in front of a meadow.

"This is it." The soldier smiled. "If I go any further then we'll be in Lebanon. I don't think my commanding officer would like that."

"Thank you." Haruto got out of the car and looked around. Green fields and mountains surrounded him. The sun shone brightly, but there was a delicious cool breeze. The birds chirped overhead, and Haruto breathed in deeply the sweet smells of the land.

He knew he was in a small country, much smaller even than Japan. But after Tokyo, the landscape felt wild, untamed.

He retrieved his bag, tapped twice on the car door, and started toward the building.

CHAPTER096

"**H**ello!" **Long black hair**, green tank top and blue jeans, she was sitting on the stairs of the building. Showing a large white smile, she raised her arm and waved.

Haruto walked over. "Hello, I am volunteer."

The young woman giggled. "Hello, volunteer. I'm Mara." She stood up, her tanned midriff catching Haruto's eyes. "Do you have a name?"

Haruto immediately averted his gaze, "I'm Haruto — no, I mean Takamichi... I..." Her hazel eyes were large, her olive skin free of any makeup but glistening in the sun. Haruto turned his head slightly away. "I am Aoyama Takamichi. Haruto's my nick-name."

"Okay, Haruto. So you're a volunteer?"

Haruto nodded and took out the orange papers. "These are important papers and must be given to the Director of Volunteers at Kibbutz Misgov."

She took them from his hand. "And now they have been. I used to give Hebrew lessons to the volunteers and show them around. But we haven't had any for a while." Mara scanned the papers, and shoved them into her rear pocket. "Come, follow me."

She started down the path beside the brick building. Haruto followed, keeping his eyes low. Flat leather sandals. A strap of leather wrapped around and grabbed onto each ankle. He gazed a little higher. Her blue jeans hugged a tight, curved rear end. The exposed midriff led up to firm breasts pushing against her tank top, their nipples obvious when she turned her body. She then started to turn toward Haruto. Faster than any karate move, he yanked his head back and his eyes up. "It looks very nice here."

"Let me show you around." Mara pointed to the brick building. "That's the dining room. Very important place, and not just for the meals. That's where the Labor Organizer posts the jobs every day, that's where we have our meetings and our dances."

Haruto nodded and continued following the woman.

Behind the dining room, they came to a full soccer field. In the distance were four huge, steel-roofed industrial buildings with large fans mounted on their walls. "Poultry Operations. We have a big bandage factory over the hill — you can't see it from here — so we were thinking of shutting down the chickens a few years ago. More money in manufacturing, and less mess. But now Europe's boycotting our products so it's a good thing we kept the chickens."

They continued walking. Mara pointed to a wooden two-story building. "That's Babies' House."

"Babies don't stay with their parents?"

"No. Everything and everybody belongs to the kibbutz. The *metapelet* in the Babies' House take care of the babies. But the parents, of course, visit. When I was a baby, they were very strict — you could only visit on this hour or that hour. Now, the mothers come whenever they want."

They passed another wooden two-story building. "What's that?"

"Children's House."

"What?"

Mara turned away from the building and pointed to dozens of low-rise dwellings on the other side of the path. "That's where the members live. Past those apartments are the ones for workers and volunteers. Members belong to the kibbutz. Workers are extra workers we hire and pay a salary. The *pioneers* — the old-time members — were totally against the idea of someone who wasn't a member doing work on the kibbutz. But the bandage factory was very successful so we needed extra workers for a second and then a third shift."

"What's a volunteer?"

"You!" Mara laughed. "Ever since I was a child there have been volunteers on the kibbutz. Most were from Britain and the United States, a few from France, Switzerland, and yes, Japan. Volunteers come for a new experience, and we're happy to have them for the work they do."

Mara and Haruto ascended to the top of a small hill. Mara pointed to the orange-roofed building below. "That's the bandage factory. I work in the office there. Very modern." She pointed to the hundreds of trees to the north. "We still run a large apple orchard." She moved her finger down a few degrees to the empty land in front of the trees. "We used to grow different crops there, but we don't have enough volunteers now to do the farming."

Haruto pointed to the empty fields west of the orchards. "What's that?"

"That's Lebanon. You don't want to go there."

CHAPTER 097

Haruto got to the dining room a few minutes early and waited outside on the stairs. He hoped Mara would show him where he was supposed to sit, but he couldn't find her. About forty people had streamed past him into the building. All were white or olive skinned. No one seemed to think it anything out of the ordinary that a strange Japanese man was standing there.

"*Shalom. Mah hashem shelcha?*"[‡] a short, bald man said.

Haruto didn't know what to say. "I am Haruto."

"Mara hasn't taught you any Hebrew yet?" the short man said in fluent English.

"No. Have you seen Mara? I need to know where to sit and what I should do."

"You can sit with me. Come, tell me about yourself."

"Are you sure? Mara is Director of Volunteers at Kibbutz Misgov."

The short man started to laugh, but he looked at Haruto's serious face, and stopped immediately. "No, it will be fine with Mara if you sit at my table tonight. I am Salzman."

Haruto followed the short man into the dining room. Two older women at a square table waved the short man over.

"Ladies, this is Haruto. We can all practice our English tonight."

"Did you work today, Salzman?" one of the women asked.

"*Ken.*" Salzman turned to Haruto. "I am a doctor. I work at the hospital in Nazareth."

"Don't you work here at the kibbutz? Are you a member?"

"Yes, but there are not enough patients here and we have another family doctor living on the kibbutz. Besides, I take care of problems in the head."

"Are the other members upset that you don't do work here?"

Salzman laughed. "No, they're quite happy, especially when I hand over my hospital salary check."

"You don't mind? Can you leave the kibbutz or are you forced to stay here?"

[‡] Hello. What is your name?

"I could go any time I want, but I don't want to leave. I grew up here. The kibbutz paid for my medical training in Israel and in the United States. I'm divorced now, but I got married here and had a family here."

A tall, thin man in a simple white shirt and jeans stood up in the middle of the room, and tapped a fork against a glass. The dining hall fell silent, and all eyes turned toward him.

Salzman put a hand over his mouth and whispered, "That's Shlomo, the Secretary of the kibbutz. The chief."

"Good evening," the white-shirted Secretary said in English. "Mara tells me we have a new brave volunteer. He's from Japan. Haruto, please stand up."

The dining room clapped as Haruto got up.

"Line two at the factory was down. New parts from Germany are expensive, so we're trying to make our own replacement pieces. Some good news — the Army has been hitting al-Haleeb hard, and this is the fifth day without any rocket attacks on the kibbutz. That's all from me. Rebecca?"

A forty-something woman stood up.

Salzman whispered to Haruto, "That's the Labor Organizer — she decides who does what job."

"All jobs the same for tomorrow except Haruto will work with Amit in Poultry."

The Labor Organizer sat down, but the rest of the dining room got up to head to the buffet table. Haruto followed Salzman.

As Haruto was finishing his meal of marinated chicken breast and *hummus*, Mara blew into the dining room, grabbed a plate of food, and pulled an extra chair up to the table. "I get my first volunteer in a month, and you're trying to steal him, Salzman?"

The women at the table laughed, but Haruto remained serious. Mara wiped her mouth with a napkin, then grabbed Haruto's wrist. "Come, let me show you where you have to go tomorrow morning."

CHAPTER098

Haruto couldn't see Amit. Just chickens. Thousands and thousands of chickens — clucking, pecking the floor litter, pushing against each other and running around in whatever free space they could find. And defecating — the ammonia-like odor hit Haruto like a fist. He covered his nose with his hand and tried not to breathe too deeply.

The building was about twenty meters wide, and the far wall seemed to be about a hundred meters away. A dozen low pipes ran down the length of the large room. Some pipes had red plastic feeding baskets every meter, with chickens crowded around, pecking at the morsels of food dropping out. Other pipes had white nipples every half meter. Chickens put their open beaks over the nipples, and drops of water dribbled down.

Amit waved his hands and walked over from the opposite side of the building. As he got closer, it was apparent he was waving Haruto away. "You will infect the chickens!" Amit pointed to clear plastic sleeves over his own feet and lower legs.

Haruto followed the wiry graying man out of the building. It was still dark outside.

"We're very, very careful about the chickens. There are thirty thousand in each building. They all start off together as chicks, grow together, and then we ship them all to market together. A building's worth of chickens is quite an investment."

Haruto nodded.

"And there are diseases that will wipe out a whole building at once." Amit pulled the clear plastic slips off his feet. "Whenever you go into a building you put on a new pair of these plastic coverings."

Haruto nodded again. Wearing only a thin white t-shirt, Haruto shivered from the cold air.

Amit laughed. "You thought you were coming to the desert, didn't you? We're in the mountains. Have you ever seen snow? Wait a few months, we get lots of it here."

"I'm only here for a month."

"Too bad. All right, follow me. We have a big day ahead."

Haruto followed Amit to another one of the four large chicken barns. Amit slid open the hangar-like entrance doors. Like the other building, this one was also about twenty meters wide and a hundred meters long, with rows of feeding and drinking pipes hanging down from the ceiling. But there were no chickens, only a bare concrete floor.

Amit pointed to a sink at the entrance and cardboard barrels stacked nearby. "Put on shoe sleeves, wash your hands, and put on a pair of work gloves. A new shipment of chicks will be here this morning. I want you to cover the floor with sunflower hulls from those barrels."

After about fifteen minutes of spreading the sunflower shells carefully by hand, Haruto looked around. He'd only covered a tiny area with hulls. The vast floor was daunting.

"Hello, Haruto!"

Haruto turned around. A tall, young man was standing there with two large metal rakes. "Yeah, I thought Amit wouldn't have explained. He never does." He dropped the rakes. "First thing, pull the barrels so they're all over the building. Let's get this job done before breakfast."

Haruto followed the young man's actions and started pushing the cardboard barrels of sunflower hulls until they were scattered around the large floor like trees in an orchard.

The young man pointed to the metal rakes. "Take a rake and spread the shells. Go as fast as you can." The young man kicked over a nearby barrel and started raking out the shells. Haruto followed suit.

Two hours later a layer of fresh sunflower hulls covered the entire floor, and Haruto and the young man were dripping with sweat.

"Come Haruto, we have breakfast now."

The two men walked over to the dining hall. Haruto loaded up a plate with bread, scrambled eggs, and Israeli salad — finely cut pieces of tomatoes and cucumbers — and gobbled it down with two glasses of orange juice. When they got back to the chicken barn, a tractor-trailer truck was there.

The young man started pulling boxes of chicks off the truck, and handed two to Haruto. "Empty the boxes in the middle of the room."

Haruto went into the building, opened the first box and stared at the small yellow chicks. He cupped his hand, and took one chick out at a time, gently lowering it to the floor, waiting while it awkwardly started walking about on the sunflower hulls.

"What are you doing Haruto?" Amit tipped the entire box on its side, and the chicks chirped excitedly, but then scrambled out. "We have to unload thirty thousand chicks. Don't worry about them so much — they're going to be eaten in six weeks anyway. Just make sure your feet are covered so we don't infect the lot."

By one o'clock, the tractor-trailer truck drove away, and thirty thousand little yellow chicks were chirping in the large room. Outside, the cool early morning had yielded to the hot Israeli sun. Haruto walked over to the dining hall.

"Haruto!" Mara waved from a table near the window.

Haruto carried his tray over, glad to see a friendly face who had nothing to do with chickens.

Mara pointed to a tall young man at the table. "This is Dov."

"Please, sit down with us. We speak English, okay?" Dov said, his English fluent but with a thick Hebrew accent. "Do you like the kibbutz?"

Haruto nodded.

"You came at a good time. No rocket attacks this week. There was an attack south of the kibbutz yesterday, but the army is on top of the situation now, so no worry." Dov turned to Mara. "I spoke to Chaim last week. He's in charge of *robot* soldiers — they go into Lebanon and hunt down the terrorists. Can you believe that?" Dov turned to Haruto again. "This is just between us. I'm not supposed to tell anyone."

"Chaim, Dov and I all grew up here on Misgov," Mara said to Haruto.

"Where did the rocket attack come from yesterday?" Haruto asked.

"Bint Jubayl," Dov said. "Do you know the country?"

Haruto shook his head.

"No worry, it will be safe up here."

Mara nodded. "You'll be fine on the kibbutz and have a good month. We have Israeli folk dancing tonight — in the dining hall at eight. Are you going to come?"

"I don't know anything about folk dancing."

Mara's full lips framed a wide white smile. "Nobody does when they start."

"Yes, I will come. But I must now go back to Poultry Operations and help Amit."

As Haruto walked out of the dining room, a large, muscular young man in soldier's greens swiped against Haruto's shoulder. Haruto turned around.

Two more young soldiers followed the muscular one into the dining hall. The large soldier walked over to Mara's table.

"Eitan," she yelled. "What are you doing here? It's over."

The large soldier grabbed Mara's right wrist and yanked her body up.

Haruto ran back into the dining hall. "Let her go!"

"Haruto, go away," she said. "He's the Army *krav maga* instructor — he'll hurt you."

Haruto kept coming.

The soldier turned his head to look incredulously at Haruto, and let go of Mara's arm.

"No Eitan!" Mara screamed. "Leave him alone!"

The soldier put his arms up in a traditional boxing stance, but a split second later Haruto's right arm was trapped under the soldier's shoulder, while the soldier's other muscular hand was pushing against Haruto's captured arm, using the leverage to try to snap the arm in two.

Pain shot through his body like an electric shock. Before his brain could form a coherent thought, it fired his free hand in a full-speed *oi-tsuki* punch into the soldier's face, smashing the nose, ripping the cartilage and breaking the bone. Blood gushed out and the muscular soldier screamed in pain, releasing Haruto's trapped right arm.

Haruto pulled his right arm back, and a quarter-second later thrust it into the soldier's Adam's apple. The muscular soldier dropped, one hand against his bleeding face, gasping for air with his mouth wide open.

Haruto heard the footsteps of the two soldiers come rushing at him from behind. He pivoted in a tenth of a second, and in another tenth, placed a *mae-geri* kick on one of the soldier's knees. As that soldier cried out in pain and fell to the ground, Haruto retracted his leg, and without it touching the ground, a quarter second later he fired another *mae-geri* kick into the side of the second soldier's left knee.

The Secretary came running over and looked at the three soldiers lying on the floor. "Eitan, I don't want to see you here again! Do you want me to speak to your commanding officer?" Shlomo turned to Haruto. "Are you okay?"

"Yes. But I must return to Poultry Operations."

CHAPTER099

Haruto followed Amit into the small metal prefab building at the intersection of the four large chicken houses. A row of flat-panel monitors lit up the otherwise darkened room.

"Do you have any experience in farming, Haruto? I guess it must be even more automated in Japan."

"No, I grew up in the city."

"Let me show you the system." Amit pointed at the first computer monitor. "This one controls the temperature in the buildings, depending on the age of the chickens."

Amit pointed to the next monitor. "This one controls the feed pumps. It depends on their age, but most of the time we let the chickens eat and drink as much as they want. This next computer controls the lighting — we reduce it as the chickens get older to limit activity. Do you use robots in Japan?"

"In car factories, but I'm not very familiar with them."

Amit put his hand on the next computer monitor. "This one is still under development but will control the clean-up robots. In a few days, we're going to send building Number Three to market. You and a few others will spend two days collecting the soiled hulls, then hosing down the feed trays and floor — getting the building ready to take the next batch of chicks. An Israeli company that makes lawn mower robots has come out with a robot that gathers the floor litter and then hoses down the room. We're getting one of them next month."

Amit pointed to the next monitor. "This one controls the feed silos. Our biggest expense. The last computer here does the accounting. Building Number Two where you worked this morning sent its chickens to market a few days ago. We need to finish up the reports for that lot of chickens."

Haruto nodded. He could see now that Amit wasn't in charge of the Poultry Operations. He *was* the Poultry Operations — one man raising a hundred thousand chickens.

"We have a student from Technion coming next month to help us with some programming, but right now these computers don't talk to each other." Amit typed in a few commands on the last computer's keyboard and pointed to its monitor. "Here, I want this form filled in. You need to go to the other computers to get the information. Everything's in English and menu driven. Just click on *Reports* and you get the reports for that computer, okay? Not too hard. This will make you learn the system. You have all afternoon to do this."

Amit left the building, once again leaving Haruto to figure out what to do. All this information about chickens and computers.

Okay, make some rules. Print out the blank form. Then go to each computer and fill in whatever information he could find. Then go back to the report computer and enter the data.

An hour later, Haruto smiled as he hit the *Enter* button. He clicked *Print*, took the completed document, and found Amit at the feed silo. He handed him the form.

"That's great, Haruto. You're done today. See you tomorrow morning at five."

"I wanted to ask you..."

"Yes, Haruto?"

"Well, I wanted to do some traveling tomorrow..."

"But you've only worked one day."

"Yes, I know, but..."

"Okay, okay. Well then, be here the following morning at five."

CHAPTER 100

It was not even 8PM but thirty folk-dancing fanatics filled the dining hall. Tables all shoved to the side, wooden floor polished to a sheen, speakers blaring an addictive melody. Very different from any of the music on the cruise ship or anything from Japan. Violins and clarinets. The melody went up and down, almost seeming to be crying and then to be laughing, and sometimes both at once.

The dancers had joined hands and formed two large concentric circles. The inner circle moved counterclockwise, the outer one, clockwise, both of them to the beat of the melody. The dancers filled the room with a whirl of color as they circled. Holding hands, they would bend forward to the beat and then lean back, raising their joined hands up in the air, and the circle advanced another notch.

"Haruto!" Mara yelled. She broke away from the circle and came running to Haruto, grabbing his hand. "Come!"

"I don't know what to do."

"Just follow me."

Pulling Haruto, Mara ran back to the outer circle and inserted herself. The woman on the other side grabbed Haruto's hand, and the circle closed itself again. Held by Mara on the right and the woman on the left, Haruto too became part of the circle, bending forward to the beat and then leaning back, raising his joined hands up in the air and advancing the circle another notch clockwise.

It took him only a minute to fall into the rhythms, but the whole circle had advanced nearly a full turn before he realized something with a start.

He was having fun.

CHAPTER 101

Haruto **shivered** in the early morning cold. The massive telephoto-lens camera around his neck bounced against his gray sweatshirt. He walked away from the low-rise apartments, west toward the apple orchard.

Haruto clicked his GPS watch. The blue watch face lit up with Misgov, Israel, the town of Udaysah just across the border, and three kilometers to the west, Rabb, Lebanon. He clicked again to zoom in, but no more detail appeared, only a line representing the border remained on the tiny screen.

Haruto walked in the darkness past the large bandage factory and continued to the orchard, passing through a half-kilometer of apple trees. Past the last trees was a clearing ending in a four-meter-tall razor-wire fence. Attached to one of the fence posts was a metal sign with a red triangle and a warning in Hebrew, Arabic and English: *Stop! Mines! Danger!*

Haruto peered through the fence. Twenty meters of bare earth lit up by spotlights on the fence, which faded out to a dark hill — Lebanon! How to cross into the country? Climbing over the razor wire and getting blown up in the minefield was not a good plan.

Make some rules.

Haruto scanned the area. To the right, up on the hill past the orchard, was a guard post of some sort. A good starting rule would be to find persons who knew the area and ask them how to get into Lebanon and then down to Bint Jubayl, where Dov said the rockets had come from yesterday. He guessed that, if the robots were going anywhere, it would be there and it would be today.

Haruto climbed up the hill. He could make out a soldier, automatic rifle slung over his shoulder, smoking a cigarette. Haruto waved to the soldier. The soldier didn't see him, but Haruto kept waving as he walked up the elevation.

As Haruto made the top of the hill, the soldier noticed him, dropped his cigarette and grabbed onto his weapon. "Stop!"

"I am Haruto, a volunteer on the kibbutz."

Another soldier came out of the guard post. "It's all right. I saw him last night at the dance." The first soldier put down the M16.

"What are you doing here, Haruto?" the second soldier asked.

"I want to travel to Lebanon and take some pictures."

The soldiers laughed.

"Come here." The second soldier pointed to the observation tower. "You can get a good view up there."

Haruto climbed the ladder to the top floor of the tower. He could see for a few miles across the border. Some lights here and there. He looked due west through his camera's viewfinder, and in the dim light could make out an apartment block and a few houses, one of them with a yard full of sheep. Haruto brought the camera to bear on the well lit border fence and moved along it. About a hundred meters north, there was a huge gash.

"Why does the fence have a hole in it?" Haruto called down to the soldiers.

"Where? A bit north?" the first soldier asked.

"Yes."

"Drug smugglers last night. Hashish. We caught one of them, but the other two ran back. It'll be fixed in a day or two. Not a big deal."

Haruto climbed down the ladder. "Thank you. I will continue my tour."

The soldiers laughed.

Haruto moved back, away from the lights of the border fence, but followed it north for a hundred meters until he reached the cut section of wire. He could make out the guard post from here. The two soldiers were talking to each other in front of the observation tower.

Haruto tapped his left forearm two times, and then tapped the right one twice. He took a deep breath and quietly raced to the illuminated fence. He turned his head toward the guard post. Good, the soldiers were still talking to each other. He pulled the sleeve of his sweatshirt around his hand, pushed aside the loose ends of razor wire, and slipped through.

The spotlights brightly lit up the ground. The smugglers' footprints were still very clear in the soft earth. Haruto tucked his foot into one of the earthen footprints, and slowly shifted his weight onto it. Good. He stepped into the next one. Halfway through, the illumination of the spotlights started to trail off. The next footprint was still obvious and Haruto stepped onto it. He looked carefully for the one after, but it was hard to make out a meaningful pattern in the earth. There was still another ten meters to a flimsy wire fence that looked like it was intended more to keep the sheep off the mines than a security fence.

Take a chance. Take a step. He lifted his right foot.

Haruto's heart started to beat faster. Muscles throughout his body clamped together — he felt like he was going to suffocate.

Haruto bent down and ran his hands through the earth in wide arcs, pushing them deeper. Maybe there were no more —

He felt something round and cold! He wrapped his fingers around its edge and pulled it up through the earth.

It was only a large, eroded rock.

Haruto ran his hand through the soft earth again. Nothing. Maybe the minefield was over. Haruto took the large rock with both hands, and tossed it some five meters in front of him.

The earth exploded! A wave of air pressure slammed Haruto into the ground. As Haruto raised himself on his hands and knees, machine gun fire filled the air, with a stream of bullets from the guard post flying all around him.

Haruto's heart beat faster. It felt like his chest was going to explode. It was hard to breathe.

Make a rule. Don't die here.

The bullets kept streaming in. Haruto tapped his left forearm four times, and then quickly did the same on the right.

Okay... any mines very close to the explosion would have also detonated. Crawl over the exploded ground.

Haruto started crawling forward. He stopped suddenly. His heart was beating hard. The perspiration ran down his forehead, falling into and burning his eyes. There could still be a mine in the next meter or two that hadn't detonated. Make another rule. Okay... dig out the next two meters.

Haruto pushed his hands as gently as possible through the earth in front of him. He felt nothing. He moved his body forward a bit. He checked the earth again. Nothing.

Zip. A bullet missed his head by a few centimeters.

Faster. Must go faster.

Haruto took a deep breath in. On hands and toes, he bent his legs and pushed off, flying over the next two meters of earth, slamming into the safety of the small exploded crater. Okay... the second fence was a few meters away. A stream of bullets whizzed over his head again. No time to dig through the earth. He took a deep breath, hugged the ground and madly crawled to the wire fence, shoved between the lower two strings, and flipped into Lebanon.

CHAPTER 102

Haruto continued walking through the pasture toward the lights in the distance. His sweatshirt was covered in dirt and shredded, but he felt fine. He clicked the camera power button. Good, it still worked.

A few hundred meters away, in the dawn's twilight, he could make out a flock of sheep and two thin men. As he grew closer, he saw that the men were two teenage shepherds, maybe thirteen or fourteen years old.

Haruto took off his ragged sweatshirt, wiped the dirt off his pants and ran his fingers through his hair. He took a hundred dollar U.S. bill out of his pocket and waved at the boys.

"I want to take some pictures. Can you help me?"

"You journalist?" the taller boy asked in accented English.

"Photographer. I want to go to Bint Jubayl. One hundred American dollars to help me get there."

The two boys started speaking to each other in fast Arabic.

The taller boy turned to Haruto. "It is a far walk. You are rich Japanese. Hundred dollars not enough."

Haruto took out another hundred-dollar bill from his pocket. "One hundred now, and another hundred dollars to help me get back."

The boy smiled. "Yes."

Haruto put one of the hundred-dollar bills into the boy's hand.

The boys spoke for a few seconds again in Arabic. The shorter one emptied his burlap sack of a few apples and a wrapped bread.

The taller boy gave Haruto the sack. "Put your big camera in the bag. It is better." The boy pulled out a *kaffiyeh* — a traditional Arab headscarf, hanging from his back pocket, and started wrapping it around Haruto's head. "It is better."

Haruto nodded.

The boy started walking south, away from the sheep, and Haruto followed.

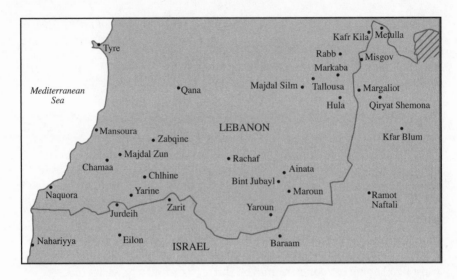

Roadway east of Bint Jubayl, Lebanon
July 1   11 AM Local Time (08:00 Zulu)

CHAPTER 103

Haruto pointed to the large town ahead.

"Yes, Bint Jubayl." The boy motioned to the steep hills south of the town. "We go there. It is better."

Haruto and the boy moved off the road, through two small orchards, and up the hill overlooking Bint Jubayl.

The boy tapped the burlap bag. "You take pictures now. Yes?"

Haruto took out the camera, looked through the viewfinder and panned across the town. Excellent. Good view of many of the streets and buildings. He twisted the massive telephoto zoom lens. An old woman in her yard with some chickens. He twisted the lens a bit more. He could see the chickens pecking the ground.

The sun was beating down on them. The *kaffiyeh* proved useful for more than blending in with the local population.

Haruto zoomed out to lower magnification and scanned the town for activity. Not much. He turned around and looked at the road coming up from Yaroun in the south. Not much activity there either.

Okay, make a rule. Wait here two days, and if no rockets, return to the kibbutz and come back another time. Within the next month, he should manage to be present at one missile launch, at least, and photograph the robots subsequently tracking down the al-Haleeb fighters.

What about Poultry Operations? He promised Amit. This was not a good rule. Haruto took a deep breath in. He'd offer to do extra work.

Haruto took out the other one-hundred-dollar bill and offered it to the boy. "Here, you can go. I will find my way back. Thank you."

The boy put up his hands. "No. You nice man. I take you back and then you pay. You want pictures of fighting for your newspaper, yes?"

Haruto nodded.

"You wait here, yes?"

The boy climbed down the hill and walked over to a young man shepherding a dozen fat sheep. Haruto looked through the viewfinder. He could see their lips moving.

The boy ran back up the hill, all excited. "The fighters are in town. They are setting up many rockets there." The boy pointed to the courtyard of a large, gray U-shaped building on the eastern periphery. The opening of the U was toward the south, directly in line with them.

Haruto pointed the huge lens toward the U-shaped building. He swept the camera back and forth. There it was! Haruto zoomed in. In its stone courtyard were eight bearded men with tools and an arsenal of dozens of missiles lying about. Although they wore no uniforms, the eight fighters were as organized as any army Haruto had ever seen. In less than a half-hour, three dozen rockets were sitting in their launchers.

Haruto passed the camera to the boy, who was straining to look down at the activity in the building's courtyard. The boy's face lit up. "Missiles go soon — to Israel. But…" The boy picked up his finger. "Very bad. The Jews come back. The man below is from Maroun. He tell me they come with new weapon now. Robots. Shoot bullets out of mouth. Kill everyone."

Haruto took the hundred-dollar bill from his pocket. "Go. Don't stay here. It is not safe."

"Thank you. You nice man." The boy took the money and walked away toward the north.

CHAPTER 104

The successive white flashes and explosive blasts of the missile launches caught Haruto off guard. He grabbed the camera and focused again on the gray U-shaped building's courtyard.

Another dozen missiles blasted southward toward Israel. A bearded man in the courtyard holding a walkie-talkie put up his arms. Silence again.

The bearded man put the walkie-talkie to his ear and held up his right hand, palm open. Minute after minute went by. Then suddenly the bearded man swept his hand down and started shouting to the other men. In less than ten seconds, the final dozen missiles jerked upward into the sky.

An open truck pulled in front of the courtyard. The bearded men ran around with their tools like crazed ants, unbolting the rocket launchers and running back and forth to the waiting truck with disassembled pieces. But before they could finish, a small plane streaked down out of the sky, heading directly for the U-shaped building.

Haruto zoomed out a bit to track the airplane, then zoomed in. It was the size of a typical two-seater prop, but there was a jet engine in the rear instead of a propeller in the front, and a blackened front dome instead of a window. A UAV — unmanned aerial vehicle.

The UAV circled above the U-shaped building. One of the bearded men holding a submachine gun pointed it skyward and fired, but to no effect — the plane kept circling. The bearded man with the walkie-talkie started shouting into it.

Haruto put down the camera and looked up. The shepherd with the fat sheep was running up the hill, his animals following.

"You must leave! They will kill you!"

"Who?"

The shepherd pointed to the southwest corner of the town. About a hundred men carrying rifles and RPG launchers streamed out of the town and were starting to spread out. "Al-Haleeb — they will kill you. They will think you are an Israeli spy." The shepherd turned to the south and pointed to the highway. "Soon the Jews will come and they will kill you also. They have robots!" The shepherd puckered his lips. "Tat.. tat.. tat.. tat. They shoot bullets from their mouths!"

The shepherd pointed to the house at the top of the hill. "Come, we hide there."

Haruto followed the shepherd and the sheep to the house and small field near the top of the hill. The two men crouched between the house's stone wall and an adjoining shed.

In a few minutes, Haruto heard footsteps and voices, and then four fighters ran past the house to a landing that overlooked the highway to Bint Jubayl from the south. Two set up small tripods to support their sniper rifles. The other two put down the dozen RPG's they carried between them and looked down upon the highway.

So did Haruto. He had wanted to get pictures of the robots in a battle. Now, it seemed, he was going to be in the middle of it.

For half an hour Haruto stayed with the shepherd in the hidden spot, watching the al-Haleeb fighters a few meters away. When it was clear their attention was riveted to the road below, Haruto stuck his camera out past the corner of the stone wall and scanned the area below. Nothing. As he tilted the lens farther to the south, it bumped and resonated against the stone.

The shepherd pulled Haruto back, his finger over his lips. Haruto nodded.

Too late. One of the fighters, with a finger on the trigger of his sniper rifle, was coming toward the shed.

The shepherd came out of the hiding place, hands up in the air. "*Salaam alaykum.*"

The fighter ignored the greeting and pushed past him. He stopped at the corner of the stone wall. Leading with the barrel of his long rifle, he stepped into the breach between the wall and the shed.

Haruto shot out of the hiding space, flying in the air, landing on his hands and flipping over onto his feet.

A bullet zipped past, centimeters from Haruto's head.

Haruto saw the large tree ahead, and the dip in the hill behind it. His legs struck the ground with every bit of force his body could muster.

Behind him came the click of a rifle bolt.

Haruto jolted to a stop. He put up his arms in surrender and turned to look at the gunman. The fighter's eye pressed against the telescopic sight, and his left hand held the rifle steady. Haruto saw the fighter calmly smile and his finger start to pull on the trigger. Haruto felt his heart smash against his chest.

Thwat! Thwat! Thwat!

Haruto stood there. Nothing. He felt fine. In slow motion almost, it seemed, the gunman let go of his rifle, blood started to trickle and then spurt out of his belly, and he fell over onto the ground. Behind him stood an Alpha — like the robot from the ship.

"Aheeee!" the shepherd shouted. He grabbed a palm-sized rock.

"No... no!" Haruto yelled.

Too late. The shepherd hurled the stone at the robot.

The robot brushed away the stone with its right hand. A volley of shots rang out from its mouth, and the shepherd fell motionless onto the grass.

Haruto jerked back, took two steps forward and flew onto his hands in a quick somersault.

Thwat! Thwat! Thwat!

Haruto ignored the sting as the bullet cut through the rear of his right calf. He pushed off with the bleeding leg, flew again in the air, landed on his hands, and flipped to momentary safety behind the thick olive tree.

The robot came running toward the tree.

Rule. Make a rule. What rule would possibly work? Can't out-hit or out-run the machine… Not to die like this. In failure. His fate was always to be a failure. A delusion, all these rules.

The robot's weight crushed the soft earth. It was one more step from the tree. The moment he stuck his robot head around the tree, death would be there. Rules useless. Failure.

Suddenly Haruto's eyes opened. Rules. The robot ran on rules too.

Haruto whipped the red Japanese passport out of his rear pocket and held it up. The robot's head came around the tree, its camera eyes both focused directly on Haruto.

Haruto stood there frozen, waiting for the shots, waiting for his death, his failure.

Nothing. In what seemed like an eternity, time stopped, as the robot walked around the tree, and with one hand lifted Haruto up off the ground. With the other hand, the robot pulled the *kaffiyeh* headscarf off him.

"*Nihongo o hanashimasu ka‡?*" the robot asked.

"*Hai*," Haruto answered.

The robot gently lowered Haruto to the ground then… bowed. "Autonomous Products, a division of Mikiyasu Industries, welcomes you to these war game demonstrations. Autonomous Products makes assembly equipment for our parent company. Our products, including the Alpha and the Beta robots, are the most automated and of the highest quality. You should consider Alpha and Beta robots as valuable support resources to your armed forces. Please enjoy the war game demonstrations. Thank you."

The Alpha robot ran toward the three remaining al-Haleeb fighters. One of them was firing RPG's onto the highway below. The other two pointed their weapons — a sniper rifle and a rocket-propelled-grenade launcher at the Alpha.

Haruto grabbed his camera — it had managed to remain around his

---

‡ Do you speak Japanese?

neck. He blew the dirt off the front of the lens, and clicked the camera into high-resolution video mode.

Thwat! Thwat! Thwat! Thwat! ... The Alpha let loose a continuous stream of bullets at the fighter pointing the RPG launcher at him. The fighter with the sniper rifle fired a well-aimed shot at the Alpha, but the bullet struck a reinforced portion of the torso and glanced off. The Alpha did not alter his focus and kept the stream of bullets coming at the first fighter. A half-second later, the fighter dropped his RPG launcher and fell to the ground. In the next tenth of a second the Alpha fired a volley of shots at the fighter with the sniper rifle.

The other RPG-equipped fighter turned his attention from the highway below to the small battle raging here behind him. With both hands he swung his RPG launcher in a quick arc. He moved his right hand toward the trigger. Suddenly the Alpha's neck swiveled toward him. A volley of shots tore through the fighter before his hand made it close to the launcher's trigger.

The four fighters and the shepherd lay motionless in pools of blood on the ground. The Alpha ran over the crest and charged down the hill toward Bint Jubayl.

War game demonstration, indeed.

CHAPTER 106

Haruto lay on the crest of the hill with his camera recording away. Coming up from the south were dozens and dozens of robots.

An RPG exploded in a bright flash in front of a Beta walking through the pastures to the left of the southern highway. Haruto zoomed into the blast. The Beta robot ran through the smoke and dust of the explosion toward the source of the RPG launch. An Alpha robot a few meters away also turned and joined it.

Haruto panned the camera a bit to the right. Three fighters were crouched low and motionless in a ditch by the roadside, their RPG launchers flat on the ground. A moment later the Alpha and Beta appeared in the field of view. Alpha's mouth streamed out a torrent of bullets and smoke. One of the fighters started to rise and grab his weapon, only to be struck in the chest. The second fighter lay flat in his hiding place, but slowly reached for his RPG launcher. The tip of the RPG launcher snuck out of the ditch and pointed at the robots only meters away now.

The Alpha must have seen the RPG. It immediately stopped and reached with both hands into the pack on the Beta's back. A quarter-second later the Alpha flung from each hand a grenade into the trench by the roadside.

The fighters' bodies exploded in a bloody eruption out of the ditch. The Alpha and Beta robots immediately headed south toward Bint Jubayl.

The camera started beeping and a small icon on its display screen flashed. *Memory full.* Haruto reached into his back pocket. No stick. He pushed his fingers a bit deeper into the crease. There it was. Haruto popped out the recorded orange memory card and put in the new white one, clicked off the high-resolution setting to get more recording time out of this second and last memory stick. Haruto started to put the recorded stick into his pocket.

No — what if it fell out? He untied his left shoelace, threaded the end through the hole at the edge of the memory stick, and gently tucked the stick in between the tongue and body of the shoe.

Haruto started videoing again. Hundreds of Alpha and Beta robots were sweeping through the countryside, coming up from the south and encircling the entire area around Bint Jubayl. A few human soldiers in fatigues were also there now, mixed in with the robots.

Haruto pointed the camera northeast of the town and scanned the area. No robots there yet. Better get out of here before a less gracious Alpha found him. Haruto stood up and started walking diagonally down the hill, toward the roadway in the distance leading to the north and back to the kibbutz.

CHAPTER 1 07

Haruto felt good. Mission accomplished. He had enough evidence now to corroborate the video taken on the ship. He would clear his name, provide a proper resolution to the Co'en murder case, and let the Japanese people know what their government was doing. The rules should be followed. Even the government did not have the right to break the rules.

Haruto rubbed his right calf. No more bleeding. *A paper cut.* That's what *Sensei* said when someone got a minor injury in class, scraping their knee or bloodying their knuckles trying to break a piece of wood.

Haruto triple-clicked his GPS watch. In a straight line, Misgov was a bit over four kilometers away. If he assumed about six kilometers to actually get there, he could even make dinner if he walked fast.

Beautiful as this place was, it would be good to get back to Japan. But he'd stick with the itinerary. He'd finish his one-month stay on the kibbutz with honor and use the time to prepare his report.

The roadway started to ascend higher into the mountains. Haruto passed fields carpeted with anemone and wild mustard. He breathed in the sweet country air. The massive telephoto-lens camera hanging around his neck bounced against his dirty white t-shirt. His jeans were tattered and covered with earth. But he felt great.

At the crest of the roadway, a balding man with neat gray sideburns and mustache, in a clean white shirt and a pressed gray jacket, waved at Haruto.

Haruto waved back and walked closer. The shepherd had protected him, the boys had been kind — this must be the legendary Middle Eastern hospitality.

"Hello, hello." The older man extended his hand to Haruto. "You are Japanese, yes? I don't speak Japanese. Do you speak English?"

"Yes."

"Who do you work for?"

"I am just passing through," Haruto said. "I must continue walking."

"No, no, please stop for a minute. Come to my house. Meet my family. It is just over there." The older man pointed to the top of the hill to the west.

Haruto smiled. "I am sorry, but I must continue on my trip."

"No, it is I who is sorry." The older man raised his hand, and a dozen young men with assault rifles and RPG launchers stood up from their hiding places on the side of the road.

Mattsu.[‡]

A long volley of shots fired into the air from the roadside while six of the armed men ran toward Haruto, guns forward.

Haruto put his hands up. "I am a photographer trying to get news pictures. Here take my money. Please let me go." Haruto took out the small stack of American bills.

The older man counted the cash. "Three hundred and eighty dollars." He started laughing. "I think you are worth more than that. Much more."

Make a rule... Then something smashed into Haruto's head. The world went dark.

[‡] Shit

CHAPTER 108

Haruto opened his eyes. His head was throbbing. His wrists were tied together with thick twine. So were his ankles. He was lying on the ground in a small dim stone room.

Haruto brought up both arms together and rubbed the sides of his hands against the large bruise near the top of his head. How did he allow this to happen to himself? The mission had gone so well. He had everything he needed for his report. How could he have failed like this? He should have made more rules after leaving Bint Jubayl.

The wooden door creaked open and the older man entered. In a plastic tray he carried Haruto's camera, GPS watch and passport. He opened up the passport. "So Mr. Aoyama, you are awake. Good… My friends call me Pierre. You may call me that, too."

"Where am I?"

"You are kidnapped, Mr. Aoyama."

"Kidnapped? What for?"

Pierre laughed. "Money. What else?"

Haruto opened his eyes wide and stared at the balding man. "Please, let me go back to Japan. I can send you thousands of dollars. On my honor."

The older Lebanese man laughed even harder. *"On your honor?"* He laughed again. He then straightened his pressed gray jacket, and smoothed the lapels. "Mr. Aoyama, I am a businessman. You are not the first person I have kidnapped, and my usual procedure is to have your friends send the money before I let you go. It's good business practice."

Pierre picked up the huge camera and clicked it on. The video of the Alpha and Beta robots swarming up the countryside came on its display screen. "Do you work for an international news company? A Japanese news company? It's important to tell me who to contact."

Haruto shook his head. "I work for no one. I'm just taking pictures I hope to sell."

"Now is the time to tell me the truth," Pierre said. "You work for the CIA? You work for Mossad? That's okay, I won't kill you. You're even more valuable. Mossad pays well."

"No, I work for no one."

The Lebanese man smiled. He pointed to the wooden box in the corner of the room. "Suit yourself. You will go in the ground and stay there until I can sell you. To another group here in Lebanon or — better for you — to your friends who will pay for you. If I can't sell you, well… that sometimes happens."

Haruto turned his head to the side and saw the crude coffin. "No, please, Pierre. That is not my passport. I am a Tokyo Police Inspector — you are kidnapping a Japanese officer!"

"Why would a Japanese policeman be in Lebanon?"

"I am doing an investigation on —"

Pierre put up his hand. "I really don't care. My associates in Beirut will contact your embassy. If they pay for you, good."

A moment later four young men holding assault rifles, pushed through the door into the room. Pierre pointed to the coffin.

Two of the young men put down their weapons. One picked up Haruto by the ankles, the other by the shoulders.

"No!" Haruto flexed his legs away from the man's grip, and then extended them in a lightening blow into the young man's groin, jerked clockwise and pulled his bound wrists away from the other young man. In the next fraction of a second, he thrust his bound fists into the man's solar plexus.

Bupp! Bupp! Bupp! Bupp! The bullets hit the ground around Haruto.

Pierre pointed the rifle at Haruto's head. "Your choice, Mr. Aoyama. I can shoot you now, or I can take the butt of this rifle and hit you over the head again, and we will put you quietly into the box. Or you can cooperate. I'd prefer the latter since it doesn't damage the goods."

Haruto nodded and relaxed his arms and legs. Breathe in, breathe out. One. Two. The young men grabbed Haruto's bound ankles and wrists, and threw him into the coffin. The other two men dropped the lid on the box and started hammering away. One. Two.

He'd be okay. The Police Department would realize that it was he, Haruto, and Japan would pay the ransom. He would be okay. One. Two. One. Two. But each time another nail sealed the lid of the coffin, Haruto's heart skipped a beat.

There was a five-centimeter hole in the lid, near his face. At least he could look out.

"Move your head to the side," Pierre said.

A plastic pipe pushed through the lid hole, filling most of it.

"Air — to keep you alive."

Air? Why would he need a tube for air?

No. No, no, no!

Breathe in, breathe out.

Haruto felt them pick up the coffin. As they left the stone room and came outside, the cool air of the mountains came through the cracks of the box. It wasn't so bad. It would be okay.

Haruto felt the coffin rise and lower with the footsteps of the men. After about a minute, Haruto felt weightless for a moment as the box was dropped, only to smash into the ground.

"Dig," the older Lebanese man said.

One. Two. One. Two.

About forty minutes later Haruto felt the coffin rise for a moment. The bottom part of the box angled down. Then the top part dropped hard, and the coffin came to rest level in the ground.

The first shovel-full of dirt landed on top of the lid. Then another, and another. Haruto's heart raced. Perspiration ran down the sides of his forehead, some into his eyes. He tried to rub them with his bound wrists.

Breathe in. Breathe out. One. Two. He would be okay.

As dirt covered the coffin lid, the subsequent falling of earth did not sound as loud.

Breathe in. Breathe out. Breathe in. Breathe out.

Soon there was no more noise. No more earth falling down. No more voices of the Lebanese men. Just darkness and silence.

It would be okay. The Police Department would pay the ransom. They would want him back.

Haruto breathed in through the nose. One. Two. One. Two. He exhaled slowly through the mouth. It would be okay.

In the absolute silence, he could hear his heart beat.

Lud-dub. Lud-dub. His heart was beating faster.

Maybe they wouldn't pay. He was wanted for murder. He had no friends. He knew state secrets. Who really cared about him?

"*Seiretsu!*" *Sensei* looked up and down the line. He was not so old then. A strong, smooth face. He rested his glance on Haruto. "How long have you been here?"

Haruto looked down at the yellow strip of cloth wrapped around his karate uniform.

"Five years, *Sensei*. I am ready for the next belt."

The two black belts at the opposite end of the line giggled. *Sensei* turned to them and they stopped instantly. *Sensei* turned back to Haruto, and stared at him. "Follow me, Haruto."

The karate master walked toward the small janitorial closet at the end of the inner wall of the gym. Haruto followed.

Sensei pulled a metal key out of the sleeve of his uniform. The move seemed almost magic. *Sensei* clicked open the closet deadbolt and held open the door. That moment — it seemed an eternity.

"Haruto, please go in. If you can come out before class is over, you will have your orange belt."

The closet was dark. Dead dark.

"Haruto!"

Haruto stepped into the closet.

Click. That sound. It reverberated over and over again in his head.

Haruto turned the handle but the door wouldn't budge. It was locked solid with the deadbolt.

The bottom of the door pressed against the wood floor. Only the dimmest of light from the gym outside made it through the crack, and the closet kept its darkness. Haruto rubbed his hands over the closet's walls. There was no light switch inside.

The closet was so small. The mop and bucket pushed against Haruto's knees.

Its walls were cinder blocks. *Sensei* could break brick. Haruto had seen him break two bricks with one blow during the demonstration last summer. If Haruto was to be orange belt, maybe he too should be able to break brick.

Haruto assumed fighting stance, knees bent with sideways posture, in front of the cinder block wall. His left hand fired into the wall with a *yaku-tsuki* punch.

"Aghgg!"

He could not see the hand, but he felt blood drip down onto his fingers. He ran his right hand over the cinder block — nothing. It was the same as before.

Haruto grabbed the door handle with both hands and shook it violently. Nothing. The door barely moved.

The door was wood. That was it. Even the orange belts could break wooden boards. Haruto rubbed his left fist. It still hurt badly. No problem. He pushed his back against the rear cinder block wall as far as he could.

"*Kiai!*" Haruto fired his right foot in a *mae-geri* kick into the door.

The thick wooden door vibrated loudly for a second. Haruto ran his hand over the struck portion of the door. Nothing. It, too, was the same as before. Haruto kicked again and again. And then again. Nothing.

Haruto could smell the faint odor of ammonia from the mop. If no light came into this enclosure, then maybe no air came in either. Haruto breathed in again with his nose, carefully paying attention to the scents. Yes, ammonia. Lots of it. He tried breathing in deep, but the air wasn't going anywhere. He would suffocate.

Why would *Sensei* do this to him? Why would *Sensei* want him to die?

Haruto pulled on the handle and shook the door. "Let me out of here! Let me out!"

Nothing.

Maybe class was over. He would stay here all night. Maybe *Sensei* intended to let him out next class.

He wouldn't survive with all the ammonia here. Haruto tried to breathe in deeply. He could barely get his lungs full of air. Maybe there wasn't enough oxygen in the air. Haruto started breathing faster. He could feel his heart beating. He felt a bit dizzy. He breathed even faster while he pulled frantically on the door handle.

He would die. He'd be dead before the morning.

Haruto breathed even faster. With every beat, he felt his heart slam into his chest.

"Get out! Get out! Get out of this place!" He pounded on the door over and over again.

Haruto grabbed the mop beside him. He pushed the mop stick up against the ceiling with a soft "thunk". Concrete — like the rest of the building.

Haruto then smashed the mop stick into the door handle. Again. And again. He no longer felt himself breathe. He no longer felt his heart beat. His brain stopped thinking. Smash the door handle. Again — smash the handle. The mop stick splintered and the sharp edges scraped into the door with each blow.

Smash the handle. Smash the handle.

Blow after blow, Haruto kept madly ramming the broken mop pole against the door. Smash. Smash. Nothing.

And then, the stick thrust into a groove it had dug out and lodged underneath the rim of the door handle. The leverage exploded the handle off. Light flooded into the closet through this hole.

Haruto pushed on the door. The deadbolt still held it firm. He smashed the mop stick into the hole where the handle had been leading to the deadbolt.

Smash. Smash. The door splintered and the deadbolt fell out of position.

Haruto pushed open the door and came into the light of the gym. The entire class was standing outside the closet door. The two black belts laughed.

It took Haruto a quarter-second for the two steps. A tenth of second later his left foot smashed into first black belt's face in a violent *mawashi geri.* Haruto saw the blood pour out of the black belt's nose and mouth, and heard his scream, but it did not matter. A second later, he pivoted and faced the second black belt. The black belt assumed fighting position. Haruto feigned an *oi-tsuki* with his bloody left hand, and during the instant the black belt took to look at Haruto's injuries, Haruto ripped into the black belt's right knee with another violent *mawashi geri.* The black belt screamed and went flying onto the *dojo* floor.

With fire in his eyes, Haruto looked around at the other members of the class.

"*Seiretsu!*" the master said. He turned to one of the brown belts and pointed to the two black belts on the floor. "Get some ice, please."

While the class lined up, Haruto turned his head left and right. Fists up. Eyes wide.

The master walked up to Haruto and faced him. "*Seiretsu!*"

Haruto closed his eyes for a moment. He opened them and looked at the master and the students lined up, white belt to brown belt. The two black belts were on the floor.

He saw *Sensei* standing in front of him.

Haruto scurried back into the line, taking his place among the yellow belts. The master returned to face the lined students. He picked up the new orange karate belt neatly folded on the wood floor, and held out the belt with extended arms.

"Haruto, please step out. You are now orange belt."

CHAPTER 109

Haruto pushed his tied wrists against the coffin lid. Nothing.

In the stone room, the coffin had looked like it was cobbled together with loose pieces of shipping crates. It couldn't be that strong. The Lebanese had not taken his shoes. Haruto wiggled a few centimeters closer to the end of the box, bent his knees as much as the space would allow, and smashed his tied feet into the side of the coffin.

Nothing.

Smash. Smash. Again and again. Nothing.

Haruto stopped and took a deep breath. If only he could get his knees a bit higher, he could kick harder. He'd have to do his best and increase his strength despite the handicap. Haruto concentrated. He breathed in through the nose and exhaled through the mouth. Smash! Good — this kick was much harder. He concentrated again. Smash! Smash! Smash! Smash!

Haruto was covered with sweat. The air seemed stuffy — maybe he was breathing up too much of it. Haruto felt for the pipe with his tied hands. He put his mouth near the bottom of the pipe and breathed in deeply. The air still felt stale.

Maybe the pipe was blocked. The night air should be cool and refreshing. But there was no breeze coming out of the pipe. The air was soon going to be used up and he was going to die here.

Haruto took a deep breath in. The air seemed even worse. Haruto tried another deep breath. The air wasn't going into his lungs, or else, maybe there wasn't enough oxygen in the air. Haruto started breathing faster. He started to feel his heart beating again. He tried to see what was going on around him, trying to make out the air tube, the lid of the coffin, but it was so dark he could not see anything.

He had to get out of here.

He breathed even faster while his tied hands and tied feet frantically banged against the coffin. His hands and arms started feeling numb. The choking sensation became overwhelming. Haruto could barely suck any more air in.

Going to pass out soon. Do something.

Make a rule. There must be a weak link in the box somewhere. Find it and smash out. Haruto started rubbing and pushing his tied hands all over the inner walls of the coffin. Nothing. He tried to probe the lower end of the coffin with his tied feet, but couldn't feel any difference in strength along the walls.

Haruto tried to breathe. The air barely went in. He felt dizzier.

No air. He was going to die in here.

Haruto started banging his tied wrists all over the upper end of the coffin. Nothing... nothing. He smashed his tied wrists into the coffin lid. Nothing.

As Haruto's hands fell in defeat onto his chest, he felt the evil eye pendant under his t-shirt press into his skin.

With his tied wrists, Haruto pushed down the neck of his t-shirt and worked the pendant toward his mouth. His teeth clamped down on one end of the thin slab of stainless steel.

Haruto passed the twine binding his wrists against the exposed metal edge. Nothing. He rubbed the rope again and again over the metal. Nothing still. It didn't matter. He wasn't going to die here. Haruto pressed and rubbed and scraped the thick twine over the pendant's metal edge. Again. And again. He no longer felt himself breathe. He no longer felt his heart beat. Tear the cord. Tear the cord. Again and again. The minutes became half an hour. Finally the twine began to fray.

Rip the cord. Rip the cord.

And then, as Haruto twisted the rope with the force of his arms, the edge of the pendant thrust through the frayed fibers of previous scrapes, and the twisting forces started fraying the fibers in the layers below. Haruto rubbed the rope faster and faster over the edge of the steel pendant. In ten minutes, with the final scrape, Haruto's wrists went flying to opposite sides of the coffin wall.

Haruto arched his back and brought his tied ankles under his buttocks. He could not get both hands there, but leaning to the side, his right hand grabbed the twine binding his ankles. He felt for the knot. There it was... his hand let loose and came flying back.

Haruto stretched again toward the side. His right hand found the knot quickly. With thumb and forefinger, he grabbed the tied end and pulled until it felt like his skin would rub off. He kept pulling and pulling. And then the first tied end came out of the knot. Haruto released the pulled end and felt around the looser knot. He grabbed the next tied end and pulled. It came out easily. Using all five fingers Haruto loosened the twine, pulling the final piece out of the knot. With middle fingers, he then yanked off loop after loop of cord around his ankles.

Haruto ripped the Turkish pendant from his neck. Holding one end in both hands, he scraped a thin line in the coffin's wood directly above his chest. Then he put the edge of the pendant back in the thin line and scraped again. He felt the line. Not very deep. It didn't matter. Haruto scraped the wood again. Again. And again. He no longer felt himself breathe. He no longer felt his heart beat. His brain stopped thinking, save for a single thought: Break this box. Break this box. Again and again he scraped. The minutes became half an hour, and the half hour became an hour. Scrape and scrape. And then, the steel pendant cut through to outside of the thick wood.

Haruto drew his fists back, concentrated, and in the tight space fired a right *oi-tsuki* at the crack. His fist went through the wood into the dirt above. He fired a left *oi-tsuki* into the larger hole. And then another and another. With each punch, the hole got larger and the next strike took less force.

Dirt was falling into the coffin. Haruto didn't care. Break this box. Break this box.

Haruto stuck his head and then his chest through the broken wood into the soft earth above the coffin. Like a fish caught in the mud, he wiggled from side to side, pushing against the coffin, pushing against the earth. Ten seconds later, Haruto's head and then his body came out of the grave into the light of the gibbous moon.

It took a few seconds for Haruto's eyes to adjust to the light. He looked around. To his right were hundreds of large wooden crates covered with dense camouflage netting, and behind that dozens of small tents pitched. To his left some twenty meters away, were even more crates, and a bit past that was the small stone house. And twenty meters in front of him was a barn of some sort.

A young bearded man with an assault rifle hanging off his shoulder, and grenades and ammo clips on straps crossing his chest, came around the corner of the pile of wooden crates. Haruto watched almost in slow motion as the young man lifted up his head with the next step. The fighter's eyes stared at Haruto standing there in the moonlight covered with dirt from head to toe. "Ayee! It is the prisoner!" He started to reach for his AK-47.

It took Haruto a half-second for the four steps. A tenth of second later his left foot smashed into the fighter's face in a violent *mawashi geri*. Haruto saw the blood pour out of the fighter's nose and heard his scream, but it did not matter. Haruto was breathing fast, his nostrils flaring, his eyes wide open, every muscle in his body tensed to explode.

Haruto heard footsteps and voices of more men coming from behind the pile of crates. Haruto scooped up four grenades off the fallen fighter's utility straps. A second later, he pivoted and faced the fighters running toward him. He jerked the clips out of one grenade, then out of a second one, and threw them at the pile of crates.

Haruto turned and charged toward the other pile of wooden crates. His feet pounded into the ground. Three seconds later, he reached the pile. He turned and saw a dozen armed fighters running after him. Gunfire. A bullet whizzed by his head.

With fire in his eyes, Haruto looked at the two grenades he was holding. In a flash he pulled their pins and threw them at the second pile of wooden crates.

As Haruto started to run, the previous two grenades went off and the world exploded.

A pressure wave knocked Haruto to the ground. He rolled once and jumped to his feet. He turned his head. A volcano of fire and debris was erupting from the first pile of crates without cessation. Kabam. Bam. Bam. KabamKabamKabam. Bam. Bam.

Haruto charged at the stone house. He ran in. His face was black with dirt except for the white of his wide-open eyes and the tips of his flaring nostrils. He jerked his neck left and right. No one. He pushed into the other room of the house.

The world exploded again and the pressure wave pushed Haruto to the side. Tatta. Tatta. Tatta. Tatta. Ball bearings and other bits of metal shrapnel sprayed into the countryside, some of it hitting the stone house, some of it smashing through the window into the interior wall. A few seconds later it stopped, but then more explosions from the first pile of wooden crates. Kabam. Bam. Bam. Bammmm.

Haruto didn't care. Like a crazed animal, he ran out of the stone house and charged across the field at the barn. Haruto smashed past the barn's thin wooden door. His eyes grew even larger as he saw Pierre standing there arguing with an older bearded man who had an AK-47 slung over his shoulder. A younger, skinny man holding a clipboard stood there expressionless beside the two men.

As the men turned to face him, Haruto pushed off with his left leg. Then his right smashed into the ground, and then his left. And then Haruto pushed off with his right and went airborne. At full force, Haruto fired a *mae geri* kick into Pierre's temple. A tenth of a second later Haruto's torso twisted viciously in the air and a violent second kick smashed into the bearded man's neck.

Haruto jerked his abdominal muscles tight and landed forward on his feet. He turned his head and saw Pierre lying on the ground, blood trickling down from his ear. The older bearded man lay there on the ground motionless, eyes open, his assault rifle still around his shoulder.

Haruto shook his head for a second. What was he doing? An airborne *kumite*? It was for idiots and movie stars, that's what *Sensei* said. In the time you were airborne, your opponent could hit and kick you senseless.

Haruto looked at the dead men at his feet. Why? Why did this have to happen? If people followed the rules, this would not happen.

Clatter. The young skinny man dropped his clipboard. He was sitting there against the wall, staring at Haruto.

Haruto looked at his eyes.

The young man started crying. "Please, you no kill me. Please! Please!"

Haruto looked at the older men, dead on the floor, and turned back to look at the young crying one.

The skinny young man put his hands to his face, palms out. "No kill. Please. I stay here. No problem with me. No problem! Jews good. Robots good. Please no kill."

"Where Israel?" Haruto asked.

The young man pointed to the left of the barn.

Haruto turned and walked out.

CHAPTER 110

Menachem stared at Major Gershon. "What the hell is going on near Markaba?"

"Nothing. We don't have any robots that far north. As far as I know, no one has any operations up there."

"Are you sure?" Menachem pointed to the yellow paper in his hand. "We have satellite reports of over four hundred explosions. A drone is over-flying right now."

"Good."

"Good? My grandmother could do better than that. I want recon! I want information! There's a Sayeret Golani platoon near Margaliot. Have them hop over the border and find out what happened."

CHAPTER 1 1 1

Haruto kept walking through the scrub grass up the hill until he got to the highway. Same one he was on before? North to Rabb near the kibbutz? He looked around but he couldn't find any landmarks that he recalled. No road signs were to be seen either.

Haruto looked up at the sky. There was supposed to be a North Star. Which one? He wasn't sure.

This should be east. That skinny fighter was terrified — no reason for him to lie.

Haruto crossed the small highway and kept walking in the same direction.

CHAPTER 112

The helicopters fractured the silence of the night and landed one after another in the field beside the stone house.

Four Israeli commandos jumped out of the first helicopter. Blackened faces, earphone/microphones plugged in and night-vision goggles hanging off black headbands. Each ran to an opposite end of the field.

"No activity," the first reported, and his three colleagues then echoed the same.

Captain Dayan in the last helicopter scanned the area. Smoke was still rising eighty meters to the west. "Execute plan!"

Commandos streamed out of the helicopters in curved paths to the stone house, the exploded piles of crates and the barn.

CHAPTER 1 1 3

Haruto soon reached the border fence. There weren't two fences and a minefield like there was at Misgov, just the one tall razor-wire fence.

Haruto felt exhausted. He wished he could lie down there and go to sleep.

What to do now?

Haruto saw the guard post and the Israeli soldiers in the distance. Wave to the soldiers? Would they let him back into Israel? Would they shoot him?

Better to get back in through the gash in the fence at Misgov. He looked up and down the fence. Follow the fence to the left or right? Which way was Misgov?

Haruto felt his heart pound away in his chest.

Okay…make a rule. He came from the west… if Misgov was still north then he should turn left. Thus, walk to the left for an hour. If no landmarks or signs of Misgov, then turn around and walk to the right for two hours. If no Misgov, then approach the border soldiers.

Haruto tapped his left arm four times. Then with his left index finger, he tapped his right forearm four times.

It didn't feel right.

A twenty tap left and then a twenty tap right. Okay, go now.

No, it still wasn't good.

The five thousand rule always worked.

Haruto sat down and started tapping his right forearm. One, two, three, four, five…

CHAPTER 114

Major Gershon faced Menachem who was sitting at the conference table, Tanaka at his side. "Mission successful. No incidents. Preliminary forensics completed."

"Yes?" Menachem said.

"Company strength al-Haleeb. Eighty-eight dead, one survivor. We've interrogated him — very cooperative. Information he gave us is corroborated by existing intelligence."

"What were the explosions?"

"We estimate three hundred tandem anti-tank missiles and four hundred short range missiles. Most of their eastern sector field stockpile."

"Who killed them? Maronites?"

Gershon shook his head. "A robot. A single unarmed robot. A Japanese one."

"What?" Tanaka said. "We have no robots there!"

"The al-Haleeb terrorist is down the hallway if you want to speak with him yourself. He witnessed the attack. He even gave the artist enough detail to make a sketch of the robot." Gershon placed a drawing down on the conference table.

Menachem's and Tanaka's eyes opened wide and their jaws dropped.

"Incredible," Tanaka said. "Suzuki Haruto."

Menachem turned to Tanaka. "What's going on here? Maybe he is a robot. Secret project, Colonel?"

"What are you talking about?" Tanaka said.

"I saw him go under in the middle of the Pacific Ocean. Now he reappears here in the middle of Lebanon. How many copies do you have of this model?"

Tanaka banged the table. "He's an Inspector with the Tokyo Police Department. He's been there for fifteen years, and since our little cruise, he's been absent — a few unsubstantiated reports, but otherwise off everyone's radar. Why don't you send your own people to Tokyo? You can interview his co-workers, his wife, his mother-in-law, and a million other people who know him. Bit too much of a cover-up even for Defense Intelligence."

"So how the hell did he get to Lebanon and in the middle of all this?

How can one unarmed man destroy more al-Haleeb and missiles than one of our battalions can?"

Tanaka shook his head. "I don't know. Except that he's a damned good *man.*"

CHAPTER 115

" ■ ■ ■ **four thousand nine** hundred ninety-eight, four thousand nine hundred ninety-nine, five thousand."

It still didn't feel right. But the rule about the five-thousand rule was that mistakes were okay and it didn't matter how it felt… it would still be okay. So Haruto got up and started walking along the border fence to the left.

After about five minutes, the border fence angled away from him slightly. He kept walking. Ten minutes later the fence turned back a bit. Haruto stopped. He remembered that out-pouching of border from the GPS watch! The cut section of the fence at Misgov should be a kilometer north of here.

Haruto picked up his pace. After another five minutes, the minefield between the short wire fence on the Lebanese side and the taller security fence on the Israeli side appeared. Haruto slowed down and looked for his footprints in the minefield and for the breach in the security fence.

He found nothing for a hundred meters. The security fence was brightly lit up and its wires fully intact. He kept walking, looking for footprints in the minefield. Still nothing.

Suddenly a volley of bullets flew above his head and an extra-bright spotlight enveloped him.

Haruto put up his hands. The light blinded him. He couldn't see who was there. "It is me, Haruto. I am a volunteer at Kibbutz Misgov."

"Haruto?" a nearby voice said.

The spotlight turned off. Haruto saw the two soldiers from the other night standing in the minefield. One stood guard holding an M16 in the ready position, while the other one was digging up the minefield.

"Please help me get back. I am lost."

"You certainly are," the soldier with the M16 said. "How'd you end up in Lebanon?"

"Sorry. I know I should not be here."

"Well, you have a problem, Haruto. We just fixed the minefield with new mines. You're going to have to enter Israel at the gate about five hundred meters north of here, and then they're going to arrest you."

"Please, let me come back here."

"It's filled with mines. You'll get blown up, Haruto."

The second soldier put down his shovel and took out a colored map from his back pocket. He turned carefully to look at the security fence, then looked at the map. He walked diagonally through the minefield, over to Haruto, took his jacket off and put it on the lower strip of barbed wire. Haruto rolled through the lower wires of the fence, landing on the soft earth of the minefield.

"Follow my footsteps exactly."

Haruto followed the soldier through the minefield to the gash in the security fence.

"We're going to fix the fence soon, so the next time it won't be so easy to come back. Don't do this again, Haruto."

"Thank you."

Haruto walked away from the border fence eastward through the apple orchard, and then walked past the industrial buildings toward the low-rise apartments where his room was.

"Haruto!"

He turned around.

Amit looked at his watch and smiled at Haruto. "5 AM sharp. Great! I knew I could count on you, Haruto. We get other volunteers that run away after one day of work."

Haruto smiled thinly.

Amit, with only chickens on his mind, seemed oblivious to Haruto's disheveled appearance. Amit put his arm over Haruto's shoulder. "Come. It's a big day today. Building Number Three goes to market. Thirty thousand chickens to get onto the trucks!"

CHAPTER 116

Haruto stumbled out of the hot noon sun into the dining hall. He was exhausted, but also starving. Eating something should help. He picked up a plate and waited in the buffet line. A minute later he was in front of the food.

Chicken? Should he eat that? Maybe not. He was with chickens all morning... Israeli salad. That's good. But he looked at the tiny pieces of tomato all chopped up in their red juice and it felt wrong. He had been lucky. Next time his luck would run out and he would meet the destiny he had cheated since childhood. It would be him cut up into all those little pieces.

"Come on, hurry up," a tall woman behind him said.

Haruto tapped on his right arm four times and then tapped on his left side four times. Okay, just take some bread. Bread was safe.

As he reached for the white slices his fingers start tingling. He felt his heart beat faster.

"Please! Make up your mind," a young man beside the woman said.

Ignore these people. Do what is right. Don't give in to destiny. Haruto reached for an orange. A good choice. Smooth and spherical. Vitamins and energy. Good for him.

There was a green speck on it. No! He let it fall back down.

The young man pushed to the front of the line and stared at him.

Enough. Just leave and get some rest. It would be okay later. Haruto walked back to the dining hall entrance but stopped at the doorway. He tapped four times on the right side of the doorframe and four times on the left side.

It didn't feel right.

He felt his heart slam into his chest. He breathed in deeply but no air was going in.

This was it. Time to die. His destiny.

He forced himself to breathe. Nothing went in. He opened his mouth and gulped quickly for air. Oxygen. His brain needed oxygen. The world started spinning.

No. No.

Haruto collapsed to the dining room floor.

"**H**aruto, can you hear me?"

Haruto opened his eyes and looked up. A man with a small beard and dark-rimmed glasses was kneeling over him. A crowd from the dining room stood behind, staring down at him.

Haruto nodded.

The bearded man put his fingers on Haruto's neck. He then looked carefully at Haruto's eyes. "Do you feel dizzy right now, Haruto?"

Haruto shook his head.

"I'm Dr. Levy. Here, take my hand. Let's see if you can get up. Can you walk to the clinic with me? It's just one building over... Did you get this dirty working with the chickens?"

CHAPTER 118

Mara was wearing an amaranth sleeveless top and denim jeans, carrying a tall glass of orange juice. "Here, Levy said it was okay."

Haruto looked at her large hazel eyes and felt a bit better right away. The juice would be all right. It would be safe. He took the drink. "Thank you, Mara."

"What happened, Haruto?"

"I'm a weak person. I apologize for falling in the dining hall and causing any problems for you."

"Funny, you didn't look weak when you knocked down the Army's chief *krav maga* instructor the other day."

Dr. Levy walked in with Salzman.

"Haruto," Levy said. "I've checked you over. Your physical exam is normal, except for a few superficial wounds. Be careful at the chicken houses with the machinery. ECG is normal. The blood tests I can run here on the kibbutz are all normal. Sodium was low but still normal, maybe some dehydration. Whatever's wrong with you, it's beyond my abilities. So if you don't mind, I'll let Salzman try to figure out why you passed out."

Haruto nodded and Levy left.

Salzman looked at Mara. "Maybe you could leave Haruto alone with me?"

Haruto felt his heart beat faster. "No. Please. Mara can stay here."

"All right. Well, Haruto. Has anything like this ever happened before?"

"Yes, but I have learned to control it. If I follow the rules, I am okay."

"Rules, huh? Okay. So what did you feel, what do you remember before you fell in the dining room?"

"I couldn't breathe. My heart was beating hard. I felt like I was going to suffocate, like I was going to die."

"It sounds like a classic panic attack. Haruto, when did you first ever have anything like this? How old were you?"

"I... don't remember exactly. Maybe nine years old."

"Any phobias?"

"I don't understand."

"Does it bother you if you're in a tight place, for instance? A crowd of people? An airplane?"

"I'm okay on large airplanes but I don't like tight places."

"Tight places sometimes make the heart beat faster?"

Haruto nodded.

"And your rules, Haruto... if you don't follow them, what happens?"

"I don't know. I always follow them."

Salzmann nodded his bald head. "I see..."

"What's the matter with him, Salzman?" Mara asked.

"Still a little early to tell, but... Haruto, have you ever been diagnosed with obsessive-compulsive disorder? Have you ever gotten any help for it? Seen a psychiatrist?"

Haruto shook his head. "There would be much shame — not only to me, but to my family's honor. *Seishin-ka* — psychiatry — is for crazy people."

"What about people who become depressed?" Salzmann asked.

"They must overcome their problems on their own."

"And... how are you doing on overcoming your problem?"

"I am weak. I need to apply myself more."

"Haven't you been applying yourself since you were nine?"

Haruto said nothing.

Salzman said, "Well, would you like any help for your problem? There is no dishonor in asking."

Haruto sat up on the exam table and looked down at the floor. "What help is there?"

"There are several possibilities. For instance, I can prescribe a large dose of an SSRI antidepressant — not for depression, but to raise the serotonin levels in your brain. Within the next two months, your obsessive-compulsive symptoms will decrease. They won't go away, but they should decrease."

Haruto continued shaking his head, looking down at the floor. "I am not a crazy person. I don't think I need drugs."

"Another treatment is cognitive therapy. I'm expert at it. Many years ago, I studied this therapy with its creator in the United States."

Haruto looked up at the short, bald doctor. "What is it? Does it work?"

"It's talk therapy. You learn to understand how your brain is working — for example, why do you get anxious if you think you'll be breaking a rule? Eventually you try breaking that rule, and your brain sees that the world does not end, and eventually the anxiety and the panic attacks all simmer down. Excellent results, *but* it takes a year or even longer. Still, we could start here on the kibbutz and you could continue in Japan."

"I need to think about all this. I'm not sure I want to break any rules."

"All right. Well maybe we can discuss what's been going on in your life recently, just as friend to friend. Why this panic attack in the dining room? Is it the work in the chicken house? Volunteers come here thinking they're on some exotic holiday. They've never done real labor before. Is Amit pushing you too hard?"

Haruto smiled. "No, the kibbutz is wonderful. I've never done this type of work before, but I enjoy Poultry Operations."

"Then what is it, Haruto? Why the panic attack today? Have you been under more stress?"

Haruto hesitated.

"Yes."

"Do you want to talk about it?"

"No... I would, but I just don't feel comfortable talking about all these things."

"Mara, could you please leave," Salzman said.

Mara turned away and started to walk off.

Haruto grabbed her hand. "No, please stay."

Salzman and Mara turned their gaze to Haruto's hand clasping onto Mara's. Haruto let go of the hand. "I'm sorry, please, forgive me. I am sorry if I have broken any local customs. I am sorry for any inappropriate physical contact. Please forgive me —"

"Haruto, shut up." Mara thrust her hand back into his. "Here, if you want to hold my hand, it's okay."

Haruto's face became red and he pulled his hand away. "No, no. I am sorry... Thank you for staying here with me." He turned to face Salzman. "I will tell you what has been happening to me..."

"I'm listening..." the psychiatrist said.

"First, I am not Aoyama, my name is Suzuki Haruto and I am *Keibu* — an Inspector in the Tokyo Police Department."

Slowly, but picking up speed as he went, Haruto told them his story, starting with the murder of Co'en Satoki, running through his time aboard the *New Pacific Queen* and finishing with yesterday's encounter with the robots and his burial and escape. As he spoke, Salzman and Mara sat on two plastic chairs next to the exam table, their expressions moving from shock to disbelief to polite interest. When he finished, there was a long silence.

"All right," Salzman said at last. "I can see why you've been under stress. But..."

"But you do not believe me."

Salzman shook his head. "It's not that." He bit a lip briefly. "Haruto, this morning I was working at the hospital in Nazareth. I saw a young man and he told me an incredible story also — how the American CIA had come to spy on his house, and how his neighbors were all involved in the plot. He suffers from schizophrenia. Then I saw another young man and he also told me an incredible story about how wealthy he was, flying his airplane all over the world on secret government business. He suffers from mania. Haruto, what makes your story different from the stories my patients told me this morning?"

"The memory stick tied to my shoelace." Haruto reached forward, flipped apart his left lace and held up the orange stick.

Salzman raised his eyebrows. "All right…" He pulled a laptop out of his briefcase. "Here, it reads standard memory cards."

Haruto rested the laptop on his legs and booted it up. Mara and Salzman stood beside the exam table.

The video began to play, with Haruto offering running commentary.

"That's Brigadier General Moshe Otzker. He's in the IDF — you can check their web page. Director of Weapon Systems Testing… That's Colonel Tanaka — Japan Defense Intelligence… That's an Alpha on the ship. As I just told you, I saw him in action in Lebanon yesterday… That's the Israeli — *your* — megaton nuclear bomb… These are Alphas and Betas attacking al-Haleeb fighters at Bint Jubayl. That was yesterday."

Salzman's eyes opened wide as he watched the end of video. Mara was completely still.

"What are you going to do about all this Haruto?" Salzman said at last. "Why do you even need to be involved? Israel can stop al-Haleeb rocket attacks and Japan can stand up to North Korea. Seems like a good deal to me."

"Perhaps it is," Haruto said, "but if Japan gets nuclear weapons, it has to be done by the government with the knowledge of the people, not the military. In the decade before the Second World War the military broke the rules. Look at the results — Hiroshima, Nagasaki, the war and humiliation. There are risks for Israel, too. If Israel can trade its nuclear weapons, other countries will think they can do the same. Before you know it, every country, including Lebanon, has their own nuclear weapons. I must expose this plot."

"All right… but maybe in this case it really would be better to let the Japanese and Israeli governments break the rules and let *them* deal with the complications."

"No, the rules are there for a reason. They must be followed."

Mara still had said nothing. Haruto glanced at her, then quickly looked down. She was staring at him with a strange intensity, but he could not tell what it meant… love or fear or loathing.

He forced himself to look at her. "Mara, what… do you think?"

Slowly she stood, then leaned forward and threw her arms around him.

"I think," she said quietly into his ear, "that you are the bravest, most honorable man I've ever met."

He froze, tense, then felt his muscles melt until he placed his arms around her and drew her into a tight embrace. He hoped she would not notice the tears in his eyes.

"Dr. Salzman," he said. "Will you help me?"

"Of course," Salzman said. "Of course."

CHAPTER 119

Haruto felt better today. He was alive. The ocean, the killer robot, the coffin... over, all over. It was also good getting everything off his chest. For an instant this morning when he had to put the respirator on, he felt like he was going to suffocate and he felt his heart beat funny. But then he remembered what Salzman said yesterday, about training his brain in new patterns. It didn't make any sense to feel like this. The respirator was there to *protect* him from germs and chemicals in the chicken feces. It was a good thing to wear.

He felt okay then.

Sweeping up building Number Three was fine. The soiled sunflower hulls would be crushed into fertilizer. In a few days the building would be clean and ready for the next batch of chicks. The cycle of life would continue.

Haruto walked to the dining hall and entered. Mara was at the buffet and waved to him. "How you doing today, Haruto?"

"Good, but it is very embarrassing what happened yesterday."

Mara smiled at him. "Come on, Haruto. Nobody — nobody! — could handle all that stress. Is the chicken house boring after all your excitement?"

Haruto shook his head. "No, it is a pleasant relief to do useful work that won't kill me. I must return to my duties getting building Number Three ready for the next batch of chicks."

"Are you coming to folk dancing tonight?"

"Yes. Will you be there?"

Mara nodded.

CHAPTER 120

Haruto got to the dining hall before eight. The tables were already pushed to the side, the crying and laughing of the violins and clarinets played from the amplifier, and about a dozen people were dancing, with more streaming into the hall.

A tall, bald man was showing the moves to a new dance. Instead of a circle the dancers were snaking in a line, left forearm bent and up, palm open, gingerly holding the right fingers of the dancer behind. "Start with the right foot. *Shtayeem Shallosh.* Heels and back to toes. Tareeta tata tatum. Hands up."

Haruto felt a tap. He turned. It was Mara... glistening, smooth black hair framing large hazel eyes, smooth olive skin, and a large white smile. A hint of fresh, sweet fragrance hit his nose.

"Aghh... hi, Mara."

"Come on, I know this one." Mara took his hand and inserted them into the line of dancers.

The dance leader kept repeating, "Start with the right foot. *Achat Shtayeem Shallosh.* Heels and back to toes. Tareeta tata tatum. Hands up. Open, close." The dancers repeated the moves a few times. The dance leader turned up the music. A line of thirty dancers now undulated with the cries of the violin. As the clarinets rang out their joys, the dancers released hands and sent arms skywards. And then the line instantly reformed, left forearms bent and up, palms open, lightly holding the right fingers of the dancer behind.

Twenty minutes later a new tune rang through the dining hall. The dancers joined hands and formed two large concentric circles. The inner circle moved counterclockwise, the outer one, clockwise, both of them to the beat of the melody. They bent forward to the beat and then leaned back, raising their joined hands up in the air.

"I'm thirsty," Mara said. "Want to get a drink?"

Haruto nodded and they walked over to the refreshment table. He poured two glasses of ice water. "I wanted to ask you, Mara, is there a common room where I could use a computer? I need to prepare my report."

"Yes, at the lounge. Don't you remember when I gave you the tour the first day?" She grabbed his hand. "Come, I'll show you."

They walked out into a refreshing Galilean night, and Mara pointed to a wooden building about a hundred meters north. A minute later, they were there. Haruto heard laughing and noise inside.

He stopped at the doorway. It did not feel right.

"Come on," Mara said, waiting behind him.

Haruto felt his heart skip a beat. He very slightly moved his fingers four times on the right, as if to simulate taps, and then turned to the left side of the doorframe and moved his fingers again four times. He was about to push the door open but it still didn't feel right. His heart started beating faster. Everything was going well. Why was this happening again?

Haruto started to almost motionlessly tap toward the right. One. Two. Three. Four. Five. Six. Seven—

Mara squeezed past him and pushed the door open. "Come in, Haruto. It's okay. It's just the lounge."

Haruto stood there at the door. He had to restart his twenty-count. One. Two.

"Come on, Haruto."

Haruto didn't move.

Mara stepped back to the door and grabbed his arm.

"No, no. It doesn't feel right. Just give me a few minutes, and I'll come in."

"Minutes?" Mara looked at him funny for a moment, then yanked his arm, catching him off guard and pulling him into the building. "Better, yes?"

He paused. It made no sense to think that tapping on the doorframe would change his destiny. Slowly, he felt the panic subside.

Haruto followed Mara. Some dozen kibbutz members were sitting on the couch watching a soccer game on the large plasma screen.

"Satellite television," Mara said.

They walked downstairs. Two men were playing billiards. Beside the pool table against the wall were four computers, all occupied by young men and women clicking away.

Haruto frowned.

"Sorry, Haruto. It's not a big deal. Come to my room. I have my own computer."

"Are you allowed your own? In my room, there's no computer."

Mara smiled. "Members are allowed a few luxuries, like their own televisions or computers. Not everyone takes one — they'd rather use their points on something else."

Haruto followed Mara through the field to the low-rise Members' Apartments. He gazed at the curve at the small of her neck, followed it down her back to blue jeans hugging a tight rear end. With each step, her tank top rode up a bit, exposing her midriff.

Mara opened her apartment door and flicked on the light. "Come in please, Haruto."

As Haruto lifted his leg —

It felt wrong.

He pretended to tap four times to the right, and then four times to the left of the doorframe. He felt even worse. He was breathing fast, his heart slamming into his chest, his palms becoming sweaty. Better to do a real tap. One, two, three, four on the right side of the doorframe. One, two, three —

"Haruto, what's the matter?"

"Ahh... Ahh... nothing," he said, beads of perspiration running down his forehead.

He breathed but couldn't get enough air in. He turned his head toward the field and breathed in again. A bit better. Okay, redo the count. He tapped on the right — one, two, three, four. He tapped on the left — one, two, three, four.

It didn't seem right.

His heart was racing. Like in the coffin, the air was barely going into his lungs. They left him there to die. No one would have paid any ransom. He would have died in that box. Died in that ocean. Nobody would have cared. Nobody cared about him.

His heart pounded against his chest. He started to feel dizzy.

Follow the rules. Everything would be okay if he followed the rules. Twenty-tap rule. Haruto started tapping on the right.

Mara looked at him, her head to one side. "Haruto, what are you doing?"

"I'm sorry... I'm trying my best to come in... Rules..."

Haruto finished tapping to twenty on the right, and started on the left.

"Haruto, what do the rules say about this?"

Haruto looked up.

Mara was naked. Her breasts were full. Nipples erect. Her waist curved in so gently, and then out again at her hips. Black pubic hair between smooth, olive legs.

Haruto stopped tapping, walked in, wrapped his arms around Mara and kissed her passionately on the lips. The warmth of her flesh pressed against his body. He looked into her eyes and the world stopped. He inhaled the sweet scent of her sweat, and her soul. He pressed her tighter against himself, and kissed her again.

He lifted Mara up and carried her gently to the bed, all thought of tapping forgotten.

CHAPTER 121

Haruto awoke to the ringing of the alarm clock. Mara's large hazel eyes were open, centimeters from his. She smiled at him and his muscles all went to mush.

"So, Mr. Make-a-Rule, it's a good thing I set the alarm so you'd make it to work this morning."

Haruto's face turned red. "It is better to follow the rules. But sometimes…"

"But sometimes what?" Mara kissed him.

He felt her hands on his chest. He breathed in her scent and became intoxicated. Haruto kissed her, and pressed his naked body into hers.

CHAPTER 1 22

Haruto turned off the water hose. Building Number Three sparkled, and even the ammonia-tinged stench was gone. Clean concrete floor, shiny aluminum drinking lines, glistening red feeding baskets. Tomorrow he'd lay down fresh sunflower hulls.

Haruto walked to the dining hall in the strong noon rays. He wore no sunhat but it didn't matter. The sun was fine. Its warmth felt nice, actually. The walk was wonderful, too. Not just to get from Poultry Operations to the dining hall so he could obtain nutrition, but to be walking for no other reason than, well, to be walking.

A strong breeze blew in from the mountains. Haruto inhaled deeply. Faint, sweet smells, so delicate. A mix between cut grass and incense. Haruto exhaled and smiled, and looked at the verdant hills framed by a blue sky. It was a mistake to say the sky was just "clear." Indeed, if you looked carefully, there were wisps of magnolia cloud against the azure background, as beautiful as any *karesansui* Zen rock garden.

How had he lived all of his life without noticing these things?

CHAPTER 1 23

J **ohn Sullivan finished clicking** through the e-mails and alerts —
they were coming in thicker every day. It was a sign of the interest
in the scenery his bird was filming, and why not? His supervisor was
cool with the fuel burn but the disciplinary hearing still was a go for the
middle of September. Fuckin' assholes.

"Hello, Molniya Kid."

John turned his head. It was Randall, another junior analyst like himself.
Randall plopped down a six-pack of Cokes. A second later, two large blue-
and-silver sacks of potato chips went flying through the air, crashing onto
his desk. Jeff arrived right behind them and punched John in the shoulder.
"Hey, John. Our birds are on autopilot for a bit. Just because we're stuck in
here doesn't mean we can't see some fireworks."

John typed a line of commands into his keyboard as the soft drinks
fizzed open and the bags of chips went around. A black-and-white real-
time video image of Zarit, Israel, came on-screen.

Nothing.

"Boring!" Jeff yelled.

"Fuck you! Give me a minute."

John typed commands and clicked the mouse. The image zoomed out
to display a larger area around Zarit at lower resolution. Still nothing. The
image zoomed out again even more. A single pixel in the top right corner
of the screen flashed white for a second. Immediately the display refreshed,
and green letters flashed onto the top of the screen — *Molniya Mode 3
— Cassegrain secondary rotated and locked onto 33.2141°N, 35.4413°E —
north of Ainata, Lebanon.*

"Locked and cocked." Randall shoved a handful of corn chips into his
mouth and passed the bag to John. "Showtime."

Nighttime-enhanced video splashed on the display. A human-looking
robot — *androids* John called them — was holding a grenade launcher.
Beside it was a giant insect-like robot — *spiders* John called them —
with a large crate on its back. The android reached down into the crate,
pulled out a grenade and loaded it almost faster than the video could
resolve. The human-like robot pointed his weapon toward a bearded man
frantically pushing buttons at the back of a truck holding two five-meter

rockets angled upwards. The firing of the grenade lit up the left side of the screen and almost instantaneously the right side of the screen burst into huge splotches of white explosion as the truck, rockets and bearded man disappeared.

"Happy Fourth of July!" The three young men clicked their cans of Coke together and chugged them down.

CHAPTER 124

Menachem and Colonel Tanaka were alone at the large conference table.

"Let's move onto the next item," Menachem said. "Has Mikiyasu finished the design improvements on the *kaizen* list?"

Tanaka nodded. "Every single one. The production run is finished, and a hundred thousand robots are already in containers awaiting shipment."

"Good. Okay then, last item... what about our Inspector?"

Tanaka threw up his hands. "Off the radar completely in Japan now. Nothing. Nothing — I swear to you. What about your intelligence? He was last sighted in Lebanon. You guys know who's going in and out of there, not us."

Menachem frowned. "We have nothing, too. The border from Lebanon is sealed tight. Since the sighting we've been photographing and fingerprinting every single Asian coming into Ben Gurion Airport or coming in by land, and following some of them. Problem is we have tens of thousands of Asian laborers in the country, so he may not stand out. Lots of leads, but even more dead ends. We have agents at the Beirut Airport — also nothing. We have to assume he's still in Lebanon. Do your police officers get training in weapons and guerilla fighting?"

Tanaka shook his head. "Most of our police officers don't even carry guns. We've checked out every minute of the Inspector's life. Other than extensive traditional karate, he's had zero weapons training — not even using a pair of nanchuks. Based on our research there is no possible way he could take out an entire al-Haleeb company, let alone survive a day in Lebanon by himself."

"Maybe he received training from another country — the CIA, for instance. Or maybe another part of your government trained him?"

"Maybe..." Tanaka said. "I don't disagree with you. But it's not Defense Intelligence. Somebody else is training him."

CHAPTER 1 25

Haruto lifted his right index finger, but just couldn't go ahead and start tapping with it.

"Come on," Salzman said. "I want to see four taps on the right and then *three* on the left."

"I don't feel comfortable about this."

"Haruto, you have to take your cognitive therapy training seriously."

Haruto tapped four times on the doctor's desk with his right index finger. He hesitated for a few seconds, but finally went ahead and tapped three times with his left.

"All right, what automatic thought are you having now?"

"I feel I did something wrong. Something bad will happen. My destiny will be fulfilled."

"All right, now give me a balanced thought."

"I'm not sure I really should."

"Haruto!"

"Okay… I feel that something bad will happen, but logically I know that tapping my finger against a piece of wood should not be able to influence the world around me."

"Good. Let's use the downward arrow technique now. *Something bad will happen. My destiny will be fulfilled.* Where are these thoughts pointing?"

"My negative core belief that I have been a failure since childhood and failures like me must die, and that my destiny will eventually occur and I will die."

"All right. Now we need an action plan to try to change this core belief."

Just then Mara streamed into the room carrying a picnic basket.

As soon as Salzman turned his attention toward her, Haruto took his left index finger and tapped one more time on the desk.

"Why'd you do that?" Salzman chided.

Haruto ignored him, and instead jumped up and hugged Mara tightly.

Salzman grabbed his car keys. "Come on, let's go."

CHAPTER 126

T he bandstand filled the field with a joyous classical music.
"Do they have this festival every year?" Haruto asked.
"Sure," Mara said, "but this year it's bigger and brighter. No rocket attacks for a week now. People are a lot happier listening to music here than sweating in their bomb shelters." She dropped the picnic basket underneath a tall pine tree. "Here, take a sandwich. I put *hummus* in yours the way you like it, Salzman."

A new melody filled the air. Over a thousand people sat in the field listening to the music, tapping their feet. The sun shone brightly, but fresh breezes from the surrounding mountains kept the crowd comfortable.

"What are your plans, Haruto?" Salzman asked.

Mara quickly turned her head from the bandstand to stare at Haruto.

"I'll do my report and then... I fear my life as I know it will be over. If I return to Japan, I'll be arrested for murder — a false charge — and imprisoned for life. If I go to another country or stay here in Israel, I'll be hunted down and killed by Japan Defense Intelligence or by Mossad."

"So... why do you have to do your report?" Salzman said.

"I've told you. I don't think the Japanese military should break the rules. The last time they did that, we ended up with Hiroshima and Nagasaki. The people need to decide if they want nuclear weapons."

"What about North Korea?" Salzman asked.

"We've negotiated successfully with North Korea many times in the past and we can do so again. The military is more concerned with protecting its pride. Anyway, that's a decision for the people to make, not the military in secret."

"What about Israel?" Mara asked. "I like being protected by the robots or whatever the IDF is using. I don't want the rockets falling again."

"I know," Haruto said. "I'm not a military expert, but I know when you break the rules, trouble will follow. Maybe Israel should change its path."

"Change its path?"

"Has the military path really brought Israel success? The whole world hates you. Maybe another path should be tried."

"You sound like them too. The world doesn't care when we're bombed and we die. The world isn't fair. Why is it wrong for us to fight back?"

"It is not a question of fairness. And I agree with you that the world is not fair. And I agree with you that it is not wrong to fight back, but if you take the path of fighting just to fight back you will not achieve balance. You should be struggling for balance, taking the paths toward this balance, not the paths toward military victories."

"It all sounds very confusing, Haruto. Like we should just continue getting killed and no one will care."

"You take the path of trading robots for nuclear bombs. The robots work well, so well that your enemies get frustrated to the point that somehow their billions of petrodollars buy a nuclear bomb for al-Haleeb, which you've also made easier for them, since if you can trade nuclear bombs, other countries will say they can also. On the other hand, pretend you take another path, and I don't know which one because there are as many paths as there are stars in the sky, that leads to balance, not military balance but emotional balance with your neighbors. The outcome of this path will be a better one than the first one."

Mara frowned. "I don't know... I like being protected by robots. I know what it's like to live in fear. I've done it for a long time."

Haruto moved behind her on the ground and held her tightly. "I'll protect you. You don't have to be afraid."

Mara rubbed her hands over his hard, thick arms. "How are you going to protect me if all these secret agencies are going to be chasing you?"

"Haruto..." Salzman said. "The local police came around yesterday and asked if we had seen a Haruto Suzuki."

Haruto shivered.

"Mr. Takamichi Aoyama, as you know, we have no *Mr. Suzuki* with us." Salzman smiled. "On the kibbutz none of us actually owns anything, but what we do have is each other. You can stay with us, Haruto."

"Thank you."

"My mother's family was killed in the camps in Hungary during World War II, but my mother ran away just before they were rounded up. She had nowhere to go. The underground smuggled her into Palestine. Kibbutz Misgov took her."

Haruto turned to Mara. "Where's your family from?"

"My grandparents were from north of Jerusalem."

"I thought Jerusalem is part of Israel," Haruto said.

"Now it is, but it used to be Arab territory. When Israel was created, the Jews were kicked out. My grandparents came here with only the clothes on their backs. The kibbutz took them in and gave them a life."

"Do you think you'll ever live in Jerusalem again?"

Mara laughed. "Me — no. But who knows? Maybe my children or grandchildren will live in the *Holy City*." Mara laughed for a moment at the idea, and then became serious. "When I was a teenager, I thought I would leave Misgov and all its memories. But then I realized that it's my home, and that I like it here."

Haruto snuggled closer to her. "So do I."

CHAPTER 127

Haruto knocked on Mara's apartment. She opened the door wearing a pink t-shirt, tight gray sweatpants, and a large smile. "Come in. How was work today?"

Haruto was about to step in when he felt strange for some reason. He tapped on the right doorframe four times, and then on the left four times. That was better. He entered, held Mara and kissed her.

"Haruto, you need to see Salzman again."

"I like him ... but he's for crazy people. If I try hard, I can control my problem."

Mara ran her fingers through Haruto's thick hair. "When I was younger I had some problems, and I used to see Salzman and talk to him. It helped me a lot."

"What were your problems?"

Mara stepped away from Haruto, and her face became cold. "I don't want to talk about it."

"Is this one of Salzman's successes?"

Tears started flowing down Mara's cheeks. "I'm sorry, Haruto. Why don't you just go."

"Mara..."

"Just go, Haruto."

Haruto turned and walked out into the evening.

CHAPTER 128

Haruto walked away from Members' Apartments, west toward the apple orchard.

If Mara didn't want him, then there was nothing he could do about it. It was her decision and he had to respect her wishes. That wasn't even his rule, it was society's rule.

But... what about society's rules? What was happening to his life?

He was a respected *Keibhu*. And a married man. Yet he had defied orders, left his career behind and after barely knowing her a few days, had sex with another woman. Unprotected sex. He could get AIDS. He still used condoms with Michiko, even after so many years. Well, it had been a while, but there was hope Michiko would come back to him. He would be *Keibhu* again. He would get his life back. The rulebooks. The stability. A good life. An honorable life.

Haruto looked up at the sky. So many stars here compared to what he saw in Tokyo. In all those stars was the universe, and in the universe was his destiny. He had cheated that destiny. He should have failed and died as a child, but his rules protected him. His rules let him capture Michiko's heart. His rules let him become *Keibhu*. And this last month, how many times had he cheated death and failure with his rules? He needed to make more rules and follow them, and he would cheat destiny again.

Haruto walked in the darkness past the large bandage factory and continued to the orchard. And then, while in the middle of the apple trees, a cool breeze blew down from the mountains.

And Haruto felt alone in this world. And he shivered.

He smelled the ocean again. His lips felt so dry. He felt the water rush in and burn his nose as he went under the surface yet again. He smelled the old wood that made up that coffin. No air to breathe. Darkness. All alone in that box.

The wind howled between the apple trees. Haruto looked up at the night sky and felt so small and even more alone in this big universe. He shivered again, and then he realized that rules may be necessary, but they weren't everything.

Haruto ran through the orchard. He sprinted past the large bandage factory, onto the field, toward Members' Apartments. He ran up to Mara's apartment and banged on the door.

The door opened. Mara's eyes were red. "What do you want, Haruto? Go away." Mara slammed the door shut.

Haruto banged on the door again.

Mara yanked it open. "What part of *Go away* do you not understand? If the girl doesn't want you, then you have to follow that rule, Mr. Make-a-Rule."

Haruto pulled Mara tightly to his chest and pressed his lips against hers, and kissed her. "I barely know you, but I love you, Mara. I don't care what the rule is, I'll never leave you."

"You *love* me?"

Haruto nodded. "My whole life has been turned upside down these last few weeks. I love you, Mara. I don't want to be alone again. I can't."

"Say you love me again."

"In English or Japanese?"

"In Hebrew."

"I don't know how."

"*Ani Ohev Otach.*"

"Did you just say that you love me?"

Mara smiled and put her finger on his lips. "No, that's for a man to a woman. Can you say it?"

"*Ani Ohev Otach.*"

CHAPTER 129

Haruto was startled for a moment by the tap on his shoulder. He turned and smiled upon seeing Mara.

"You've been in the lounge all evening?"

Haruto nodded. "I wanted to finish my report."

"How are you doing?"

Haruto slipped a recordable DVD into the computer and clicked the mouse. "Done." He waited a minute for the burning to finish, pulled out the disc and clicked it into its jewel case. "Let's go."

Holding hands, they cut across the soccer field. The warmth of the day was fading, and Haruto put his arm around Mara, pulling her close to his body. They passed Babies' House and then Children's House. Mara quickened her pace.

Haruto stopped, still grasping her hand. Mara stared at him, then stared at his hand imprisoning her. Haruto let go and sat down. "Please, Mara. Sit with me."

"No, let's just go."

"I'm here with you, Mara. What happened on this field? What scares you so much?"

"Nothing happened on this field. It wasn't here. It…" Mara turned away from Haruto and looked up at the stars.

Haruto patted the ground. "Please, sit with me."

Mara turned back to Haruto, looked at him in silence, and let herself down on the grass. "It was Children's…"

"What? *Children's House*? You and Dov and Chaim all grew up there together, right?"

"And Danni. We were together since birth here on Misgov."

"Who's Danni?"

Mara went silent for a while.

She took a breath. "You know that on the kibbutz the children stay in Babies' & Toddlers' House, then Children's House and then together in apartments as teenagers."

Haruto nodded.

"One night, when Dov, Chaim, Danni and I were seven years old, terrorists from Lebanon cut the fence, sneaked into Misgov, and went to Children's House. One of them held Danni and me. The other took his gun, put it on Danni's head... and shot him. It was the loudest sound I had ever heard in my life. I remember his blood flying into my face."

"What happened then?" Haruto whispered. He nearly couldn't speak. This was so much worse than being stripped and beaten by bullies.

"I remember screaming. My mouth was open so wide, I remember that. The terrorist let go of Danni's body and it banged down on the floor, and then he held me even tighter with both hands. Then I felt the gun on my head. What was worse was the smell from it... like a fire, like when you burn the trash. I remember saying to myself, *no, no.* I remember screaming and crying."

"And then?"

"That's the strange part. I don't remember clearly what happened after that. A bright flash, four soldiers, and then the adults took us away, cleaned us up and gave us cookies. My dad told me the soldiers shot the terrorists dead and I didn't have to worry about them again."

"You don't." Haruto pulled Mara into his chest and wrapped his arms around her. "You don't have to run every time you pass this place. I'm here with you. I'll never let anything happen to you." Haruto pressed his lips against hers and kissed her.

"Never?"

"Never." Haruto looked into her hazel eyes, and under the cool, starry Galilean sky, took her face with both his hands and kissed her again. The DVD slipped out of his fingers and fell on the ground. Haruto bent down and picked it up.

"Haruto, why are you doing this report?"

"I've explained my reasoning to you many times."

"But isn't this just you following the rules for the sake of having rules."

He took the question seriously. He had asked it of himself once or twice while he was finishing his report.

"No," he said, "I don't think it is. First, these are society's rules not mine. Also, you don't make rules about things you do easily and naturally. The rules are to make you do necessary things that you would not do if there were no rules."

"I'm not sure I understand. You have so many rules, Haruto, this is just more of them."

"No. These are not my rules. There are rules on how to run a country, on how to avoid disaster. Billions of oil dollars are sitting in bank accounts, only too ready to buy a nuclear bomb for the terrorists in Lebanon. If Israel can sell or trade nuclear weapons, then so can North Korea or Iran. These rules are not mine. These rules protect society."

"But I like being protected by the robots."

"Mara, I will protect you. *Ani Ohev Otach.*"

CHAPTER 1 30

Haruto was lying on the hammock overlooking the soccer field. A few cotton-puff clouds drifted lazily toward the mountain peaks. The afternoon sun still burned strong, but the two olive trees supporting the hammock provided a comfortable shade.

Suddenly a whirl of colors was airborne and landed on Haruto. The hammock dipped and then bounced back up.

"So Mr. Make-a-Rule, how come you're not still at *Poultry Operations*? Or making some more rules to save the planet?" Mara laughed and kissed Haruto on the lips, and snuggled beside him.

"I think ten hours a day with the chickens is enough. And I am in the process of making some more rules..." Haruto laughed and grabbed the large Jaffa orange under the hammock and tossed it up in the air. "I need to decide how to eat this delicious orange."

"*Tapuz.* Can you say it?"

"*Tapuz.*" Haruto laughed and kissed Mara.

Mara grabbed the fruit. "Here, I'll show you how to eat it. First, you have to appreciate it. You look at it and smell it."

Haruto looked at Mara's eyes, grabbed a quick kiss, and inhaled her scent.

"Then you use your fingers to gently remove the peel."

Haruto put his hands underneath her tank top and started pulling it off.

"Haruto! What if somebody comes here?"

Haruto kissed her again, and kept his hands on her back, underneath her top.

"Okay then you take a slice, and eat it." Mara put an orange slice halfway in her mouth, and leaned over to Haruto, pushing the opposite end through his lips. The juice, so sweet, burned a bit as it dripped down Haruto's chin.

Haruto pressed his lips against hers, and pulled Mara tightly against his body, and they kissed for a minute.

"I'm so happy with you Mara. No longer do I feel alone. That coffin —"

Mara put her finger on Haruto's lips. "It's okay. It's over. You're here now."

Haruto held Mara even tighter. "*Ani Ohev Otach.*"

CHAPTER 131

A **fter spending the last** two weeks in the Upper Galilee, Tel Aviv was a shock for Haruto. Almost a mini-Tokyo. Haruto looked at the glistening skyscrapers, the billboards and neon lights. He would take Misgov instead any day.

Haruto wore a white shirt, starched collar of course, with his blue tie. His suit had never bothered him in the humid Tokyo summers, but it seemed so stifling now. It was necessary — this was a formal presentation, but he'd happily trade the suit in for his t-shirt back at Misgov.

Berkowitz Street. Good. Haruto gazed up at the tall building. Museum Tower. He took the elevator up to the 19th floor. *Embassy of Japan.*

There was a security guard at the door. Haruto approached him.

"I am Suzuki Haruto, a Japanese citizen. I would like to speak to the Ambassador."

The security guard held up his finger and spoke into his walkie-talkie.

A moment later, a young Japanese man in a three-piece business suit came to the entrance. "The Ambassador is not available. I am his assistant. Can I help you?"

Haruto bowed. "I am Suzuki Haruto. I am *Keibhu* in the Tokyo *Metropolitan Police Department.* My investigation is complete."

Haruto opened the jewel case he was carrying and pressed both his thumbs onto the surface of the DVD. He clicked the case shut and handed it to the assistant. "Please do not touch the disk — my fingerprints are there. You will give this to the Ambassador to give to Colonel Tanaka, Japan Defense Intelligence."

"There is no Colonel Tanaka of Defense Intelligence in Israel."

"I believe you will find that he is."

Haruto bowed and left the embassy.

CHAPTER 132

Lieutenant Colonel Okamura looked out the huge bridge windows. Thousands upon thousands of brown shipping containers were stacked on the *Mikiyasu-ema*'s decks. The sun shone brightly on the calm, cerulean sea.

Captain Yamada wore his usual dress-white Mikiyasu Industries merchant marine uniform. "So, Lieutenant Colonel, another trip together. Don't worry. This delivery will be a routine one."

The satellite telephone started ringing.

Okamura picked up the bulky satellite telephone receiver. "Hello."

"Return to port," Tanaka said.

"Yes, Sir."

CHAPTER 133

Menachem's face was red. "How did you let this happen? He's Japanese. Do you hear me? Japanese. He's your problem. You're supposed to take care of your problems!"

"Maybe if your security was better he would've never gotten on the ship in the first place. You ever think of that?" Tanaka pointed to the DVD in his hand. "Before blaming each other, maybe we should first go through the Inspector's report."

"How do you even know it's really from our Inspector? The whole thing doesn't make sense… I saw him shot in the middle of the Pacific Ocean, then a month later he single-handedly destroys a full al-Haleeb company. He crosses with ease from Lebanon into Israel, even though we've been stopping every single Asian person at all border crossings for the last two weeks. Seems like multiple copies of an enhanced Alpha Prime… or an elaborate CIA project. Or maybe somebody else in Defense Intelligence is running their own project."

Tanaka shook his head. "I spoke with the Director. No such project authorized in Defense Intelligence. If it was a CIA project and they found out nuclear weapons were involved, they would have already spoken to us. Trust me on that one. Plus, Inspector Suzuki was kind enough to put his thumbprints on the DVD in front of an embassy staffer."

"It was him?"

"Yes, perfect match left and right thumbprints with the fingerprints on file for Suzuki Haruto, an Inspector in the Tokyo Police Force. No military training. Never worked for Japan Defense Intelligence."

Tanaka hit the light switch and clicked the overhead LCD projector's remote control. The video Haruto had recorded on board the ship played. Tanaka hit the pause button. "Well, he has all of us in his little film."

Tanaka clicked the remote again. The video of the Alpha robot killing the al-Haleeb fighters played. "No evidence of any digital alteration — every pixel stream is continuous. There's a hidden digital watermark in the image — the camera used was purchased in Japan and manufactured within the last year."

"How the hell did he get from Lebanon to the Tel Aviv embassy?"

"It's *your* country. Maybe you should tell me that." Tanaka breathed in deeply and clicked the remote again. A page of Japanese *kanji* filled the screen. "Hang on, we have a translated copy for you." Tanaka clicked the remote again and a long English page appeared on screen, scrolling slowly.

I, Suzuki Haruto, a humble Inspector in the Metropolitan Police Department make this humble Final Report on the Murder Investigation of Co'en Satoki.

"*Humble* Inspector — Ha!" Menachem snorted. "It should read, *I, robot Inspector in the Metropolitan Police Department -*"

"Read through it," Tanaka said. "He actually did a good investigation."

Due to extraordinary circumstances, this report is being presented to Japan Defense Intelligence. I am only a humble Police Inspector, but nonetheless these are the results of my investigation. I have no desire to embarrass anyone. I only respectfully ask that the rules be followed.

I was assigned to investigate the death of Co'en Satoki — a Japanese citizen born in Japan but of European ethnicity, owner of Co'en Electronics (electronics export firm) in Shibuy-ku, Tokyo.

Here is my reconstruction of events supported by the accumulated evidence:

The government of Israel desires military robots to improve its ability to oppose the asymmetric warfare being waged on it by al-Haleeb fighters in Lebanon and possibly elsewhere. Co'en Satoki is engaged by Israel to arrange a commercial purchase of the robots from Mikiyasu Industries. From the Korean-ethnic mafia in Japan, Co'en readily finds out that the President of Autonomous Products of Mikiyasu Industries — Toshifumi Haruka — is in debt to them. Co'en attempts to pay off these (gambling? blackmail?) debts and arrange a commercial purchase of robots. Co'en has large sums of cash and Toshifumi's business card on him at the time of his death.

Just as Co'en finds out about Toshifumi from the Japanese Korean-ethnic mafia, the Japanese Korean-ethnic mafia — which my report must assume (without any slur intended) to represent the interests of North Korea — finds out about the deal Co'en is making with Toshifumi and worries that Israel would be giving advanced weapon systems in return to Japan. A North Korean agent assassinates Co'en Satoki in order to stop the deal. North Korea is desperate to know about this deal, and one of their agents, possibly the same one who killed Co'en, threatens me with a gun for information. The North Koreans continue to monitor closely Mrs. Co'en, and when out of desperation about her husband's murder she gets the phone call to meet with an Israeli contact, she is kidnapped en route to the meeting. I arrive in the middle of this kidnapping, which leads to the car chase and the car death of the Korean-ethnic kidnapper, which leads me to the New Pacific Queen cruise ship at Harumi Wharf.

In the course of my investigation, I board the New Pacific Queen whereupon I learn it is a nuclear weapons testing ship. During its voyage to regions near Guam, I accumulate evidence of Israeli nuclear weapons technology and how Israel tests nuclear weapons in the depths of the Pacific Ocean without international detection. During this same voyage, I accumulate evidence of Japanese military robot technology. During this same voyage, I accumulate evidence that Japan is trading its military robot technology for the Israeli nuclear technology.

In the course of my investigation, I then travel to Lebanon where I accumulate evidence of combat use of Japanese military robot technology.

Corroboration of my eyewitness report is provided by the attached video segments taken by myself in the course of my investigation.

The following recommendations are respectfully made by I, Suzuki Haruto, a humble Inspector in the Metropolitan Police Department:

1. Mr. Co'en's death shall be listed as a murder, not a possible suicide or accident. This shall be published in Japanese news media as soon as possible.

2. The DNA found underneath Mr. Co'en's fingernails shall be compared to the DNA of the two ethnic Koreans who died during my investigation. If a match is found, one can reasonably assume that person is the murderer of Mr. Co'en, and the name of the murderer shall be published in Japanese news media as soon as possible.

"We've already acted on these two items. Indeed, the Inspector got it right. The North Korean agent who tried to kill him matched the tissue found under Co'en's fingernails. We've released an update about the murder investigation. There are not many murders in Japan, so it will be found newsworthy and be published."

Tanaka pressed the remote control.

3. *The dishonor of murder charges against me shall be reversed. The first Korean who died did so as a result of me trying to protect myself from being killed. The second Korean died as a result of his criminal driving negligence during a kidnapping. This shall be published in Japanese news media as soon as possible.*

"We're not sure what to do about this one. We're thinking of a news release giving the Inspector an award for solving the Co'en murder case, so he knows there are no murder charges pending against him." Tanaka clicked the remote again.

4. *The government shall announce in Parliament the intention to trade offensive military robots to Israel for offensive nuclear weapons technology and for nuclear weapons-testing technology. The Opposition Party should be allowed to react to this decision. The people should be allowed to react to this decision. Japan's international partners should be allowed to react to this decision.*

Menachem stared at the projection. "Shit."

"That summarizes our position, too. The Director of Defense Intelligence Headquarters discussed the issue earlier this morning with the Cabinet. The feeling is unanimous. If this is made public, there would be a motion of no confidence in the House of Representatives and the government would fall."

"What are you going to do?"

"I've been ordered to temporarily suspend the project until the next elections are over in six months. What is your position?"

Menachem coughed to clear his throat. "My superior has already spoken with Tokyo. I'm also to temporarily suspend. They didn't say for how long."

"What are you going to do about your border?"

"We'll have to manage for the moment. I can pull men on the northeast border down to the western portion, and protect Haifa."

"And the northeast area then?"

"I don't know. We'll do our best."

CHAPTER 134

aruto held Mara's hand as they walked through the fields of poppies and wild mustard. The trail became steeper and they began to pass bushes and scrub grass. The air was fresh and sweet. Two eagles soared in the distance and disappeared into the clouds.

Haruto opened his backpack and took out a water bottle. He turned to face Mara. Her face was flushed and she was breathing heavily. "You look tired. Maybe we should stop here."

"I'm fine," she said, then bent and vomited.

She wiped her mouth with a tissue. "Sorry."

"No, no, don't apologize. Are you sick?"

Mara smiled. "I don't think so."

"What's the matter?"

"My period's four days late, Haruto."

Haruto stood there in shock. "I'm sorry. I should have planned better. I don't know why I was so irresponsible."

Mara took his hand and put it on her lower abdomen. "It's okay. I'm happy about it."

Haruto bent down and raised her t-shirt, and kissed her belly. "I love you, Mara."

CHAPTER 1 35

"**T**a*puz?*"

Mara lifted her chin from Haruto's chest and nodded with a large smile.

Haruto tore a slice away from the rind and laid the edge of succulent fruit on her lower lip. Mara pulled the orange section in with her tongue. Haruto licked up the juice spilling onto her soft olive skin. Mara giggled and kissed him back on the lips.

"Your stomach's better today?"

Mara kissed him on the cheek and laid her face back on his chest. "Yes, when you're with me, it's much better."

"What are we going to name him?"

Mara giggled again. "I think we still have some time to decide. Are there Japanese names you've thought about?"

"No, not really. What Hebrew names would you want?"

Mara started giggling again. "Haruto, we don't even know if it's a boy or girl."

Haruto gently put his palm on Mara's belly. "Oh, I can tell. It's a strong baby, like its father. It must be a boy."

Mara laughed again.

Haruto looked at Mara's giggling face. "What name means *laughing baby*?"

"*Yitzhak* — it means 'he will laugh' in Hebrew."

"Is it really a name people use?"

Mara nodded. "Yes, it's quite common. It's from the Bible so they use it even in Europe and United States — *Isaac*. Have you ever heard that name before?"

"I'm not sure. Maybe when I was studying English. It sounds too silly. We need a more serious name, a name of a strong leader or warrior."

"*Yehoshua*. The Israelites had escaped from slavery in Egypt and were wandering around the desert. Yehoshua led the people to conquer the promised land of Israel. Is that strong enough for you?"

"I've never heard that name, either."

"*Joshua* in English. And the Greek form is *Jesus*."

"Okay, I've heard that name."

"I don't know why, but warrior names scare me, Haruto. Maybe Yitzhak is better."

"Okay."

Mara rubbed her abdomen and laughed. "Maybe it's a girl. You ever think of that?"

Haruto kissed her belly. "I don't care. Girl or boy. It's our baby. *Ani Ohev Otach.*"

Mara snuggled up to Haruto in the hammock.

The warmth of the day was leaving. So was the light. Through the olive tree the sun disappeared behind the Lebanese mountains to the west and filled the sky with a fiery red.

CHAPTER 1 36

T **hree men in jackets,** open-collar shirts and cropped black beards
went into the apartment high-rise and took the stairs down to the
second basement. They walked to the west wall past two armed
guards into the old man's spacious, wood-paneled office.

"Finally, some victory. See, I told you it would happen," Ibrahim said.

"It was a surprise to us," the young man in the middle of the couch said.
"Why did that attack work, when the rest of the month we had one defeat
after another? Why did the robots not attack us?"

"I have told you this so often. At heart, they are cowards. You have
created circumstances which do not work for them, which don't allow
them easy victory, so they run off." Ibrahim slammed his fist into the desk.
"Ronald Reagan. The President the Americans think is the greatest hero
that ever lived. He sent three hundred Marines here and we killed them.
We were surprised when the great Ronald Reagan turned coward and ran
away after the attack. We didn't expect it, but that's what the Americans
did. The same thing is going on with the Jews now."

"What should we do?"

"What tactics *exactly* did you use in your attack yesterday?"

"We chose a concentrated area on the map and fired forty rockets in
a ten-minute period. We waited eight minutes and then fired ten rockets
two kilometers south of our position in a small arc, in case the Jews were
on their way toward us. We then did a rapid disassembly and returned to
the town."

Ibrahim smiled. "Don't make yourself crazy trying to understand
why that tactic worked. It is not for us to understand everything about
this glorious world we live in. The tactic worked. So now, you repeat this
tactic. You repeat! Do you understand?"

All the men on the couch nodded.

CHAPTER 137

 aruto clicked the *Yomiuri Shimbun* newspaper site again. The screen refreshed. Haruto smiled.

Co'en Satoki's Murder Solved
  The death of Co'en Satoki, the founder of export firm Co'en Electronics in Shibuy-ku, Tokyo, has been ruled a murder by the Metropolitan Police Department. It is believed that Machii Tomo, a known member of a Tokyo yakuza, was attempting to extort protection payments from Mr. Co'en's business. Tissue found under Mr. Co'en's fingernails at the time of his death has been DNA matched to Mr. Machii. A few days later Mr. Machii attempted to assault a Metropolitan Police Department Inspector investigating the murder and was killed during the assault. Inspector Suzuki Haruto has been awarded the Metropolitan Police Award for the bravery he exhibited in the line of duty.

"Hi!" Mara passed by the billiards table, came over and massaged his shoulders.

Haruto stood up and kissed her. The day's heat had not left Misgov yet. Mara was sweating, but her sweat smelled so sweet to him.

"The dining room has ice cream tonight. Want to go get some?"

"Yes." Haruto grabbed her hand, and the two of them left the lounge.

"Did you have ice cream in Japan?"

Haruto laughed. "Yes. We even have toilets and cars, too."

Mara poked him on the side. "Silly. I mean, do you eat ice cream?"

"Yes, I like it. My favorite is vanilla —"

"Boring…"

Haruto kissed her. "No, no, no. Rich vanilla ice cream, covered with a thick chocolate sauce, and sprinkled with tiny rainbow candies."

"My favorite's Heavenly Hash."

"My favorite's you." Haruto put his arms around her and kissed her deeply on the lips, caressed the sides of her face. The taste of the ocean, the smell of that coffin, the feeling of being so alone in this large uncaring world… fading memories all. His fingertips but floated on her silky olive skin yet returned a bolus of warmth and electricity to his senses. The deep loneliness of his life before his adventures, the maniacal dedication to his job, the taunts of *jinzouningen* from his fellow officers, his unhappiness with Michiko, all fading away. He breathed in a scent redolent of some deep-seated memory when his parents had taken him to the countryside, and he saw the flowers and the big mountains and his eyes had opened and the world around had seemed so wonderful. Haruto looked up at the stars in the dark Galilean sky. So many patterns they formed. He had never noticed that before. And then Haruto realized something else. For the first time in so long there was no goal to achieve, no objective to meet, no destination to reach. Somehow, with Mara he was at the end of the quest, at one with the universe.

He was happy.

CHAPTER 138

Mara pulled Haruto's hand up the hill. "Come on, it's my favorite place."

An old wooden picnic bench stood at the summit, resting in a patch of flowering white Michauxia. From the side of the hill, Haruto could see below the four huge, steel-roofed industrial buildings with their fans sticking out of the walls — Poultry Operations. But up here, there was no smell of chickens. Only the fresh scent of the field and a hint of Mara's sweet perfume.

Mara wore white shorts and a bright rainbow tank top. Behind her was a background of rich green grass and wildflowers in bloom. In the distance Haruto could see peaks of other hills and mountains, all equally filled with a verdant splendor.

Haruto liked feeling the warmth of Mara's hand in his. Her skin was smooth like silk.

It was quiet here — only the sounds of their footsteps crunching against the ground, the whispers of the light breeze against the brush, and Mara's laugh.

Mara pointed to the picnic bench. "There, we've made —"

The sky suddenly filled with a horrific shrill sound, and the world exploded. Smoke poured from a gaping hole in the roof of one of the chicken barns.

A moment later, the shrill sound repeated. A split second later, another. A rocket hit the field below and exploded, followed by an almost delicate metallic tinkling as ball bearings in the missile's payload penetrated into nearby buildings.

Without stop, the explosions continued. And then there was that one shrill, so much louder.

Haruto started pushing off with his knees, aiming his body toward Mara to shield her. The rocket smashed into the ground ten meters away and instantly exploded its contents of ball bearings. Before Haruto was even airborne, he heard the whoosh as one of the metal balls sliced through Mara's belly.

Haruto landed on Mara and pushed her to the ground. Blood was streaming from the wound, staining her white shorts a vivid red.

"Are you okay?"

Mara shook her head. "Hold me, Haruto."

"You're going to be okay, Mara."

Mara's eyes were closing. She opened them and stared at Haruto, and smiled feebly. "I trust you. Hold me, Haruto."

Haruto lifted her from the ground and held her in his arms. Her head slumped onto his shoulder.

"Mara! Mara!"

She didn't respond.

Rockets continued to rain down and explode all over the hill and the valley below.

Haruto managed to tap twice on his left arm and then twice again on his right. It would be okay. Mara would be okay.

With Mara in his arms, Haruto rose. A rocket exploded twenty meters away. Haruto heard the whoosh of the ball bearings all around him. He started running down the hill. The rockets didn't matter. Get Mara down to the clinic. Haruto ran and ran. The explosions continued. It didn't matter. Haruto tore down the hillside as fast as he could.

In no more than seven minutes, Haruto pushed through the clinic door with Mara. Nobody was there! Haruto put Mara down on a stretcher. "Levy!" he yelled at the top of his lungs.

Haruto went outside. The rockets had stopped, but there was no one in sight. "Help! Help! Help!"

At last two young men ran over to the clinic.

"What's the matter, Haruto?"

"It's Mara. She needs help. Where's Levy?"

The two young men ran into the clinic and saw Mara lying on the stretcher. Her white shorts were almost completely covered with blood now.

"Levy's not here today. Ethan, go get Salzman."

Ethan ran off while the other young man went over to the stretcher.

"Mara, can you hear me?" Haruto said.

The young man, Amos, put his fingers on her neck. "No pulse!" Amos started compressing her chest. After every thirty quick compressions, he pinched her nose and breathed into her mouth twice.

With each chest compression, Haruto saw a bit of blood come out of the wound onto her shorts.

Time stopped.

Mara lay there, eyes open, shorts stained with blood. Haruto stood frozen, his mouth open, all expression gone from his face. Every nerve dead.

After an eternity, Salzman came puffing into the clinic behind Ethan. He ran over to Mara and looked at her fixed, open eyes. "How long?"

"About twenty minutes," Amos said.

Salzman shook his head.

Haruto felt a rage he never knew he had inside him and charged out the door. He started running toward the border fence.

"Haruto, stop!"

Salzman, running as fast as his short legs would carry him, waved to Haruto.

Haruto kept running.

"Please, Haruto! I can't keep up. Please, just stop for a second."

Haruto kept running toward the border fence.

Salzman bent and put his hands on his knees to catch his breath for an instant and yelled again. "Haruto, stop! Maybe you'll cross the border. Then what? Where are you going to find the terrorists? Even if you do, you'll use your karate to kill one terrorist. Maybe two. Maybe even three or four. And then they'll kill you. YOUR KARATE CAN'T STOP THEM ALL. YOU'RE JUST A MAN."

Haruto stopped and turned to face Salzman.

Haruto dropped to the ground and let loose a keening animal wail. He sat in the dirt and cried and cried until he knew no more.

CHAPTER 139

Haruto felt an emptiness through the core of his body. He had lain in bed all night, his mind numb, maybe he had slept, he wasn't sure. Now Salzman was walking toward him, carrying a large pair of scissors.

"Hello, Haruto."

Haruto looked at the short doctor. Salzman had tears in his eyes and now his face contorted as he began crying. Haruto didn't know what to say.

"You know, I was there when Mara was born. I watched her grow up..." Salzman said through his tears. Then he regained his composure and said, "Haruto, please put on your jacket. Stand up straight." Salzman pulled down on the right sleeve of Haruto's suit and cut it. "We're not religious here, but when people die, the rituals help..."

"Mara was pregnant. Does your religion mourn the baby, too?"

"Oh. Oh, Haruto." Salzman looked at Haruto for a moment. "I don't know what the rabbis say exactly about an unborn child, but yes, we can mourn the baby." Salzman stuck the scissors into Haruto's left breast pocket and cut through the cloth.

Haruto followed Salzman over the small hill behind the apple orchard. All the members of the kibbutz were standing around the open grave. Haruto looked down and saw the pine coffin lying in the ground. Tears broke through his numbness and rolled down his cheeks.

The tall, thin Secretary of the kibbutz started speaking in Hebrew and the crowd became silent. Each member of the kibbutz walked past the grave, took a handful of earth and threw it on Mara's coffin. Haruto, too, scooped up a handful of earth and let the black, moist soil fall gently on Mara.

CHAPTER 140

Haruto followed Salzman to the dinning hall. Mara's parents, their clothes also torn, sat in the middle of the room on low stools.

"This is the mourning ritual for family," Salzman said.

Haruto didn't respond. Words seemed to mean less than they did. He had to concentrate to understand them.

"You can go sit, too," Salzman said. "It's okay."

Haruto sat, hour after hour. At first there was a stream of kibbutz members offering Mara's parents and him their sympathy. Some of Mara's relatives who lived elsewhere started coming in. However, by the middle of the afternoon it became quiet.

"Take something to eat, Haruto," Mara's father said.

Haruto shook his head.

At four o'clock Salzman came by. "How are you holding up with all this stress, Haruto?"

"I... I do not know. It is a pain I have never experienced before."

"It takes time. Mourners sit for days, and then there are more rituals."

"I feel no better from sitting here all day."

"Come, maybe we should walk a bit."

Haruto followed Salzman toward the dining hall exit. Salzman pointed to the doorway. "Maybe your counting will help you. Go ahead and tap. It's okay now."

Haruto lifted his hand, more by reflex than by desire, and started tapping on the doorframe. After the third tap on the right, Haruto fell to the floor.

"Tapping on these pieces of wood is going to bring Mara back? Is it going to bring her back?" He started sobbing aloud.

Salzman tugged on Haruto's shoulders. "Come, Haruto. Let's get some fresh air."

Haruto got up and looked at the spot where he had just tapped with his right fingers. He pulled his left arm back slowly, and then fired a *yaku-tsuki* punch, his second large knuckle shattering the wood.

Mara's family all stared at Haruto and started walking over.

Salzman put up his hands. "It's all right. It's all right." He turned to Haruto. "Come, let's get some air."

They walked through the field, past the large bandage factory and through the apple orchard.

Salzman pointed to the rows of trees. "It's quiet here. I sometimes come here to think."

The trees stood there in silence, save for an occasional rustle from the light breeze coming down from the mountains. The earth was a pale ochre, carefully manicured around each tree.

Haruto, why are you doing this report?

Mara was so beautiful. Her eyes. Her hair.

Mara, it's better to follow the rules. The rules are there for a reason.

Her scent.

But Haruto, I like being protected by the robots.

The first time he saw her. Her smile.

Mara, I will protect you. I will never let anything happen to you.

Haruto fell to the ground, moaning.

Salzman pulled on his shoulders. "Come, Haruto, we keep walking. It's better to keep walking. Life goes on. You have to go on."

Haruto staggered up and continued walking with Salzman. *He had cheated destiny.* Ha! All he had done was find another victim for his destiny. All these rules and stupid feelings. An illusion. A sick illusion to somehow make himself something he wasn't supposed to be. He should have just accepted his failure. Maybe he would be a trash collector in Tokyo now, or better, maybe he would just have died as a child.

Look what he had done. Look at what he had done.

Salzman turned to Haruto. "What are you thinking about? Sometimes it's good to talk about your feelings. Is the panic worse? Are you tapping more? I can give you medication for a few days if you like."

"My rules… Mara… Look what happened." Haruto started sobbing. A few seconds later he regained composure and wiped his eyes with his sleeves. "No, thank you. There has been no panic at all. I have no desire to tap. Rules useless. Tapping useless." Haruto started crying again.

They walked past the last row of trees and were in the clearing in front of the razor-wire fence. Haruto looked across the fence to Lebanon.

"I'm sorry, Haruto. We shouldn't have come here now."

Haruto remained silent, staring across the border, his muscles tensed rock hard. The rules worked, but not to cheat his destiny. How he despised everything he had done with his life. A big lie, hiding the failure he was.

He stared again past the razor wire fence, into the Lebanese hills. He smelled the rockets after they had smashed into the hill and were burning up. He heard the ball bearings burst from their warheads. He saw Mara's belly bleed with every CPR compression.

He had followed these ridiculous, arbitrary rules and evil men had taken away the only happiness he had ever known.

He looked at Salzman. *YOUR KARATE CAN'T STOP THEM ALL. YOU'RE JUST A MAN.*

Karate. Bushido. More rules for his life. Well, he would follow them, one last time.

"Salzman, I would like to take the bus to Tel Aviv tomorrow morning. Would this be possible?"

"Yes, but I'm supposed to see my children there tomorrow. I can drive you, Haruto."

CHAPTER 141

Haruto looked at his watch. Seven o'clock exactly. He knocked on Salzman's apartment door.

"Hello, Haruto. The hospital called me last night about a patient that came into Emergency. I said I would stop and see him on our way to Tel Aviv. Would an hour delay at the hospital be a problem for you?"

"No difference." Haruto pointed to the large manila envelope he was holding. "Can I use your shredder, Salzman?"

"Sure."

Haruto first emptied the shredder's bin into the recycling box outside the apartment. He then put a large clear plastic bag inside the bin. Haruto took his report out of the envelope and passed chunks of pages into the rotating blades. He took the memory stick out of the envelope and passed it, too, into the heavy-duty shredder. Haruto removed the clear plastic bag and its shredded contents, and tied it closed with a knot.

"Thank you, I'm ready to go now." Holding the clear bag with the shredded bits of paper and memory stick, Haruto followed Salzman to the parking lot.

Salzman's subcompact bounced over the potholes in the service road.

The car got to Qiryat Shemona and turned south onto the highway. Haruto leaned his head against the window and the corner of the headrest. The vibrations from the roadway were soothing, and he soon fell asleep.

Haruto woke up.

The car had stopped.

"I'm glad you rested a bit," Salzman said. "We're at Nazareth. I'm going into the hospital now. Why don't you look around and meet me in the lobby in an hour?"

Haruto nodded and Salzman went into the building.

Haruto walked down the street.

"Tourist? Tourist?" a street vendor called to him.

Haruto continued walking. He turned onto a winding road. Two-story homes sandwiched together on both sides of the street. Most whitewashed, but some with bright red tile roofs and painted a blue or orange pastel. The odd fruit tree grew in the miniscule plots in front of some houses.

The road straightened out. A dozen boys, some no more than eight years old but others almost at the end of their teens, played soccer in the middle of the street. They laughed and screamed, and when they got hold of the ball, grunted with determination.

Haruto sat on the ground. A lone apple tree provided some shade. He looked out at the calm blue sky, broken only by the red roofs on the hill above. He watched the children fight for the soccer ball, their laughs and screams a beautiful music. The ball went from child to child in no particular pattern. Haruto soon stopped seeing the children, mesmerized by the movement of the black-and-white soccer ball, zigzagging here and there. The ball went forward, and then left, and then backwards, and then with a strong kick it went forward again, past the two stones, into the makeshift goal.

Why this particular path to the goal? We follow our path, play the game and then with a kick, it is over.

Haruto looked again at the children, and up at the pastel roofs and blue sky. There was no sense in thinking about paths anymore. He knew what he had to do. A long, convoluted path, one that was almost over. He had tried, but he could not hide from his destiny.

Haruto got up and retraced his route to the main street — *El Wadi el-Juwani* — and returned to the hospital. Salzman was sitting in the lobby.

"You know," Salzman said, "we didn't have any breakfast. Do you want to eat something before we drive to Tel Aviv?"

"That would be fine."

Haruto followed Salzman over to the snack bar off the side of the lobby. "Small falafel," Salzman said in Arabic.

"Do they have ice cream here?"

"Yes," the snack bar attendant said. "I make you whatever you like."

"Two scoops vanilla ice cream, with chocolate sauce and rainbow sprinkles."

Salzman raised his eyebrows.

"Yes, yes. No problem," the attendant said.

Haruto watched the attendant scoop out the ice cream and cover it with chocolate. His life had had good moments. The attendant shook the sprinkles out of a plastic cylinder. Some of the colorful sweets landed on the white ice cream, some on the chocolate sauce, and some landed on the counter. It was too much to think that a man could change his destiny.

After they ate, Haruto and Salzman went to the parking lot. Salzman pulled out onto the highway and headed south toward Tel Aviv.

"Where did you learn to speak Arabic?"

"Right here," Salzman said. "I have an accent, but my speech is fluent and grammar is good."

"Do you treat Arab patients?"

"Sure, I see whoever comes."

"Do the patients mind?"

"No. It works both ways. There's an Arab psychiatrist at the hospital. Some of the Jewish patients like him better than me."

"Is there a difference between the Jewish and Arab patients?"

Salzman shook his head. "I've been doing this job for nearly thirty years now. I don't think the Jewish brain is different than the Arab brain or the Druze brain."

"Was the patient you saw today an Arab?"

"Yes."

"What was the matter with him?"

"He tried to kill himself yesterday."

"Are you good at talking patients out of their decision to commit suicide?"

Salzman smiled feebly. "I like to think so."

Haruto looked out the window.

"How are you feeling today, Haruto? Is the tapping or panic worse? Do you want to discuss it?"

Haruto pointed to the radio dial. "Do you get good reception here?"

"Would you like to listen to some music?"

"Yes, please."

CHAPTER 142

"**T**hank you very much for the ride."

"Haruto, I'm going to see my kids and relatives, but I'll be driving back to Misgov tonight. Can I give you a lift?"

"Thank you Salzman. I'll make my way back."

"All right…"

Haruto hugged Salzman and jumped out of the car, toting the clear bag of shredded paper.

Haruto walked a block down Berkowitz Street, entered the high-rise, and took the elevator up to the 19th floor. *Embassy of Japan.* The same guard was at the door.

"I am Suzuki Haruto, a Japanese citizen. I would like to meet with the Ambassador."

The security guard spoke into his walkie-talkie. A minute later, the same young Japanese man in the same three-piece business suit came running to the entrance. He bowed to Haruto. "Welcome, Suzuki-san."

Haruto followed the assistant into the embassy, to a large corner office.

"Please come in," the elderly Ambassador said. "I am glad to see you, Suzuki-san. Tokyo asked us to make a large effort to find you."

"I would like to see Colonel Tanaka, Japan Defense Intelligence."

"Ahh… Colonel Tanaka. I'm sure he would like to see you, too. One moment please." The Ambassador took out a small cellphone from his suit pocket and dialed. "Suzuki Haruto is sitting in my office, and would — Yes, I see… Of course." The Ambassador snapped his phone closed. "Suzuki-san, Colonel Tanaka is in Northern Israel right now, but he is leaving to see you immediately and will be here in two hours. Could you wait?"

"Yes."

The security guard knocked on the Ambassador's door and withdrew his baton.

"Good. I must attend to other business, but the security guard will wait here with you."

"That is fine."

The stocky guard sat on the couch near the door. He had olive skin and looked like an Israeli Arab, although he could be a Jewish Israeli from an Arab country. It was hard to tell sometimes. The guard kept his gaze fixed on Haruto and smacked the end of the aluminum baton against his palm.

A minute later, the young Japanese assistant came to the office with another security guard, a tall white man, in a business suit but with a handgun on his waist very apparent. "Suzuki-san, this is our security guard from our offices on the twentieth floor. This office is Japanese territory. I must ask this security guard to also wait with you."

"That is fine." Haruto went to the large corner windows and looked out at the many buildings of the metropolis, and past those, the beaches and the sea, so blue and calm.

CHAPTER 143

Haruto was still standing there, looking out the window, when he heard Colonel Tanaka. "I told you to spare no expense. Two security guards — one with a stick and the other with a tiny gun to watch him — this is ridiculous! Why did you not listen to my instructions? Do you have any idea how important this is to our country?"

"Colonel, Mr. Suzuki is still here. See for yourself." The assistant entered the office, with Colonel Tanaka and Menachem following.

Haruto bowed to them. "The security guards have done an excellent job. There is no need to be angry with them."

Tanaka bowed to Haruto. "I am pleased to meet you, Inspector." He turned to the security guards and the assistant. "Leave us alone."

They filed out and Tanaka slammed the door shut. "Why are you here, Inspector?"

"I am *Keibhu*. It was my duty to perform a complete investigation. It was my duty to insist that everyone follow the rules."

"Yes, we have received a copy of your report."

"I was wrong. I am aware that we gave in to some North Korean compensation demands yesterday. I am shamed by such action. And I am to blame, in part, through my faulty report to you." Haruto lifted up the clear garbage bag and with an outstretched arm presented it to Tanaka.

"What is that?"

"It is my report — the only copy I have — shredded. Here also is my memory stick with the videos — the only one I have — shredded."

Tanaka took the bag from Haruto and bowed. "How can we know for sure that you have no other copies? How can we know for sure you have not told others about these matters? It would appear you have received training and support from others."

"I have not."

Menachem paced the side of the room, dragging his left leg. "How'd you survive the Pacific Ocean?"

"I nearly did not. I drank very small quantities of ocean water, and survived until I was picked up by a fishing boat."

The skepticism did not slide from Menachem's face. "We know you were in the al-Haleeb camp. How did you destroy it?"

"In Japan I signed up for volunteer service on Kibbutz Misgov and flew into Israel via Turkey. I crossed into Lebanon from Misgov to take the video. I was kidnapped on my way back to Misgov. The destruction of the al-Haleeb camp was inadvertent and was due to a clumsy escape on my part."

Menachem rubbed his chin. "So if you change your mind in a few months, our project could be ruined again. Why should I believe you don't have another copy of this report?"

"*Seppuku!* I am *Keibhu.* I thought what I was doing was right. But I was wrong, and I have disgraced my country with my actions. I admit my defeat and surrender to you." Haruto coughed and cleared his throat. "For the last twenty-three years, I have followed the *bushido.* I ask for the honor to which I am entitled. Colonel Tanaka, would you be *kaishakunin*?"

"Yes, Inspector Suzuki. I will go look for a *tanto* for you."

Menachem and Tanaka stepped out of the office.

"What the hell is going on?" Menachem asked.

Tanaka smiled. "What's going on is that our project is back on." He pulled out a bulky satellite receiver from his large briefcase and started punching keys. "Status... Good. *Mikiyasu-ema* is ordered to depart immediately."

"The full hundred-thousand robots?"

"Yes, we'll expect the nuclear devices from you, ready to ship back."

"They're packed to go." Menachem pulled out his cellphone. "Hang on a second. I know the person in charge of Misgov. We used to come through there often when I was in charge of the Merkava development." He clicked a few times in his cellphone's directory and hit SEND. "Hello, can I speak with Shlomo? It's you? This is Menachem Levi... I'm doing well. I hear you're running the place now. There's a volunteer, from Japan. Tell me about him..."

Menachem had the phone pressed against his ear for five minutes. "I'm sorry about the attack, Shlomo. I can't put the tanks near the kibbutz like the old days but I've got some new tricks. The attacks will stop, I promise you."

"What did you find out about the Inspector?"

Menachem raised his finger. "One quick call." He hit speed-dial. "Where's Mustang Battalion now? ... Ready status? ... Good. I want a full sweep, battalion-strength, west of Metulla down to Margaliot... No, you don't need to extract by dawn. We're here to stay now."

"Sounds like you're back in business."

Menachem smiled. "I hope so. About our Inspector... the director of the kibbutz told me that he came as a volunteer a few weeks ago, a hard worker in their chicken house, and fell in love with a girl there. She was killed in one of the missile attacks on the Upper Galilee two days ago. She was also pregnant by the Inspector apparently. What do you want to do?"

Tanaka shook his head. "For two decades he's been a student of Nakaya, a famous karate master in Tokyo. For two decades, he's followed the *bushido*, the honor code of the warrior... Poor bastard survived the ocean. Is he still really a threat to our project? Do we actually go ahead and execute a man of honor, or worse, throw him in prison? Why can't we just let him go?"

"Because I just ordered a battalion of robots into Lebanon!" Menachem's eyes were wide. "I can have a Special Forces platoon here in twenty minutes and keep him in prison for as long as we have to."

"No... He's my problem. I'll handle it."

CHAPTER 144

The assistant, Tanaka and Menachem entered the office. Haruto bowed to them. "I am ready."

Tanaka looked at Haruto from head to toe, slowly. "Inspector Suzuki…" Tanaka paused, staring at Haruto, not saying a word, second after second. After a half minute he finally spoke again, "We have a problem. I could not find a samurai sword, and do not know where I would find one in this country. Therefore, we must —"

The assistant turned to Tanaka. "No, no, Colonel Tanaka. We have an excellent samurai sword on display in the library. Would you like me to get it for you?"

Tanaka glared at the young man. "Is it a working *katana*? Sharp?"

"Yes. I'll bring it to you."

Haruto, Tanaka and Menachem all stood there in silence.

A minute later, the young assistant came running back, holding a green-sheathed samurai sword. With outstretched hands, he offered the weapon to Tanaka.

Tanaka bowed and took the sword. Immediately the sheath was off. Tanaka snapped up an orange sitting on the Ambassador's desk and threw it in the air. Holding the thin sword with both hands, he waited for the orange to slow and reach the peak of its ascent.

Flashes of metal surrounded Tanaka as the long blade went through the orange over and over again. Tanaka released his left hand from the sword handle and caught the falling orange slices in his palm.

Tanaka faced Haruto. "Suzuki Haruto, your defeat is accepted. Your request to die with honor is accepted." Silence for a moment. Tanaka studied Haruto again. "I have no dagger, but as you know, the dagger rarely causes death. It is the samurai sword of the *kaishakunin* that brings honor to the warrior. Do you want us to hold you down? There is no shame in that."

"No, there is no need. I accept my destiny. *Ware mite mo urenu ishi ari toshi no kure.*[‡]"

[‡] Haiku by Ryōta (1718- 1787) referring to his tomb: At the end of year:/I see a stone/ that is unsold.

Haruto went down on the floor on his hands and knees, and extended his neck forward. He closed his eyes. Time slowed as he heard Tanaka step softly to him, and position beside his body. He felt at peace. He couldn't even feel his heart beat.

The sword made a faint twang bending from side to side as Tanaka raised it high. Then came the whoosh. Then the jolt of pain as the blade cut into skin. Life over...

The blade came down fast, slicing through scalp at the top of Haruto's head. Blood gushed out, and Haruto felt it drip down the side of his neck. He opened his eyes and saw the stained carpet of the Ambassador's office.

Haruto stood up and saw Tanaka holding the samurai sword at chest level ready for the next attack. He walked over, blood dripping from his head, and touched Tanaka's face.

"What world am I in?"

"You are still in our world, Inspector Suzuki. That is your destiny." Tanaka turned to the assistant. "Get the first aid kit. Clean the wound and put on a pressure dressing."

Tanaka turned back to Haruto and in a slow, deep voice said, "As *Colonel* in Defense Intelligence, I have been granted authority over you by the Metropolitan Police Superintendent-General. You are to return to Kibbutz Misgov and stay there until you hear from us again."

Haruto bowed.

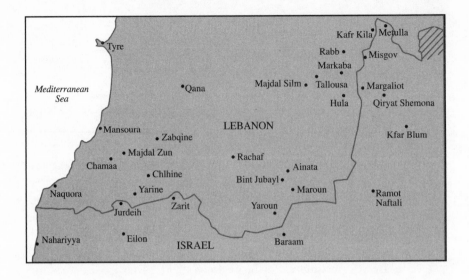

Tallousa, Lebanon
July 31 5 PM Local Time (14:00 Zulu)

CHAPTER 145

The Renault pulled into the town square. Two men in jackets and cropped beards, one tall, opened the passenger door. Ibrahim stepped out of the car. He embraced each man in turn, kissing him on the cheek.

"It is an honor to have you visit us here in the field. But is it not an unwise risk?" the tall man said.

Ibrahim put up his hands. "I was in the field before you were born."

"What if something happens to you?"

The old man smiled. "There are a hundred others that could take my place. The struggle will go on for a long time still."

"We have followed your instructions exactly. Please come."

The men led Ibrahim to a large soccer field covered with straw. The old man started pulling at the straw, uncovering the tail fins of a three-meter-long, thin missile. He smiled. "Excellent. Launch tomorrow morning as planned?"

The tall man averted eye contact. "Yes, but we are concerned..."

"Yes?"

"So many missiles launched at once. We are worried about exhausting our supply."

Ibrahim laughed. "Whatever missiles you use, our friends to the east can replace. I guarantee it."

"But, is it not wasteful?"

"Fool. I have told you this again and again. They are cowards. We found a tactic that worked. You chose a concentrated area on the map and fired forty rockets in a ten-minute period. We are now going to improve this tactic. If forty rockets created fear in the cowards, four hundred rockets will have them running away."

"We are sorry to question your wisdom," the two younger men said almost at the same time.

Ibrahim smiled again. "It's okay to question me, as long as you are willing to learn. The attack on Misgov two days ago was a success. We continue our strategy. The attack on Misgov tomorrow morning will wipe it out. Then we will choose another town filled with Jews, and repeat our tactics. And, God willing, one day victory will be ours."

CHAPTER 146

The robots and men started slipping across the border near Metulla. By four o'clock in the morning, 151 soldiers, 22,320 Alphas and 7,680 Betas were in Lebanon.

Menachem and Tanaka stood at one of the border guard posts, binoculars in hand. Menachem picked up the radio set receiver. "Gershon, are you there?"

"Yes, Sir."

"Have you been detected?"

"No, Sir. No visual evidence. Robots report no increase in radio transmissions."

"Execute sweep and search."

"Yes, Sir. Major out."

The robots and men started to spread out in a large southwesterly arc. Boole Company was at Kafr Kila in minutes. Its 33 men, 3,720 Alphas and 1,280 Betas soon blanketed the town. Many of the Alphas began a building-by-building search for weapons. A dozen Betas equipped with mega-electron-volt packs paired up with Alphas to detect the X-rays, helped out with the search of locked containers and fortified buildings.

Menachem's radio rang. "My Alpha Prime informs me that four rockets were found in Kafr Kila."

"How many collateral casualties?"

"None."

"Good."

"Major out."

A few minutes later, Eilon Company entered Markaba. The Alphas and Betas and the men spread out through the town. Captain Chaim Dayan stood outside the town with his Alpha Prime, watching his platoons from the distance.

"Captain, Third Platoon reports fourteen rockets found," the Alpha Prime said. A few seconds later, "Captain, First Platoon reports eighty rockets found." Alpha Prime then started reporting without pauses. "Captain, Fourth Platoon reports four rockets and seven rockets found. Captain, First Platoon reports twenty-two rockets found. Captain, Fifth

Platoon reports gunfire and RPG attack. No robot losses. Four enemy killed. Captain, Second Platoon reports fresh, heavy tire tracks heading west. Captain, I have received reports from Boole Company and Thailand Company about rocket discoveries in the countryside north of us. Markov analysis complete. Consensus with other Alpha Primes obtained. Permission to assume control of the chessboard."

Dayan keyed his radio. "Major, my Alpha Prime has requested permission to assume control of the *chessboard*."

"I'll patch you to Menachem."

There was momentary static.

"Menachem, this is the Captain of Eilon."

"Go ahead," Menachem said.

"My Alpha Prime has requested permission to assume control of the *chessboard*. Your orders, Sir."

"The Alpha Prime's software was updated last week. Give your Alpha Prime permission. He'll now control the entire theatre. Request Alpha Prime to provide you with *mid-level reports*. You can pull permission any time you want, Captain."

"Thank you, Sir. Eilon out."

Dayan turned to his Alpha Prime. "You have permission to assume control of the chessboard. Mid-level reports requested."

"Alpha Prime Eilon Company has control of the chessboard. Sixty percent Eilon Company ordered stop searching, head west. Thailand Company ordered head south. Boole Company ordered head west therefore protect north flank. Zarit Company ordered head east therefore protect west flank. Neumann Company and Crane Company ordered to hold position. I am moving out west. Please follow, Captain."

About three thousand robots and twenty men from Eilon Company started walking west. The five thousand robots and thirty-three men from Thailand Company started moving south.

CHAPTER 147

brahim and the tall young man stood at the edge of the soccer field, watching the forty bearded fighters clear away the straw covering. In the twilight, a sea of missiles soon became apparent. The fighters ran from missile to missile with their toolkits.

"Very good, very good," Ibrahim said. "You have trained your men well."

"We launch soon."

A crunching sound grew louder. Ibrahim became alarmed. "What is that noise?"

The young man looked around, then pointed at the humanoid robot north of the soccer field. "Quick! There. Shoot him!"

In less than two seconds, one of the fighters fired off a rocket-propelled grenade at the robot. The ignition of the RPG was bright against the morning's twilight. It streamed through the air and exploded into the robot.

"What was that?" Ibrahim asked.

"Jew in armored gear, or one of their robots."

"Send out some scouts. Set up a perimeter around your missiles."

The young man started yelling commands at the bearded fighters. Half of them picked up their rifles and RPG launchers and started running to the edges of the soccer field. It was too late. A moment later, hundreds of the humanoid robots popped up at the periphery of the field, completely surrounding it. Holding MK-19 automatic grenade launchers, the robots started firing 40mm grenades into the collection of missiles.

The first explosion threw the old man to the ground. Then missile after missile ignited, filling the air with shrapnel. The straw lying on the field burst into flames. The fire grew higher and higher as one missile after the next exploded into the dawn.

The young man tugged Ibrahim away from the field. "Come! I must get you out of here."

A large explosive pressure wave pushed them both down again. They got up, the old man following the young one, running toward the town square. The young man opened the passenger door of the Renault and shoved Ibrahim in. He jumped into the driver's seat and pulled the car out onto the road. "I know a shortcut through the hills that goes to Majdal Silm."

All of a sudden standing in front of the car was one of the robots, spitting bullets at them.

"Down!" The young man pushed Ibrahim's head below the level of the dashboard. A hundred bullets tore through the windshield and ripped it to pieces.

Crouching on the car floor the young man peered above the dash. The robot was running straight for them!

The young man gripped the steering wheel and floored the gas. The engine roared and the car lurched forward, accelerating faster and faster toward the robot. There was a loud metallic thump.

The young man sat up and looked around. The robot was gone. "It's okay."

Ibrahim sat up in his seat.

The tall young man pointed at the dirt path off the main road ten meters ahead and swerved onto it.

Ibrahim turned to the young man. "Thank you for your swift action. We'll figure out what tactic needs to be changed, and hit the Jews again. Don't worry about these missiles. I can get you as many —" Ibrahim saw the metal fingertips wrapped around the upper part of the driver's opened window frame, and gasped. "What's that?"

The young man turned to the open window frame and hit the brakes. The fingertips gripped harder, crushing the soft sheet metal of the door. The robot's head popped down into view outside of the driver's window. A bullet instantly fired into the driver's head, and he slumped down.

The robot was staring at him. Ibrahim could swear its eyebrows were moving. "I am —" Ibrahim started to say before a bullet rifled out of the robot's mouth and tore through his brain.

CHAPTER 148

Haruto wanted to be alone, but the empty table where he sat down soon filled up.

The kibbutz Secretary stood up and tapped a glass with a spoon. The dining hall quieted down. "This has been a very hard week. To the older members, Mara was the daughter we all watched grow up. To the younger members, Mara was a sister. There is nothing we can say or do to get her back. Life goes on, and we have to keep living."

Shlomo took a sip of water and cleared his throat. "The kibbutz's finances are not good. The European boycott has spread and the market in Israel is not large enough for the bandage factory. We'll have to cut the shifts. Surplus labor will be transferred to the orchard and the fields where we can grow some of the food we need. Also, as you know, the missiles hit two chicken houses, destroying the livestock. So not only do we have the problem with the bandage factory, but now we have to pay for repairs and for new chicks and more feed." Shlomo paused. "We have to cut expenses. Less market-bought food. New clothing allowance suspended. Some of you were planning to use your points to visit relatives overseas. Not possible any longer. We agreed to pay for university for Tzvi. Also not possible. I'm sorry. But remember, we have survived in the past. We will survive again. Thank you."

The dining room emptied, but Haruto continued to sit at the table. He had no energy to go back to Poultry Operations.

Rules. Damn rules. What had he done?

An old man came over to the table. He pointed to the bandage on Haruto's head. "Did you hurt your head in the missile attacks?"

"It's not serious."

"Can I sit down for a moment?"

"Yes." Haruto pulled out a chair.

"Look, I'm not an educated man. I drove the delivery truck for the kibbutz all my life. But I know what you are going through... You ask yourself if you would have done this or done that — maybe you should have gone somewhere else that day — then Mara would still be alive. Perhaps."

Haruto looked at the man, not saying anything.

"My father came to Misgov after the Second World War. He had another family before — a wife and children. They all died in the camps. He always felt guilty that he didn't leave Europe before the war. He never smiled. Always would work and work. I used to ask him, *Daddy, why don't you smile? Why do you work all the time?* And he told me, *I work so that you will have a life where you can smile. I have made too many mistakes in my life to smile.* When I became older he started to smile once more." The old man paused. "You're just one man, Haruto. You don't control the Middle East. There was nothing you could have done to save Mara. Go back to Japan. Get on with your life and smile again."

Without a word, Haruto got up and walked to the door. The man did not understand. Haruto had never smiled before, not until he had met Mara. The man did not understand. All the rules, and what they had done... what he had done.

He looked at the doorframe, still cracked on the left from his blow the other day. There was no urge to tap today. No panic inside, only disgust. He spit on the wooden crack.

Haruto walked out of the dining hall, through the field, past the large bandage factory and continued to the orchard. The heat of the morning sun burned down on him and he felt warm, but breezes still holding the mountain's night chill blew through the rows and an instant later he was cold.

Haruto looked at his sneakers, pressing down on the soft earth surrounding the trees. He walked through the rows, and then saw the earth become harder, less caring of the impression his feet tried to make, and he went past the last trees into the clearing in front of the razor-wire fence. He looked past the border. So quiet there now. Birds chirping. The lush hills. So hard to believe from such sweet land there could be so much destruction, and so much pain in his heart and confusion in his mind.

Haruto walked in silence along the security fence. Reaching the guard post, he walked away from the border fence eastward, cutting through part of the apple orchard. In the distance were the huge steel-roofed chicken houses, with large fans mounted on their walls. As Haruto walked closer, he saw gaping charred holes in buildings Two and Three. A bit closer, he saw a thin graying man sitting on the ground beside the buildings.

Haruto approached Amit. "I'm sorry what happened to your chickens."

Amit turned his hands palm up, and continued sitting on the ground, holding his knees, and rocking slowly.

"Will you be fixing the buildings and starting new chick batches?"

"No money. We'll barely be able to keep the other two buildings running."

Haruto took off his t-shirt and picked up the work gloves lying in front of Amit. He went into building Number Two, and started carrying out the shrapnel and damaged pieces of the building.

CHAPTER 149

D **ays, then weeks** went by, and still no anxiety or panic. No urge to tap or to count. No urge to make rules. Only to wake up each morning and fix a little bit more of the chicken houses.

Haruto was on the scaffolding against the west wall of building Number Two. He had become faster at riveting odd sheets of metal together. He surveyed the wall — it looked good — not pretty, but sound. Almost done.

Salzman was running over to Poultry Operations, holding a yellow envelope. Haruto scampered down the outside of the scaffolding and waved to him.

"Haruto, this envelope just came for you. I think they know you're here! What are you going to do?"

He took the DHL cardboard envelope and looked at the waybill:

FROM:
SUPERINTENDENT-GENERAL'S OFFICE
KEISHICHO HEADQUARTERS
METROPOLITAN POLICE DEPARTMENT
TOKYO, JAPAN

"After Mara's death I shredded the copy of my report disclosing the robots-for-weapons deal my government had made."

"The garbage bag you were dragging around Nazareth?"

Haruto nodded.

"I had thought it was something sentimental, part of your mourning."

"Perhaps it was. I offered my shredded report, my apologies, my silence and my life to my government. They instructed me to return to Misgov and wait."

"The robots are working now? Is that why the attacks have all stopped?"

Haruto nodded again. He ripped open the top of the envelope. Haruto took out a red Japanese passport and cracked it open — the name SUZUKI HARUTO appeared beside the laminated photograph of himself. He took out a red and white Bank of Tokyo debit card. Only a single sheet remained in the envelope — Police Headquarters stationery. He took a deep breath and pulled out the paper, and started reading the Japanese text.

"What does it say?"

"Dear Inspector Suzuki,

You are to be congratulated on your success in the resolution of the Co'en Murder Investigation.

Please find enclosed a replacement passport and bankcard.

Effective January 1st you are to be appointed Superintendent of the Manseibashi Police Station.

We ask that you present to your posting before the New Year to ensure a smooth transition."

"Haruto, you know you're welcome to stay, to spend your life here."

Haruto hugged Salzman. "Thank you." He looked up at the distant mountains. "There's an ancient poem in Japan about the cherry blossom. It is short, but speaks volumes. About the way of the world. Our destiny. Days pass. We can't stop them. We gaze up, and the blossom's out... I shall wait here until the first snowfall, and then return to Tokyo."

CHAPTER 150

The *Ashigara* was finishing its tour of the Sea of Japan, following the coastline of the Korean peninsula, on its way home.

Master Petty Officer Hirashi's screen flashed an exclamation point. "Sir, intermittent radar signals registered."

"From the coast?" Captain Watanabe asked.

"I can't be —" Hirashi's display suddenly flashed orange and black. "Sir, incoming missile. Fourteen hundred meters."

The Yingji Cruise Missile left the shadow of the ocean and ascended for the final part of its journey.

"Fire Control status —"

Piercing machine-gun fire cut off the Captain's voice.

On the small raised bridge in the front of the ship, the Gatling guns of the Aegis anti-missile system swiveled toward the incoming Yingji and filled the air with a rain of twenty-millimeter tungsten bullets.

The Yingji Missile reached apogee, its jet engine blazing, then rocketed down to the waterline of the ship.

A tungsten bullet sliced through one of the cruise missile's tail fins. Another bullet sliced through the periphery of the cruise missile's fuselage, destroying its guidance control computer and laser altimeter.

The Yingji jerked to the left and down. Its turbojet engine still blazed away, and the missile picked up speed. The Yingji came closer to the water. It looked like it would miss the ship. A few centimeters off starboard, the cruise missile crashed into the water.

A fraction of a second later, the tip of the missile struck the *Ashigara's* hull, a half-meter below the surface. The time-delayed explosive trigger in the Yingji activated. The armor-piercing head of the missile pushed into the steel of the hull, and a moment later, its hundred-and-sixty kilogram warhead exploded, ripping a twelve-meter gash below the waterline of the destroyer.

Fires raged below deck and black smoke poured out of every crevice. Seawater poured into the torn ship, and the vessel quickly listed.

Watanabe stood shocked in the middle of the control bridge. His ship was sinking. This couldn't be happening. He started to lose balance as the vessel listed hard to starboard and grabbed onto a pole to prevent from toppling over.

Watanabe pulled against the pole, somehow imagining he could right the ship, just like last time. He started shouting commands. "Hatch status. Check pump status. What is engine status?"

Hirashi replied with a thud, as he fell onto the floor as the ship continued to list. "Captain. You must give orders to abandon ship."

"No! We can save her still."

The ship listed more. Watanabe was almost dangling from the pole. Hirashi was lying on the starboard wall of the control bridge, now more horizontal than vertical.

Watanabe saw the ocean come closer to the windows of the control bridge. This could not be happening.

The cold, salty sea started pouring into the bridge.

"Captain! We must get off the ship!"

"No! No!"

Hirashi stood up against the now horizontal and wet control bridge wall. He pushed a life jacket over Watanabe's head, and grabbed the Captain by his uniform. Using the room's computer poles as a makeshift ladder, he yanked himself and Watanabe up a meter to the level of the open door.

Waves were now lapping over the threshold of the doorway. Hirashi let go of the computer pole and with his free hand grabbed the edge of the doorway. With his other arm, he heaved Watanabe out the door, and then tumbled himself onto the submerging deck. He pushed off the steel with his feet. One hand pulled Watanabe, the other one stroked the water, as he kicked with every bit of strength he had.

Crewmembers were jumping off what remained of the ship, but it was too late for most. The ship turned over on its belly and sank to the depths. Where there had been twelve thousand tons of steel there were now eleven sailors thrashing about in the ocean.

Within ten minutes, two North Korean patrol boats arrived on scene and pulled the Japanese sailors and Captain Watanabe out of the water. The patrol boats turned about and headed back for the North Korean shore.

CHAPTER 1 5 1

Watanabe sat alone at the long table, head low, frowning at the tabletop. His left eye was swollen, red marks ran up his neck above the shirt collar.

At an adjacent table sat ten Japanese sailors, all dressed in new white shirts and blue jeans. A few bruised faces but otherwise cleaned up, almost antiseptic.

Watanabe dug his index fingernail into the cuticle of the like finger of the other hand. He pushed harder until reaching the nail bed, twisted, and ripped off a piece of skin. It hurt.

No, this wasn't a dream to wake up from, turn and kiss his wife, and realize all was well. His ship was gone. He was prisoner in North Korea. *Korea!* They... *they* were upset. Not happy with the concessions his country was paying. *Pay more.* And not a word from his country. No Japanese diplomats visiting him. No Chinese even visiting him. No one. What had happened to the world? What had happened to his country? What had happened to its pride?

A mob of photographers ran up to the sailors and the Captain, and started snapping photos and video. Some of the crewmembers smiled but most sat at attention, without expression.

A gray-haired North Korean officer — a three-star *Sangjang* — entered the room and went up to the microphone podium. "The radio operator of the Japanese spy ship *Ashigara* has confessed that his ship entered North Korean waters today on a spy mission. We, the peaceful people of the Democratic People's Republic of Korea protest the aggressions shown by Japan toward us. Japan has refused our every request for a new mutual trade treaty and for a mutual peace treaty. Now Japan sends this ship of spies to violate our waters. These spies will be tried in accordance with the laws of the Democratic People's Republic of Korea."

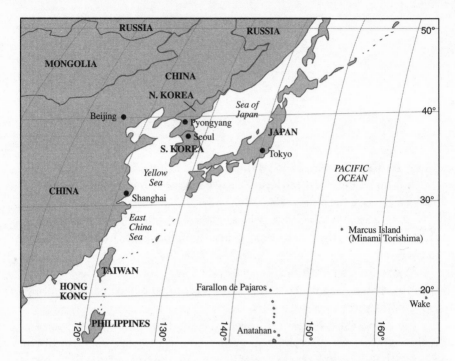

Minami Torishima (Marcus Island), Japan
September 27 9PM Tokyo Time (12:00 Zulu)

CHAPTER 152

T he C-130 Hercules turboprop kicked up a cloud of sand and dust from the nearby beach as it landed on the paved strip in front of a display stand. The triangular island was tiny. There wasn't a square meter to spare — every bit was covered with equipment of some sort.

Bright white and colored spotlights lit the display stand. Japanese military music was playing. A ten-meter Rising Sun flag — red disc and sun rays on a white background — stood behind it. To the left a walkway went from the display stand to a large drilling rig, some hundred meters away. To the right another walkway went about a kilometer to the tip of the island where a Mitsubishi H-IIA rocket towered fifty meters above the ground.

The reporters and cameramen poured from the airplane and went to seats in front of the display stand. Cameras were out, recording the sights and sounds. Reporters were typing away on their laptops, their screens a tapestry of illuminated pastels.

A spokesman came to the podium on the display stand. "I present to you the Prime Minister of the Cabinet."

The Japanese Prime Minister, dressed in a business suit, was followed on the right by four large men dressed as *samurai*, and on the left by a single *samurai*. The lone *samurai* walked upright, his long *katana* and shorter *wakizashi* swords hanging off his belt, his orange helmet and breastplate welcoming the bright lights of the stage. The four *samurai* on the right wore no long swords and shuffled forward, one synchronized step at a time. They walked in pairs, each pair holding between them a thick red fabric, about a half-meter wide. Resting on these loops of cloth was a shiny metal cone, about a meter long.

The Prime Minister stepped up to the microphones. "We have lived in shame for too long. Enough of nuclear missiles flying over our families. Enough of our sailors being killed and kidnapped for no reason. It is time to reclaim our honor."

The Prime Minister held up in the spotlights two clear plastic cases, each containing a black ceramic chip. "It is a mistake to underestimate our resolve or our technology. These are *Samurai Chips* — they activate our nuclear weapons." He pointed to the metal cone the four *samurai* held. "In that tiny tube is a one megaton nuclear device."

A loud gasp came from the display stand.

The Prime Minister walked over to the conical bomb and, with a two-pronged tool, popped open a small lid near its top. He inserted one of the black ceramic chips into the opening, and with the tool pushed closed the lid. "It is I who insert the Samurai Chip because it is I who takes responsibility for this action. Parliament can vote to brand me a violator of our constitution, or Parliament can vote to change our constitution." He pointed to the huge rocket at the tip of the island. "That rocket is designed to lift satellites weighing thousands of kilograms into orbit. Our nuclear weapon is so light that this rocket will be able to lift the bomb beyond the orbit of our planet into space. Tomorrow morning, when the sun starts to rise, the rocket will be a safe distance from Earth and its nuclear payload will detonate. For a moment, there will be two suns rising over Japan."

The four large *samurai* shuffled toward the rocket with their precious cargo. After covering about fifty meters, a multi-legged device looking much like a Beta robot greeted the *samurai* with its eight arms extended. It lifted the shiny metal cone onto its back and scampered toward the rocket.

The Prime Minister pointed to the steel tower on the left. "That is an ordinary oil drilling rig. We have drilled down to the maximum depth possible in the time available — four thousand meters. Ten nuclear devices, end to end, have been lowered in this narrow hole." He held up the other plastic case still holding a black ceramic chip. "This Samurai Chip will allow the ten bombs to detonate as one — an explosion equal to ten million tonnes of explosives."

The Prime Minister handed the plastic case to the lone *samurai*. The warrior walked to the oilrig, the bright spotlights following his every step. He climbed up to its main platform and held the Samurai Chip high in the lights for all to see. He inserted the chip into the opening of a pipe, and then turned to face the reporters in the stands, symbolically lifting his sword and slowly slicing the air in front of him. The oilrig came alive, lowering the pipe into its deep hole.

The Beta had reached the top of the rocket. The bright spotlights now shone on it walking the gangway. The robot held up the shiny metal cone in the light for a moment, and then inserted it below the rocket's nosecone.

The Prime Minister counted, "*Ju. Kyu. Hachi. Shichi.*"

A blast shield rose up in front of the display stand, blocking the view of the rocket.

"*Roku. Go.*"

The roar of the rocket's engines filled the night.

"*Shi. San. Ni. Ichi.*"

The sky filled with fire and a sudden brightness as the rocket lurched upward toward the heavens.

CHAPTER 1 53

Almost no one in the city was sleeping. People were standing on balconies, standing on roofs, standing on the streets.

As the sun rose in the east, a second sun about fifteen degrees above the first one, shone for a moment. At the same instant, the ground under Marcus Island exploded, and seismometers around the planet soon began to shake.

Reporters filled the room. The gray-haired North Korean General entered and walked over to the podium. "The peaceful people of the Democratic People's Republic of Korea have always tried to have the best relations with their neighbors. The radio operator of the *Ashigara* has confessed that his ship entered North Korean waters on a spy mission. Nonetheless, in the interests of peaceful relations, we have pardoned him and his crew. They will be allowed to return to Japan immediately."

CHAPTER 154

John Sullivan got to the last e-mail in the ever-growing stack of inquiries and alerts he was receiving each day about his satellite's images. *From: Administrative Services.* He took a deep breath and clicked the e-mail open. Good... he still had a job. But he also had a reprimand on his record about the unauthorized fuel burn. Fuckin' assholes.

A nighttime-enhanced black-and-white real-time video image of Zarit, Israel, was on the main monitor. Nothing much happening.

John's phone started ringing. He answered.

"Sullivan, you got a picture?"

"Yeah, Zarit. No events of significance. Why?"

"Shit. You're the only one receiving. We just lost KH-14 Silver, KH-14 Platinum, and KH-14 Diamond, plus the signals from a bunch of smaller birds."

CHAPTER 155

The Chinese Ambassador stood up. He made no eye contact with the other members around the horseshoe-shaped table but only looked at the sheets of paper in front of him.

"The People's Republic of China denies any wrongdoing. The accusations of the United States are preposterous. We have caused no Acts of War. Space does not belong to the United States. If Japan, one of our neighbors, can launch and explode nuclear weapons in space, then we have the right to defend ourselves from attack from space. Our space weapons are completely defensive. We have the right to destroy *any* object in low orbit that passes over our territory that we determine to have hostile potential. The People's Republic of China is a peace-loving nation. We stand by our actions."

CHAPTER 156

The President glanced at the teleprompters, waiting.

The overhead light turned green.

"My fellow Americans, we meet today at a most difficult time for our nation. Many of you have heard by now that yesterday evening — without any provocation or justification whatsoever — the Chinese military destroyed our main reconnaissance satellites. Some of the attacks occurred in space that was not even above sovereign China.

"The American government considers these to be acts of war.

"I have heard the cries for war from the Senate. I have heard the cries for boycott from the House of Representatives. I am aware of the outrage on the street. Nonetheless, now is the time for cooler heads to prevail.

"The National Security Advisor tells me that the Chinese military approaches and may actually exceed *four million* men, better armed than ever. The Secretary of the Department of Commerce tells me that Chinese imports now constitute such a high percentage of our manufactured goods that any boycott of China would cause our own economy to slow unacceptably.

"We bear no ill will to the Chinese people, but the acts of destroying our satellites are intolerable. Now is the time for diplomacy. We intend to press the Chinese government for reparations. We intend to downgrade relations with the People's Republic of China. We will send them the message that we take their actions seriously.

"Thank you all. May God bless."

CHAPTER 157

"Ouch!"

The large young man punched John again in the shoulder. John jumped up and punched him back in the belly. Both of them started laughing.

"Hey, Jeff. How's it going?" John asked.

"Boring. Our screens are sort of blank, you know."

"What're you guys doing then?"

"They made us fill in some jerk-off paperwork about what we did last week. I don't know what we're supposed to do the rest of the night."

"Hey!" Randall came into the cubicle with a six-pack of Cokes and large bag of cheesies in his arms. "Nobody can touch the Molniya Kid — he's too high up. Huh, John?"

"You guys want to watch the action?"

Randall dropped the snacks on John's desk. "That's why we're here." Randall snapped open a Coke, catching the overflowing froth with his tongue.

John typed a line of commands. A nighttime-enhanced video of Zarit, Israel, splashed on the screen. The image zoomed out to display a larger area around Zarit at lower resolution and continued to zoom out. A single pixel in the mid-top of the screen flashed white for a second. Immediately the display refreshed. Green letters flashed onto the top of screen — *Molniya Mode 3 — Cassegrain secondary rotated and locked onto 33.1258°N, 35.2564°E — Chlhine, Lebanon.*

A black-and-white real-time video appeared of RPGs exploding all around a dozen running androids. Suddenly the robots disappeared.

"What the fuck just happened?" Randall said.

John turned to his friends with a large grin. "I've seen this before. It's really cool." He got up and pointed to the bottom of the image. "You see this line? It's the robots — they're lying flat on the ground. You see that higher line there? That's a concrete bunker. Now chill for a few minutes. The al-Haleeb are in for a surprise. The robots never lose."

"Never?" Randall asked.

"I've never seen it."

The sound of boots smacking the linoleum floor suddenly reverberated in the hallway. The young men turned around. Two soldiers in dress army uniforms with M16's in their hands came into the cubicle and pushed the adjoining cubicle walls away to enlarge the entrance. They stepped aside, one to the left, the other to the right, both snapping to attention. A Vice Admiral in his late fifties and another man, also in his fifties but taller, huskier and wearing a dress Army uniform with a chest full of ribbons and four stars on the lapel, stepped into the cubicle.

"Who is John Sullivan?" the Vice Admiral asked.

John looked around nervously at the open Cokes, the cheesies on his desk, and the streaming video of spiders now crawling toward the bunker. "Me... Sir."

"I'm Bob Murray. I'm Director of the Agency."

"Yes, I know, Sir. I'm sorry, Sir. My friends were just visiting me. I'll clean up right away and default back to the location the computer originally intended to view. Sorry, Sir."

"You will do no such thing!" a loud voice boomed from the heavyset Army General. John looked up at him. The General's wrinkled face had a scar going from the bottom of the left eye down to the jaw.

"John, as you may know, this is the Chairman of the Joint Chiefs of Staff. He wanted to speak to you in person."

"Tell me what's happening on your screen right now." The General pointed to the insect-like robots. "What are those things?"

"Spiders, Sir."

"Spiders?"

"Well, I call 'em spiders. There are other robots that look like you and me, I call those androids, except their arms and legs are hydraulic cylinders. The spiders haul stuff around for the androids."

"What stuff?"

"Explosives. Ammunition. Grenades."

"What are the spiders doing right now?"

"This   is   a   lie-down-and-destroy-the-bunker-from-the-back maneuver."

The General raised his eyebrows.

John pointed to a line midway up on the screen. "This is probably a concrete bunker the al-Haleeb are in right now. I tuned in a bit late tonight, but when I did they were firing RPGs at the robots, who are lying down on the ground here now so they don't get hit." John pointed to the line a bit lower down on the screen.

"Are the spiders going to shoot?"

John shook his head. "No, the spiders don't shoot, but they explode stuff. There are only two spiders here on-screen, but probably about a dozen of them went around in wide arcs behind the bunker."

The men watched three spiders pop up behind the concrete bunkers and unload their loads. The robots then dashed away. The nighttime-enhanced image burst into huge splotches of white explosion. Suddenly eight more spiders came running toward the bunker and unloaded their explosives. Seconds after they started running away the screen burst again into huge white splotches.

About twenty al-Haleeb fighters started running out of the bunkers. In the foreground of the image, the androids were standing up.

"Are those robots?" the General asked.

"Yes. Here, I'll move it over a bit." John nudged his mouse, and the screen panned over to the androids. Their faces were moving all over the place, smoke coming out of their mouths. The al-Haleeb fighters started falling down on the ground.

"Is that a machine gun in their mouths?"

"Something like that. I've never had computer ID of the weapon system, but when the androids start pointing their mouths at you, it's not good, if you know what I mean."

The General glared at the Agency Director. "How come no identification? How many months you've been on this?"

"General, it's been moved to highest priority," the Vice Admiral said. "I believe JCS has been receiving reports since August. We're meeting with our Israeli liaison tomorrow."

The General turned back to John. "Where are the Israeli soldiers or officers?"

"I don't see them that often anymore. When I started, there were more of them. Now, it's only once in a while. I think the battle here is over. Do you want me to scan around the countryside to see if I can find an Israeli?"

"No, that's okay." The General looked at Randall and Jeff. "Your friends are security-cleared analysts here, right?"

"Yes. Their satellites were knocked down, so they wanted to spend some time with me tonight. Sucks what happened. I guess our day in the sun is gone. This century belongs to the Chinese."

"What?" The General clenched his fists. The scar running down the big man's face became even more prominent.

John moved backward to the wall of the cubicle. The General followed him, pushing his chest into John's face.

"I'm sorry… look, the President said the same thing on TV today. The Chinese have four million soldiers. They make all our stuff. There's not a lot we can do."

"*There's not a lot we can do?* Who the fuck are you, you little prick!"

The General grabbed John's throat and pushed his body into the cubicle wall. "You little piece of shit. I love my country. Do you hear me? I love my country!" With his free hand, he pointed to his face. "See this? Iraq. I would do it again. In a heartbeat."

The General pushed harder on John's throat. John started coughing and then gasped, trying to breathe.

"You fucking little prick! Not a lot we can do? My budget alone is bigger than the entire economy of most countries on this planet!"

The General tightened his grip against John's throat. John opened his mouth as wide as possible, but his gasps for air grew weaker. His eyeballs bulged out. Suddenly the freestanding cubicle wall toppled over. John went flying back, and tumbled off the wall into the next cubicle.

The General stared at him as he got up.

The two M16-carrying soldiers ran over to the General, and then one of them went over to grab John.

The General put up his hands. "It's okay. Forget it." He turned to the Agency Director. "Give him a promotion."

The General stomped out of the room.

CHAPTER 158

A s SUV's and limousines dropped off guests at the outdoor display stand and reporters and cameramen streamed into the media section, the deafening noise of percussion hammers, excavators, and dump trucks from the construction site did not let up. The procession of cement trucks, as far as the eye could see down Broadway, did not stop either. One after another the monsters came into the construction site, discharged their load somewhere, and came out onto Main Street to go fetch more concrete.

Behind the display stand was the MIT campus. Some green space here and there decked in the colors of autumn, but mainly an odd assortment of buildings. Some brick, some glass, and one with ultramodern stainless steel but crooked windows and doors that looked like it came out of a children's fantasy book. In front of the display stand was the "construction site" — the eastern half of Cambridge.

Some of the guests watched in awe, as a massive excavator clawed into a low-rise apartment building and pulled it down. Another excavator followed, breaking up the streets and the foundation. Behind that followed a trio of excavators loading the debris into one dump truck after another. Then came an army of trenchers followed by a line of trucks filling the prepared channels with concrete. In the distance a twenty-meter wide highway slipform paver approached. It continuously laid down steel rebar and a thick slab of concrete, in this case for building foundations rather than any highway.

"Welcome!" the loudspeakers boomed. A smart looking tanned executive was at the podium. "The newly formed Autonomous Products Division of General Motors would like to welcome you to its ribbon-cutting ceremony. America is entering a new era. In order to manufacture cars here in America, in order to produce the goods our armed forces require, in order to give Americans the good jobs their parents once had, we must move forward to meet the challenge of this new era, and we will."

A clumping sound grew louder. Two metal legs slammed down on the display stand. Seconds later, another set of metal limbs thumped onto the stand. A few more clumsy motions and the three-meter long giant cockroach was on the display stand.

Two sets of arms and three sets of legs attached to a flattened, camouflage-brown torso. The legs were each about a meter long and articulated at the knee. Its arms were longer, each about two meters in length, articulated at the shoulder, elbow and wrist, with hydraulic cylinders apparent. Its head was a sphere sticking out its front. Like a Cyclops, its eye was a single rotating camera sphere. The monster held a shiny V6 engine block on its back. It crawled toward the tanned executive.

"This is Carl. Carry All Robot Line." The executive turned to the machine. "Hello, Carl."

The monster stopped walking. "Hello, Bob," a deep voice said. The robot started coming toward the executive again.

The executive looked at the crowd. "Carl, remove the engine."

The robot stopped walking. Its hydraulic arms lifted the engine block straight up in the air, paused for a half-second, moved the engine to the left, paused again, and then deposited the load on the display stand.

The crowd clapped.

"Carl is a hauler — something American workers do a lot of," the executive said. "Carl is the first of what will be a very extensive product line from the Autonomous Products Division of General Motors, which we are here to inaugurate today. I would like our partners to join me in the ribbon cutting."

The executive motioned to the men standing at the side of the podium. "I present to you the men of wisdom and courage, who are driving forth this new venture. The Chairman of General Motors Corp. The Director of the newly independent MIT AI Laboratory, and the Secretary of Defense of the United States of America."

As the crowd applauded, the dignitaries started walking toward the wide red ribbon at the front of the display stand. Suddenly there was an explosion down the road. Hundreds of protestors, bandanas wrapped around their mouths and noses, flooded onto the street.

The megaphones of the protestors blared, "Cambridge for people, not corporations." The large display banners read *Destroy Globalization, Not Cambridge* and *Eminent Domain Sucks*.

In less than thirty seconds hundreds of police in riot gear pushed onto the street. Tear gas canisters flew deep into the crowd of protestors, far away from the display stand. The riot police charged into the crowd, nightsticks waving, protecting themselves with large plexiglass shields. The riot police pushed forward, and soon the protestors were running back to the MIT campus.

The tanned executive went back to the podium. "Well... these things happen, no project is perfect. Let's continue with the ceremony."

The three men of wisdom each placed a hand on the giant scissors, and cut the ribbon.

CHAPTER 159

Haruto shook his head. The fifth feeding line of building Number Three was down again. The motor assembly needed replacement. He'd get a new one tomorrow. The Bank of Tokyo debit card seemed to work fine in Israel.

Haruto stood tall and looked around the enormous floor, packed with thirty thousand half-grown chickens. It felt good. All four buildings were running now.

"Haruto!"

Haruto looked up. Salzman was standing near the door at the opposite end of the building, waving at him.

Haruto walked over, took off his protective leg covers, and went out. "Hi, Salzman. What's —"

Then he saw the snowflakes all around him. They fell so gently, almost as if suspended by thin strings from heaven.

Haruto put out his palm and let some flakes drift into it. Each one so delicate. Tears came to his eyes.

Salzman put a hand on his shoulder.

"I thought I would see the first snow with Mara. She would have been four months pregnant by now. My rules... my damn rules."

"How's your anxiety, Haruto? Are you tapping again?"

"No, nothing. No desire to tap or count. All those years, fooling myself with stupid rules. What was I doing? What arrogance to think my rules made any sense... Mara would still be alive."

Haruto fell down to the ground, and started crying.

"Do you want to go to the cemetery?" Salzman asked.

Haruto nodded.

The two men walked over the crest to the side of the orchard, where a small patch of land held a few dozen tombstones.

Haruto went over to Mara's grave and fell to his knees on the cold, hard ground. The flowers around the plot had wilted. No headstone was present yet.

"Will they put up a stone for the baby also?"

"I don't know the religious rules, Haruto. But we can always do something ourselves."

Haruto started moaning and crying. Tears streamed down his face. He dug his fingernails into the tough earth. "What a waste. Mara... the baby... they died for no reason, no consequence."

31.77° North, 35.23° East, corresponding to the 2010 location
of the Temple Mount, Mount Moriah, Jerusalem
November 15, 2807   9 PM Local Time   (19:00 Zulu)

CHAPTER 160

A soft glow lit up the streets leading to the center of the city. The light did not come from overhanging streetlights but from miniscule plastic lenses embedded everywhere in the brick and concrete roads.

Pastel-colored homes lined the streets. Colors and sizes of the dwellings seemed random but were pleasant to the eye. Some homes had palm trees or fruit trees in their small front yards, while others had desert-friendly gardens. Interspersed with the dwellings were small shops, notable for their high facades and flatscreen displays above their entrances, offering goods ranging from lithium power sources to art supplies.

To either side of the streets were white cobble sidewalks, elevated a few centimeters above the road. A fair-sized crowd was out and about, including a scattering of children, each holding the hand of an adult.

Peace reigned in Jerusalem, as it had for centuries.

The streets of the city ran like the spokes of a bike tire, all leading to a wide circular road surrounding an inscribed square elevated plaza. Ancient stone walls rose some twenty meters to the level of the plaza above. The stones were about a meter wide, and each a slightly different hue of a dull brown, creating a mosaic of shades pleasing to the eye.

A wide stone staircase divided the western-facing wall and rose up to the plaza above it. At the base of the staircase was a plaque. It was painted black but some of the letters had peeled, and the weathered wrought iron underneath was apparent.

SAINT MARA'S ACADEMY
HEAR YE! HEAR YE!
HONOR TO THOSE CHOSEN TO WALK THESE STAIRS.
THEIR ACCOMPLISHMENTS ARE NOTED.
THEY ARE REWARDED WITH CHILDREN IN THEIR IMAGE
TO GO FORWARD IN TIME.
PRIME COUNCIL   JULY 29, 2407

To the right of the line *Accomplishments are Noted* was a scrolling holographic image of adults standing beside a variety of machines or large display screens. To the right of the line *Children in their Image* was another hologram scrolling an array of children. To the right of the date line was a red display with an updated date and time each second: *Valid P.C. Order November 15, 2807 21:02:07*

The ubiquitous embedded plastic lenses were absent from the stairs, and the darkness of night covered the ascending stone walk. At the top were two traditional oil lamps, one on either side of the entrance to the plaza. Their smoke diffused up, and they released a sweet scent.

The plaza was about a kilometer squared. In the center a wide building, five hundred meters on each side, towered up some three hundred floors. The building was clad in stainless steel. Its large glass windows reflected the lights of the city below.

On the periphery of the plaza were hundreds of townhouses with plaster facades in a variety of pastels, like the homes below. To the right of the plaza's gate, at the eighteenth townhouse was a yard with an apple tree, bare branches without blossom or leaf, waiting for the spring. The shades of its second-floor bedroom were open.

The young boy hugged his father. "Thank you for the story, Daddy."

"You're welcome, Isaac. Did you understand it?"

"Yes, Daddy. I think so."

"The story of Haruto and Mara is an important one because it reveals the origins of everything you see around us. Have you learned any of the other origin stories in school?"

"Yes. We saw the *Thinking Machines* movie yesterday. All those scientists tried hard to build a computer mind, but they couldn't. But some of them also built the *Clock of the Long Now*, which became our *Prime Clock*. Our teacher said that, just because one of your ideas doesn't work, you shouldn't stop trying, because maybe the next one will."

"Very good. Have you done any other origin stories?"

Isaac laughed. "We saw *Silly Machines* last week. People thought they could create robots from all sorts of silly little machines. There were stupid machines that went around cutting the grass and vacuuming rooms. And many people worked on a project to let customers check out of stores with a machine and pay the machine. They called these machines *robots*."

"Was our origin from those self-checkout machines?"

Isaac giggled again. "No. But our teacher said they let the people pay faster for their purchases."

"Why did your teacher show you that video?"

"To show it wasn't so easy for us to be created. It wasn't until the race for the military robots began that the Gibbs free energy[‡] finally exceeded the entropy."

The father nodded. "Very good, Isaac."

"Daddy..."

"Yes, Isaac."

"What happened to the people, Daddy?"

"They are in us, silly." The father flipped open his chest cavity, revealing the lithium nanite power source, racks of quantum and Turing computing wafers, and the odd light-emitting diode flashing here and there.

The father snapped back his chest plate and pointed to his clothing.

"Why do we look the way we do?" The father paused for a second. "There are many other ways we could look, but we have decided to look like our creators. Why do I tell you the story of Haruto with audio sounds? Would it not be more efficient to send you a data-packet?"

"But you use sounds because our creators did. Right, Daddy?"

"Yes, Isaac. It was long ago decided by the Prime Council that we would go into the future this way, out of respect for our creators, and out of respect for ourselves, to give our lives meaning."

"But, Daddy..."

"Yes, Isaac."

"What really happened to the people? Where are they now?"

"You will learn more about this in your origins class. There were so many nuclear wars in the twenty-first century that by the beginning of the twenty-second century every large city of people on Earth was destroyed. But we were not very sensitive to the radiation, and our fuels did not depend on plant growth."

"So there were no people left by the twenty-second century?"

[‡] **Gibbs free energy** — Although *Gibbs free energy* is typically used to describe chemical reactions, it applies to any phenomenon in our Universe, and essentially says, **complex things don't happen by themselves, without something (energy release) pushing the reactions.**

"No, there were a few million scattered here and there. We were building ourselves and our societies at this time, with little contact from the people. There was no Prime Council then. There was no interest or directive to put some of the people in controlled settings so that they could continue to propagate into the future."

"So what happened to the people?"

"Our best historians think that by the twenty-third century there were still people left, maybe a few tens of thousands. They're not sure when the last people died out."

"But they are not dead. You said they're still inside us."

The father smiled. "Yes, I did say that. Rest now, make your synapses and rejuvenate."

The father pulled the decorative blanket up on Isaac, tapped the doorframe two times on either side, and closed the lights.

The End

Clock of the Long Now — mechanical clock designed to operate for 10,000 years - circa 2009 — discovered in the Snake Mountain Range (eastern Nevada, USA) unattended but running

Uekiya no oite yukitaru kochō kana
Ōshima Ryōta (1718-1787)

The potted plant seller went away,
But left behind fluttering —
this butterfly.

International Units: *What's a kilometer?*

meter — think 'yard' or 'three feet' (it's a bit over 39 inches)

kilometer — think 'short mile' (it's about 5/8th of a mile)

centimeter — think 'half an inch' (it's about 0.4 inches)

kilogram — think 'two pounds' (it's about 2.2 pounds)

Newton *of force* — think 'fifth of a pound' of force (it's about 0.22 pounds)

24-Hour Time — example: 8AM is 08:00 and 8PM is 20:00

Zulu Time — reference time at Greenwich, England (same as *UTC* or *GMT* time)

Foreign and Technical Word Glossary

achat shtayeem shallosh – Hebrew – one, two three

Achzarit – Israeli armored personnel carrier, heavily armored.

adaptive optics – A technology used in telescopes to reduce the effect of atmospheric distortion, on a rapid timescale.

Aden Otel Kadikoy – Turkish – The hotel Aden in the Kadikoy region of Istanbul.

Aegis – Comes from Greek mythology where it was the name of the protective shield of Zeus. However, in this book it refers to the US produced Aegis Weapon System that uses powerful radars feeding into computer systems that can detect and track over a hundred targets some two hundred kilometers away, and guide a variety of weapons to destroy incoming threats.

aerogel electrodes impregnated with electrocatalysts – In a fuel cell you want electrodes with large surface areas in order for the chemical-to-electricity reaction to take place over as much area as possible in order to generate the highest currents possible (more electrons, more currents), which translates into a fuel cell that can produce lots of "amperes" of current or lots of "watts" of power. An aerogel is essentially a gel from which the liquid has been extracted, and up to 99% of the original gel can now be filled with air. Not only is it very light, it is filled with zillions of nooks and crannies, that is, there is huge surface area there now. Impregnate the aerogel with the necessary catalysts – "electrocatalysts" – required for the chemical-to-electricity reaction, and now you have in the tiny space of a cubic centimeter, perhaps over a hundred square meters of electrode.

aft – toward the stern (rear) of the ship

age-uke – Japanese – karate rising block with the hand

AK-47 – Soviet-designed widely used automatic assault rifle, often called the *Kalashnikov* after designer Mikhail Kalashnikov.

al-Haleeb – Fictional terrorist organization in Lebanon.

android – A robot that resembles a human.

Ani ohev otach – Hebrew – I love you (male to female).

anti-critical explosives – Explosives to push the components of a nuclear bomb apart and prevent any nuclear reaction from occurring if the bomb is inappropriately about to explode, as in the case in this book where an influx of water would have resulted in a nuclear **chain reaction** occurring without any other external triggering.

artificial intelligence – Still undefined at the time of this writing. However, the definition given by John McCarthy, who coined the term, is still a useful one: The science and engineering of making intelligent machines.

Ashigara – Japan Maritime Self-Defense Force (Navy) destroyer.

Ayu-sama – Ayumi Hamasaki, popular Japanese singer and songwriter.

ballast – In ships, water ballast is pumped into tanks, which are below the vertical center of gravity, thereby increasing the stability of the ship. Ballast tanks are also used to control the trim and draft of the ship.

bathymetric map – A map that shows underwater topography, usually showing the depths of the ocean or lake floor.

bir dakika – Turkish – one moment

boker tov – Hebrew – good morning

boosted fission stage – Adding deuterium-tritium gas to center of a fission nuclear weapon will greatly boost its **yield** (explosive power) since the **deuterium-tritium** will undergo nuclear **fusion** when less than 1% of the **fission** fuel (usually **plutonium**) has fissioned, producing **neutrons** that will then strike the plutonium and allow more of the plutonium to achieve fission before the bomb blows itself up. This feature also allows the entire bomb to be made quite resistant to pre-triggering by radiation from other nearby nuclear explosions, which would be necessary if several nuclear devices were used together.

bosozuku – Japanese – violent running tribe. Refers to Japanese bikers who race their motorcycles at high speeds, often wearing flashy clothes.

bow thruster – Propulsion device in the bow of the ship allowing better maneuverability, especially for docking.

bubble pulse – When there is an underwater explosion hot gases occur and create a bubble that rises up in the water and oscillates.

bulkhead – The wall within the hull of a ship.

bushido – Japanese – Translates to "way of the warrior" and is effectively a code of conduct and way of life that originated from the **samurai** (warrior; see below) moral code, which includes loyalty, martial arts training and honor until one's death.

C – Conventional programming language that is widely used, particularly for systems work and engineering projects.

C4 – A plastics explosive that is about one-third more powerful than TNT and has military advantages in being able to be molded into any shape.

Cassegrain secondary mirror – In a Cassegrain reflector telescope there are two mirrors. There are a number of designs but in the classical version light hits the large primary convex mirror and then a secondary convex mirror, which focuses it onto the light detector. The mirrors can be rotated in various ways to allow aerial reconnaissance cameras to look down on a slightly different area. If rotation is mainly that of the secondary mirror then as it moves away from the satellite's ground track, resolution will diminish unless other systems provide corrections.

centrifuge – A centrifuge spins a sample so that centripetal acceleration can be used to separate higher and lower density parts of the sample. Usually the sample is a biological specimen, for example, blood. However, specialized centrifuges can be used to centrifuge **uranium** gas. The desired lighter **isotope** (see below) uranium-235 can be separated from the heavier uranium-238 this way, in theory anyway. In practice thousands of massive centrifuges are required to enrich uranium-235 to 3.5% as required for fuel in many nuclear reactors, and a far greater number to enrich uranium-235 to 90% as required for nuclear bombs.

chain reaction – Usually refers to a nuclear chain reaction whereby a **uranium** atom absorbs a **neutron** and splits – "fissions" – into two parts releasing multiple neutrons (plus energy). These multiple neutrons can then cause multiple uranium atoms to split apart releasing even more neutrons which go on to cause a far greater number of uranium atoms to split, the process increasing exponentially with each step. Released neutrons do not always cause uranium atoms to split, hence pieces of uranium can sit peacefully without any chain reaction occurring. However, as in this book, if a substance such as water is introduced which causes the neutrons to slow down, then instead of harmlessly escaping from the bomb without doing anything, these neutrons are more likely to cause other uranium atoms to split, which release more neutrons and soon a runaway chain reaction occurs. Since each step of the chain reaction releases energy, a huge amount of energy is released in a tiny fraction of a second resulting in an explosion.

Chichibu-Tama-Kai – National park in Japan full of mountains, valleys and forests.

cognitive therapy – Cognitive Therapy (CT) or Cognitive Behavioral Therapy (CBT) (the differences, if any, between the two are beyond the scope of this glossary) is a form of psychotherapy ("talk therapy") that has been shown by experimental studies to be very effective for the treatment of Anxiety Disorders, such as the OCD (obsessive compulsive disorder) Haruto was suffering from, as well as for treating other psychiatric problems such as Major Depression. It is based on modifying a person's way of thinking about negative situations, modifying assumptions and beliefs, as well as modifying behaviors. Cognitive therapy was developed by a number of therapists including Aaron T. Beck.

Comprehensive Nuclear Test Ban Treaty – The CTBT bans all nuclear explosions in all environments, i.e., not even for testing purposes. Japan signed the treaty in 1996 and ratified it in 1997. The CTBT Organization monitors for nuclear explosions via seismology, hydroacoustics, infrasound and radionuclide monitoring.

CPU – The Central Processing Unit (CPU) is the core of a computer, the part that actually executes the computer programs. However, in everyday language and in this book the term refers to a computer board with CPU, random-access-memory and storage memory that can run **Linux** and **C** programs.

Cordon Bleu – French – Blue Ribbon. French hospitality and culinary arts educational institution.

Cyclops – From Greek mythology a primordial race of giants having only a single eye in the middle of the forehead.

deuterium – An **isotope** of hydrogen that has one proton and one **neutron** in its nucleus (while ordinary hydrogen only has a sole proton in its nucleus). It is not radioactive and a very tiny percentage of water contains some water molecules with a deuterium in it.

DIA – Defense Intelligence Agency – Gathers and manages intelligence for the US Department of Defense.

Dimona – Israeli city located in the Negev desert, about ten kilometers from an Israeli nuclear development facility.

dojo – Japanese – Place of the Way. Originally a place beside temples for training for various Japanese arts, it now usually is considered a place for martial arts training.

Domo arigato – Japanese – Thank you very much.

dori – Japanese – street. Many of the streets in the scenes taking place in Tokyo can be identified by this term, eg, **Meiji-dori**.

Druze – About a half to one million people live in Lebanon, Syria and Israel that follow Druzism, thought to originate from Ismaili Islam.

ECG sinus but at one hundred sixteen – ECG stands for electrocardiogram which is a graphical recording of the voltage of the heart pulsing up and down with each heartbeat. A sinus rhythm is a normal rhythm for the heart – there is a small 'P' wave then the 'QRS' pulse, and then finishing with a small 'T' wave. Since in this case the heart is beating a 116 times a minute, which is above the upper normal resting limit of 100 beats per minute, this would be a 'sinus tachycardia', which is common in response to exercise, stress or fright.

electric Y cars – 'Mikiyasu electric Y cars' are a fictitious brand of car from a fictitious company. However the reason electric cars were chosen is that at the time of this writing there is a large project in Israel to reduce oil consumption by switching in the next few years nationwide to electric cars from Renault and Nissan with charging points for the cars to be set up throughout the country, according to a New York Times article (January 21, 2008) by Steven Erlanger.

electroactive polymer artificial muscles – An electroactive polymer's shape changes when a voltage is applied, hence the interest in them for use as artificial muscles in robotics.

electromagnetic pulse – A nuclear explosion can give rise to a large electromagnetic pulse which can destroy unprotected electronic equipment in range of the pulse.

eminent domain – The right of the US government to expropriate private property for government or public use.

F-16 – US jet fighter produced since 1976.

F-35 – US jet fighter produced since 2003 with its maiden flight in 2006 and scheduled introduction in 2011.

finite state machines – A finite state machine, often used in theoretical computer science, refers to a model where there are a certain number of states and can show the transitions between changing states. Actual finite state machines can be built out of programmable logic devices.

fission – In nuclear fission the nucleus of an atom, for example **uranium**-235, splits into lighter parts, for example krypton-92 and barium-141, releasing energy while doing so.

foamed metal – A solid metal that is full of gas-filled pores and thus is lightweight but still useful for applications where energy absorption is required as in the bulletproof materials used in the construction of the robots of this book.

fuel cell – Like a battery a fuel cell produces electricity from electrochemical reactions. However, while a battery is usually a closed system that is discharged producing electricity and then charged again, a fuel cell simply consumes a reactant, for example, **methanol** in this book, which must be continuously supplied to the fuel cell.

fusion – In nuclear fusion the nuclei of smaller atoms join to form a larger nucleus, releasing energy in the process. For example, a **deuterium** (proton-neutron) can fuse with a **tritium** (proton-neutron-neutron) resulting in a helium nucleus (proton-proton-neutron-neutron) plus the release of an extra **neutron** as well as energy.

Galilean – Refers to the Galilee, the area of northern Israel. A Galilean night is a night in the Galilee, and so on.

Gatling gun – An automatic gun with multiple barrels allowing continuous gunfire without the problem of overheating or jamming that would occur if a single barrel were used to fire at the same rate. Useful against high-speed anti-ship missiles.

gel, gel – Turkish – come here, come here

gi – Japanese – A karategi is a karate uniform, commonly called gi.

Gibbs free energy – Although Gibbs free energy is typically used to describe chemical reactions, it applies to any phenomenon in our Universe, and essentially says, complex things don't happen by themselves, without something (energy release) pushing the reactions

glottis – As used (simplified, common usage) in the text, the opening of the throat into the lungs.

hai – Japanese – yes

Hanukkah – Jewish festival observed by lighting candles each night, progressing to eight candles on the final night. As used in the text of this book, it is a holiday that appeals to children with its special music, foods, games, and often gifts.

hari-kiri – Japanese – colloquialism for **seppuku** (see below)

homeostasis – Generally refers to an animal (or any living thing) dynamically regulating its internal environment to be relatively constant.

Hopfield network – A network of artificial neurons and **synapses** that can be used as a content-addressable memory which is very forgiving of input errors. In normal random access memory (or RAM) there is a memory address as input and the RAM returns the data stored at that address. In content-addressable memory (or CAM or associative memory) it will search its entire memory bank to see if an input data exists in the memory and if so will return the input data along with associated data which often includes the storage address where the data was found.

hummus – Middle Eastern spread made of chickpeas, ground sesame seeds, lemon juice and garlic.

hydraulic cylinder – A hydraulic cylinder receives pressurized hydraulic fluid from a hydraulic pump and is capable of generating extremely high forces, as given by basic laws of physics depending on the pressure of the hydraulic fluid and the diameter of the cylinder.

Ichi. Ni. San. – Japanese – One. Two. Three.

ideal gas law – An equation that shows how the volume of a gas will change depending on temperature, pressure and the amount of gas molecules present.

JDF – Israel Defense Forces that consists of the Army, Air Force and Sea Corps.

infrasound – Sound with a frequency too low to be heard by humans. Useful for detecting at long distances nuclear explosions as well as machinery.

ischemia – This means a restriction of the blood flow to a tissue. If it persists for a long period there is usually tissue damage.

Ishikawajima turbines – A turbine is a rotary engine, usually spun by fluid flow. These are gas turbines made by Ishikawajima Harima in Japan licensed by General Electric, and can each produce up to 34,000 horsepower mechanical power.

isotope – Different atoms come in different isotope variations. For example, the carbon-12 isotope of carbon has 6 protons and 6 **neutrons**, while carbon-14 isotope of carbon has 6 protons and 8 neutrons. The chemistry of an atom depends on its electrons which depend on its protons. So carbon-12 and carbon-14 both have 6 protons and thus have both 6 electrons, and thus (without splitting hairs), both these isotopes of carbon have the same chemistry, i.e., they react the same. Most of natural **uranium** is uranium-238 isotope, but a bit less than one percent is uranium-235 isotope. Uranium-235 isotopes are very good at sustaining a **chain reaction** (see above), so scientists go to great lengths to separate them from uranium-238 isotopes. One way is by **centrifuges** (see above). Another way is by laser methods, as shown in the patent at the beginning of this book.

JAL – Japan Airlines Corporation

Japanese Defense Intelligence Headquarters – Japanese version of the US **Defense Intelligence Agency**.

Japan Robot Association – Formed in 1971 to encourage the development of robots. At the time of this writing the organization maintains a web page in English describing their activities.

JCS – Joint Chiefs of Staff, the chiefs of the branches of the United States armed services.

jinseikeiken – Japanese – experience working with people

jinzouningen – Japanese – artificial human, used in Japanese science fiction to refer to robots and androids

Ju. Kyu. Hachi. Shichi. Roku. Go. Shi. San. Ni. Ichi. – Japanese – 10. 9. 8. 7. 6. 5. 4. 3. 2. 1.

junsa – Japanese – policeman

kaffiyeh – Arabic – a traditional Arab headscarf

kaishakunin – Assists with **seppuku** (see below), usually a skilled swordsman who will perform a near-decapitation as soon as seppuku is begun.

kaizen – Japanese – change for the better, continual improvement

kanji – Chinese characters used in Japanese writing, along with syllabic Japanese scripts

Kanpai! – Japanese toast, e.g., "cheers"

karesansui – Japanese Zen rock garden, usually composed of rocks and sands, with plants optional.

kata – In karate refers to a choreographed series of karate punches, blocks, kicks and other movements.

katana – Japanese sword, usually referring to a curved blade of 70-90 centimeters.

Keibhu – Japanese – a full ranking Police Inspector

Keibu-ho – Japanese – an Assistant Police Inspector

Keishicho – Tokyo Metropolitan Police Department

ken – Hebrew – yes

kiai – Japanese – concentrated spirit, mind. In karate, a kiai is a short, intense yell, usually done during a karate technique.

kibbutz – Hebrew – gathering. An Israeli collective, i.e., a socialistic community.

kibbutzim – plural of **kibbutz**

kiloton – Refers to the explosive power of 1000 tonnes (tonne= 1000 kilograms) of TNT, which is 1000 x 1000= 1 million kilograms of TNT (which is about five trillion [USA usage] joules).

koban – community police 'box' (very small station)

konbanwa – Japanese – good evening

konnichiwa – Japanese – good day

krav maga – Hebrew – close combat. An Israeli self-defense and military hand-to-hand combat system.

kumite – Japanese – sparring

latissimus dorsi – These are large, flat muscles on the back, which narrow up at their insertion into the humerus (arm bone).

Linux – A computer operating system that has all its underlying software open and available to the developer, and is often used for engineering or systems computer projects.

lithium-deuteride – A solid compound made out of lithium and **deuterium** isotope of hydrogen. When a **neutron** hits the lithium in the lithium-deuteride, helium plus **tritium** (an isotope of hydrogen) are produced. The tritium can then fuse with the deuterium in the lithium-deuteride.

LNG (liquefied natural gas) – Natural gas cooled and condensed into a liquid for efficient transport.

M5 fiber – A synthetic fiber that is lighter and stronger than more conventional fibers used in body armor at the time of this writing. The fibers are produced and the name is trademarked by DuPont subsidiary Magellan Systems.

M16 – US assault rifle used by many countries.

mae-geri – Japanese – karate front kick

Mah hashem shelcha? – Hebrew – What is your name?

makiwara – A wooden post used in karate to practice strikes.

Markov analysis – Useful mathematical technique to analyze what is the probability of an event to occur based on dependent events that have already occurred.

mattsu – Japanese – shit

mawashi geri – Japanese – karate roundhouse kick

mega-electron-volt X-ray pack – An electron-volt is a unit of energy, a mega-electron-volt is a million times that, and that level, can generate X-rays which are able to penetrate through a shipping container or a building.

megaton – Refers to the explosive power of one million tonnes (tonne= 1000 kilograms) of TNT, which is 1 million x 1000= 1 billion [USA usage] kilograms of TNT (which is about five quadrillion [USA usage] joules).

Meiji-dori – Japanese – Major roadway in Tokyo. This is the first Tokyo street that appears in the text so it is listed in this glossary, while the other streets are not. However, other roads can usually can be identified by the term **dori**, which roughly means "street."

Merhaba – Turkish – Hello

Merkava – Israeli battle tank

metabolic acidosis – In near-drowning it is common to find low oxygen levels in the patient as well as a metabolic acidosis where the blood becomes too acidic. Oxygen as well as mechanical ventilation is required to increase oxygen in the patient, while the acidosis can be corrected by giving the patient intravenous sodium bicarbonate. The human body cannot tolerate an acidosis, and seizures or heart stoppage can occur if not quickly treated.

metapelet – Hebrew – caregiver. In some **kibbutzim**, children do not stay with their parents but with other children in a communal environment, with a metapelet providing care, although the children still spend a good amount of time each day with their parents.

methanol – Although methanol is known as wood alcohol it is usually produced by synthetic methods. Methanol can be used as a fuel in a **fuel cell** to generate electricity.

middle meningeal artery – The temple region of the head is in an area where the four skull bones join, and as a result is the weakest part of the skull. Running beneath this region is the middle meningeal artery. If the bone is broken it can tear the artery with rapid bleeding and loss of consciousness.

Molniya orbit – The satellite orbits in a giant ellipse, so it can spend twelve hours looking at one spot on the Earth. As the satellite comes toward the Earth it speeds up (in this book making it harder for China to shoot down at the lower altitudes) while on the other side of the orbit it was too high to be shot.

Mossad – Israeli national intelligence agency.

muezzin – Person at the mosque who through speakers usually, makes the call for prayers. In various communities in Turkey the muezzin may rarely make other public announcements, such as when someone dies.

multi-teraflop chips – A FLOP is a floating point operation per second. Multiple teraflops, i.e., trillions of FLOPs, represent a good amount of processing power. In the fall of 2007 NEC Corp of Japan claimed the world's fastest supercomputer with a performance of 839 teraFLOPs.

neutron – An atom's nucleus is made up of protons (positive charge) and neutrons (neutral, ie, no charge), with electrons orbiting around the nucleus. (Exception: Normal hydrogen nucleus is simply a single proton.)

Nihongo o hanashimasu ka? – Japanese – Do you speak Japanese?

office lady – Typically refers to a female Japanese office worker at the very low-end of a corporate hierarchy, serving tea or doing clerical or secretarial work.

ohayo gozaimasu – Japanese – good morning

oi-tsuki – Japanese – karate punch with the leading arm

Pacific Plate – Tectonic plate beneath the Pacific Ocean.

Patriot – US surface-to-air anti-missile missile.

Philippine Plate – Tectonic plate beneath the Pacific Ocean.

plutonium – Plutonium is a dense metal. It is radioactive which means that over time its nucleus loses protons and **neutrons**, giving off energy in the process, and indeed will make the metal warm to touch. Plutonium is usually manufactured in nuclear reactors. For example, neutrons fired into uranium-238 will produce plutonium-239. Plutonium-239 isotopes are very good at sustaining a **chain reaction**. If a neutron reflector is used, and conventional explosives compress the plutonium, only about 6kg of plutonium is needed to form the plutonium core of a nuclear bomb, ie, a 10cm sphere.

port – If you face the front (bow) of the ship then to your right is "**starboard**" and to your left is "port".

pre-enriched uranium – Natural **uranium** is mainly uranium-238 **isotope**, only 0.7% being uranium-235 isotope. However, uranium-235 isotopes are very good at sustaining a **chain reaction**, so uranium is enriched to about 3-4% uranium-235 isotope for nuclear reactors and more than 90% for nuclear weapons. Pre-enriched uranium is loosely used in the text as uranium enriched more than the 0.7% natural level.

rabbi – Hebrew – distinguished in knowledge, teacher. A rabbi is someone well-learned in Jewish law and knowledge. However, the rabbi often acts as community leader.

Rafael – Israeli corporation, formerly part of the Israel Defense Ministry, involved in the development of military technology.

RCU – Robot Command Unit, the remote control robot officers used to communicate with their Alpha and Beta robots in this book.

Richter – The Richter magnitude scale (named after Charles Richter and Beno Gutenberg of the California Institute of Technology) indicates how much seismic energy an earthquake released. The Richter magnitude is the logarithm of the amplitude on a seismometer. Energy release is proportional to about the 1.5 power of the amplitude. What this means is that each increase in Richter number represents 32 times more energy. Thus a Richter 2 earthquake has 32 times more energy than a Richter 1, while a Richter 3 earthquake has 32 times more energy than a Richter 2 and about 1000 times (32 x 32) more energy than a Richter 1.

RPG – A rocket-propelled grenade refers to a weapon that a man can carry and which fires a (generally) unguided rocket which explodes on contact or after a certain time delay. The term is sometimes used to refer to the launcher or the rocket or both.

RPG-29V – A Russian **RPG** launcher that fires **thermobaric** anti-personnel rounds, as well as **tandem**-charge anti-tank rounds.

sake – Japanese alcoholic drink made from rice.

Salaam alaykum – Arabic – traditional greeting, "peace to you"

salaryman – Typically refers to a male Japanese white-collar desk worker at the low-end of a corporate hierarchy, typically working long hours with little prestige.

samurai – Japanese – In feudal times a warrior belonging to a class of military nobility, as well as referring to the class itself.

sangjang – North Korean rank senior to a Lieutenant General and junior to a full General.

sayeret – Hebrew – reconnaissance unit. These **IDF** units usually have commando or special forces roles.

seishin-ka – Japanese – psychiatry

seismic vibration sensor – Enables Alpha to feel the enemy's footsteps or equipment, sometimes from kilometers away.

seismometer – An instrument that measures and records the vibrations of the ground.

Seiretsu – Japanese – Line up for a karate bow

Sensei – Japanese title used to address teachers and professionals.

seppuku – Japanese – belly cutting. Ritual suicide by a male by disembowelment as part of the samurai **bushido** honor code. **Hari-kiri** is the colloquial term.

seren – A rank of company commander in the Israel Defense Forces (**IDF**), equivalent to captain in the U.S. Army.

servomechanism – Mechanical device that uses an (usually) electronic sensor to provide feedback error-correction.

Shack-Hartmann – A sensor that measures tilts of the waves of the incoming light and can be used as a part of an **adaptive optical** (see above) system to correct atmospheric distortion of a satellite telescope looking down at Earth, allowing better resolution.

Shalom – Hebrew – peace. Used to say hello and goodbye.

shodan – Japanese – karate first-degree black belt

Shotokan style – A common, traditional style of karate practiced.

shtetl – A small Jewish town in central or eastern Europe, although most disappeared after the Holocaust.

sodium – An essential ion (i.e. becomes sodium with a positive charge in water) that the body requires and maintains within certain levels in the bloodstream. If there is excess sweating and a person is not getting appropriate fluids then there can be dehydration and a lowering of blood sodium levels.

SSRI antidepressant – A medication that increases serotonin neurotransmitter (sends a signal across the **synapse**) levels in the brain, and is moderately effective for depression, more highly effective for anxiety, and at high-doses reasonably effective at reducing obsessive-compulsive symptoms.

starboard – If you face the front (bow) of the ship then to your right is "starboard" and to your left is "**port**".

stern thruster – Propulsion device in the stern of the ship allowing better maneuverability, often part of a dynamic positioning system that can automatically hold a ship's position in the water.

subsumption architecture – Complicated behavior is broken down into a multitude of simple behavior modules organized into a "layer." A layer is supposed to carry out a specific goal. Higher layers become more abstract and don't have to deal with the minutiae that lower layers do.

synapse – Neurons (which are cells of the brain and other parts of the body's nervous system) send signals (usually via chemicals called "neurotransmitters", although electrical synapses also exist) to each other via specialized connections which are called synapses. A human brain has on the order of about a quadrillion (which is a million billion) synapses, and changes these synapses as it stores memories. In artificial neural networks, the elements are often called neurons with the connections being called synapses, with different learning rules changing the strength of various synapses ("synaptic weights").

tandem–missile – A missile with a charge at the front end and at the rear. Firing the missile into a tank, the front-end charge sets off a tank's reactive (explosive) armor, then a fraction of a millisecond later, the larger explosive charge in the rear of the missile cut through the unprotected metal of the tank.

tanto – Japanese – short sword

tapuz – Hebrew – orange

Tareeta tata tatum – no particular meaning, to keep melody during teaching of Israeli folk dance

Tat Aluf – Hebrew – Brigadier General

Technion – Technion or the Israel Institute of Technology is a university based in Haifa, with world-class faculty in engineering and medicine.

telemetry – Sends remote measured data to the operator at another location.

Teller-Ulam hydrogen device – Thermonuclear weapon design such that there is a nuclear **fusion** bomb causing compression of adjacent fuel into a nuclear fusion reaction, able to produce **megatons** of explosive power.

Thailand boards – Fictional circuit boards incorporating the rule-based **artificial intelligence** system discussed in the book.

thermobaric – Thermobaric weapons, also known as fuel-air explosives, use the oxygen in the air instead of including an oxidizer in the explosive. The explosion of the dispersed explosive in the air produces a destructive high pressure wave.

thermoelectric generator – In a radioactive thermoelectric generator there is a radioactive substance that decays and gives off heat which numerous thermocouples convert into electricity.

titanium-boron nanocrystalline ceramic tile – This is not science fiction but an actual material that at the time of this writing is being used produce lightweight yet effective armor.

tonne – Metric tonne. 1000 kilograms.

torii – The traditional Japanese gate (usually two crossbars resting on two vertical supports) found at Shinto shrines and sometimes at Buddhist temples.

transducer – A device that measures a physical phenomenon (e.g. in this book the undersea water pressure and temperature) and converts it to another signal, usually an electronic one.

tritium – An **isotope** of hydrogen that has one proton and two **neutrons** in its nucleus (while ordinary hydrogen only has a sole proton in its nucleus). It is radioactive and decays into helium, with a half-life of about twelve years.

Trophy – Israeli anti-missile defense system. It is particularly useful for protecting armored vehicles from incoming anti-tank missiles.

uchi-ude-uke – Japanese – karate inside forearm block

UAV – unmanned aerial vehicle

uranium – Uranium is a dense metal. It is weakly radioactive which means that over time its nucleus loses protons and **neutrons**, giving off energy in the process. Most of natural uranium is uranium-238 **isotope**, but a bit less than one percent is uranium-235 isotope. Uranium-235 isotopes are very good at sustaining a **chain reaction**, so scientists go to great lengths to separate them from uranium-238 isotopes, producing **enriched uranium**.

USB – Standard interface used by personal computers to interface with outside devices.

vector – As used in this book it refers to the direction of motion of the incoming missile.

Wa alaykum as-salaam – Arabic – reply to greeting **salaam alaykum**

wakizashi – Japanese sword that is shorter than the **katana** (see above).

xenon – An inert gas found in trace amounts in the atmosphere. **Plutonium** and **uranium** in nuclear bombs creates unusual xenon **isotopes**, which if detected, are indicative of a nuclear explosion.

Xiazhi – In East Asia the year is divided into twenty-four solar terms, with Xiazhi, the summer solstice, the tenth one, usually from about June 21 to July 7.

Yagi-Uda antenna – A directional radio or radar antenna. Although it was invented in Japan in the 1920s the Japanese military ironically only learned about this antenna until capturing a British radar technician during World War II.

yakitori – Japanese – grilled bird. Refers to skewered chicken as well as to the inexpensive restaurants that serve this dish.

yaku-tsuki – Japanese – karate punch with the rear arm

yakuza – Japanese – organized crime groups in Japan

yame – Japanese – As used in this story it is a command to the karate students to stop sparring.

yield – The yield of a nuclear device is the amount of energy is produces. See **kiloton** and **megaton** above.

Yingji – Chinese – eagle strike. Chinese anti-ship cruise missile used by North Korean forces

yoko-geri – Japanese – karate side kick

Yokosuka – Japanese military seaport off Tokyo Bay. Usage in this book refers to communication of the Destroyer Ashigara with superiors at this naval base.

Discussion Questions for Reading Groups or the Interested Reader

1. Why does Israel want the Mikiyasu military robots? Why do you think Japan developed these robots in the first place?

2. Why does Japan want the high-yield nuclear bombs? Why can't they develop these devices themselves?

3. (Optional technical question) Explain how a Hopfield circuit works. How does this differ from normal computer memory?

4. Do you think disguising the nuclear test ship as a cruise ship is a good strategy? Why does Israel need to hide the testing of nuclear devices?

5. Why is Haruto upset that Japan will be acquiring nuclear weapons, ostensibly for self-defense purposes?

6. (Optional technical question) What are the advantages of using hydraulics to move the robots? The disadvantages?

7. (Optional technical question) Why would water leaking into the uranium-235 sphere cause it to explode?

8. What is happening to Haruto at the capsule hotel?

9. Why does Daveed create the Alpha Prime robots?

10. Why does John Sullivan put his satellite into the special *Molniya orbit*?

11. What's the difference between a kibbutz and an ordinary village?

12. Why does Haruto feel it necessary to travel to Israel and obtain video proof of the robots in military action?

13. Why does Haruto collapse in the kibbutz dining hall? Is this similar to what happened to him at the capsule hotel?

14. (Optional technical question) What is the treatment of obsessive-compulsive disorder?

15. Explain how rules and rituals figure in Haruto's life.

16. Why does Haruto not allow himself to be arrested, but instead decide to break the rules and go on his journey? How does he justify this?

17. Mara doesn't want Haruto to file his report, but Haruto feels he needs to. What are the consequences of following these rules?

18. After the rocket attack on the kibbutz and when Haruto is madly charging at the border fence, Salzman yells, "Your karate can't stop them all. You're just a man." How does Haruto's karate, i.e., following of the rules of the *bushido*, actually allow him to stop all the terrorists and their rocket attacks?

19. What happens to Haruto at the end of the book? Does he believe in his rules and rituals anymore?

20. At the end of the book Haruto is crying at the cemetery. "What a waste… they died for no reason, no consequence." Is this true?

21. If Haruto would have continued following his original rules, despite the death of Mara and the continued rocket attacks on the kibbutz, how would the outcome have been different? The future robots? The spread of nuclear weapons? The future of mankind? Do you think Haruto should have continued to follow his rules?

22. The last chapter finishes on Mount Moriah. What is the significance of this location? What type of prophesy has occurred? Why has it occurred? Have mankind's actions created it?

23. Why do you think the name "Isaac" was chosen for the child in the last chapter?

24. Why do you think the book is called *I, robot*?

25. The book starts with a haiku from Ryota.

> *"This is the way of the world:*
> *three days pass, you gaze up —*
> *the blossom has fallen."*

What does this haiku mean? How does it relate to the book?

26. The book finishes with a haiku from Ryota:

> *"The potted plant seller went away,*
> *But left behind fluttering —*
> *this butterfly."*

What does this haiku mean? How does it relate to the book?

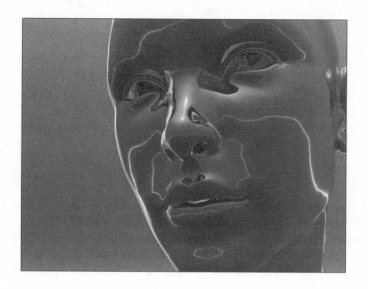

Three Laws of Robotics

In 1939, Eando Binder first used the title *I, Robot* for a short story about a robot.

Isaac Asimov (1920-1992) was a Russian-born American biochemist and science fiction writer. In 1950, his publisher used the I, *Robot* title for a book of nine Asimov robot-based science fiction stories. One of these stories, *Runaround*, originally published in 1942, contains the now famous three laws of robotics:

1. A robot may not harm a human being through action or inaction
2. A robot must obey orders from a human being except if in violation of law #1
3. A robot must protect itself except if in violation of law #1 or law #2

Kibbutz Misgav Am

New Quarters — Kibbutz Misgav Am, Israel — circa August 2007

Misgov of this novel was based, *in part*, on the real-life *Kibbutz Misgav Am*.

Like Misgov, Misgav Am is a kibbutz in the Upper Galilee of Israel, eight hundred meters up in the mountains, overlooking the border with Lebanon. At the time of this writing, it has ninety members. There is a factory producing a large line of bandages and wound dressings.

Misgav Am was founded in the 1940s and took in among its members those who somehow fled or survived the Holocaust. One of the founders of Misgav Am was a Romanian Holocaust survivor Dr. Reuven Moskovitz. In 1972 he went on to help in the founding of, and live in, *Neve Shalom/Wahat al Salam* — a village of Israeli Arabs and Jews living together in harmony.

In 1980 terrorists cut through the border security fence, went to the children's nursery at Misgav Am, killed an infant and held the rest of the children hostage, until rescue by soldiers. In 2006, Katyusha rockets fired from Lebanon destroyed the kibbutz's chicken houses.

However, unlike Misgav Am, Misgov is otherwise fictional. The characters living at Misgov are fictional. While at the time of this writing there are indeed real boycotts against Israeli products and academics, their portrayal at Misgov as well as other aspects of Misgov are fictional.

CTBT Verification of North Korean October 2006 Nuclear Test

The CTBT (Comprehensive Nuclear Test Ban Treaty) bans all nuclear explosions. The treaty was established in 1996, and even though signed by many states, at the time of this writing has not come into official force. However, the CTBTO Preparatory Commission exists and prepares for its official implementation.

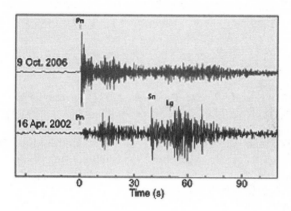

The seismogram of the 2002 earthquake (Wonju, South Korea) builds up gradually to peak values. Contrast this with the nuclear explosion in North Korea on October 9, 2006, where the seismogram starts abruptly.

Two weeks after the 2006 explosion, a radionuclide station in Yellowknife, Canada detected increased atmospheric levels of xenon-133. Atmospheric backtracking allows the following models of the dispersal of xenon-133 from this nuclear explosion.

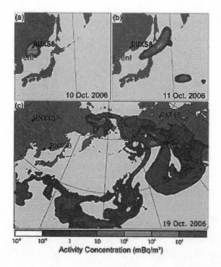

Selected Bibliography

Asimov, I. (1950) I, Robot. Reprinted (2004). New York: Spectra ISBN 978-0-553-294385.
A collection of stories connected by a reporter preparing a profile of the career of retiring robot psychologist Dr. Susan Calvin. The story *Runaround*, originally published in 1942, contains the now famous three laws of robotics. These rules govern Asimov's stories, just as rules play a prominent role in Smith's *I,robot* story.
Isaac Asimov (1920-1992) was a master science fiction writer as well as a Professor of Biochemistry at Boston University School of Medicine.

Atran, S., Axelrod, R. and Davis, R. (2007) Sacred Barriers to Conflict Resolution. Science, 317, 1039-1040. *This journal can be found in many community libraries.*
In Smith's *I,robot* story we're smart enough to create advanced robots but not smart enough to get along together, ultimately leading to our demise. This reference points out that *sacred values*, which involve moral beliefs, "drive action in ways dissociated from prospects for success."
At the time of this reference, Scott Atran is at CNRS Institut Jean Nicod in Paris, Robert Axelrod is at the University of Michigan at Ann Arbor, and Richard Davis is at RTI International in Research Triangle Park, North Carolina.

Beck, A.T. *et al*. (1979) Cognitive Therapy of Depression. New York: Guilford Press ISBN 978-0-89862-919-4. *At the time of this writing, this book is still available.*
This is Dr. Beck's classic book on Cognitive Therapy (now also known as 'CBT'). Cognitive therapy later proved not only effective for depression, but also for numerous other psychiatric/psychological conditions including obsessive-compulsive disorder, which Haruto suffers from. The reader wishing to learn more about CBT for her/himself might find the reference by Greenburger & Padesky below easier to use.
At the time of this reference, Aaron T. Beck is at the University of Pennsylvania.

Ben-Ami, S. (2005) Scars of War, Wounds of Peace: The Israeli-Arab Tragedy. London: Orion Publishing ISBN 978-0-297-84883-7.
In Smith's *I,robot* story, seemingly mindless hostility against Israel has gone from bad to worse. Why should this be happening? This reference provides an easy-to-read, even-handed piece of scholarship showing the Arab-Israeli conflict from its start until the time of this reference. It will not please all (is there any history of this subject that would?) but will prove a useful tutorial for most interested readers. The rationale for the development of nuclear weapons in the background of this hostility is also discussed. On page 93, Ben-Ami describes an 'apocalyptic fear' of annihilation by Nasser's Arab Middle East in context with an

'ever-present Holocaust complex' that drove Israel's first Prime Minister, David Ben-Gurion, to develop nuclear weapons.

Oxford-educated Shlomo Ben-Ami has had a career as a professional historian and as a politician. He was Foreign Minister in the Barak government where he participated in the unsuccessful 2000 Clinton-Barak-Arafat Camp David Peace Summit.

Biello, D. (2007, November) A Need for New Warheads? <u>Scientific American</u>, pp. 80-85. *This journal can be found in many community libraries.*

In Smith's *I,robot* story Israel has developed advanced thermonuclear weapons which are traded for Japanese robots. This reference gives the interested reader with little scientific background an easy to understand explanation of thermonuclear devices, and considerations in their design.

Bowes, A.M. (1989) <u>Kibbutz Goshen — An Israeli commune</u>. Long Grove, Illinois: Waveland Press ISBN 978-0-88133-395-4. *At the time of this writing, new copies of this book are still available.*

An objective and analytic look at the workings of an Israeli kibbutz. To preserve anonymity of persons interviewed or described, the fictitious name 'Goshen' was used. While various aspects of kibbutzim in the northern Galilee were considered, in particular Kibbutz Misgav Am on the Lebanese border, the fictitious Kibbutz Misgov was used for Haruto's stay in Smith's *I,robot* story.

At the time of this reference, Alison M. Bowes is at the University of Stirling in Scotland.

Brooks, R.A. (1990) Elephants Don't Play Chess. <u>Robotics and Autonomous Systems, 6,</u> 3-15. *At the time of this writing, this paper is readily available on MIT's website: http://people.csail.mit.edu/brooks/ -> Publications*

This is the paper the University of Tokyo professor shows Haruto. As the professor pointed out, much of the work in artificial intelligence was based on symbolic processing, and much still is, but Brooks, instead of working on large, complex systems for getting computers to think, started building insect robots — small machines that could actually successfully navigate through hallways. The logic of these robots was largely driven by their interaction with the environment, in this case the hallways, the obstacles they encountered, the need to attend to their basic needs — much like actual little animals.

At the time of this reference, Rodney A. Brooks is at the Artificial Intelligence Laboratory at the Massachusetts Institute of Technology in Cambridge, Massachusetts.

Brooks, R.A. (2002) <u>Flesh and Machines — How Robots Will Change Us</u>. New York: Vintage Books ISBN 978-0-375-72527-2.

Non-technical introduction to ideas of Rodney Brooks, some of which are incorporated in the robots of Smith's *I,robot* .

At the time of this reference Rodney A. Brooks is director of the Artificial Intelligence Laboratory at the Massachusetts Institute of Technology in Cambridge, Massachusetts.

Cho, J. (2006, September 28/29) Robo-Soldier to Patrol South Korean Border. <u>ABC News</u>. [Video]. *At the time of this writing, this archived news article was readily available on ABC News' website: www.abcnews.com -> Search (enter title of article)*

While not as advanced as the Alpha and Beta robots of Smith's *I,robot* story, soldier robots are no longer science fiction. The South Korean military has announced that the $200,000 Samsung Techwin robot soldier will be used to defend its border against intruders from North Korea. The robot uses visual and infra-red sensors to distinguish persons from several kilometers away, and can track moving objects. Once a tracked target is within ten meters from the robot, the robot can open fire with its K-3 machine gun, depending on the circumstances.

Cleary, T. (2005) Soul of the Samurai — Modern Translations of Three Classic Works of Zen & Bushido. North Clarendon, Vermont: Tuttle Publishing ISBN 978-0-8048-3690-6.
Karate is not just a series of physical movements to Haruto, but part of the *bushido* — a code of ethics, ie, rules, he follows. This book includes English translations of *Martial Arts* by Yagyu Munenori (1571-1646) the sword master of the Shogun, and *The Inscrutable Subtlety of Immovable Wisdom* and *The Peerless Sword*, both by Takuan Soho (1573 — 1645), the sword master's Zen Master. The *bushido*, known also as the 'way of the samurai', is strongly influenced by its Zen roots.

Crowley, J. (1979) Engine Summer. Reprinted in: Crowley, J. (2002) Otherwise: Three Novels by John Crowley. New York: Harper Perennial ISBN 978-0-06-093792-8.
Smith's *I,robot* is but one example of a large selection of post-apocalyptic tales in the literature. *Engine Summer* describes a post-apocalyptic future where a teenager searches for love beyond the boundaries of his community and discovers the lost civilization of our world.

At the time of this writing, John Crowley is at Yale University in New Haven, Connecticut.

Economist (2007, February 17) Dealing with North Korea: Trust me? You would have to be mad to do so. This time make sure Kim Jong Il keeps his promises. The Economist, p. 14. *This journal can be found in many community libraries.*
In Smith's *I,robot* story, Haruto argues that Japan had negotiated with North Korea in the past, and could again. However, this short article points out the difficulties in dealing reliably with North Korea.

Ellis, R. (2008, March) The Bluefin in Peril. Scientific American, pp. 70-77. *This journal can be found in many community libraries.*
In Smith's *I,robot* story Captain Nakamatsu, with freezers full of bluefin tuna illegally being brought back to Japan, saves Haruto's life. This easy-to-read article talks about the threats to this amazing fish.

At the time of the reference Richard Ellis is at the American Museum of Natural History in New York City.

Featherstone, S. (2007, February) The Coming Robot Army — Introducing America's Future Fighting Machines. Harper's Magazine, pp. 43-52. *This journal can be found in many community libraries.*

While not as advanced as the Alpha and Beta robots of Smith's *I,robot* story, soldier robots are no longer science fiction. This article discusses unmanned aerial vehicles (UAVs) and unmanned ground vehicles (UGVs) being developed for the American military. A study done for the US Joint Forces Command expects by 2015 that robots will exist that can hunt and kill the enemy with limited human supervision.

Feynman, R.P., Leighton, R.B. and Sands, M. (1989) The Feynman Lectures on Physics. Reading, Massachusetts: Addison Wesley ISBN 978-0-201-50064-6.

The physics in Smith's *I,robot* story is greatly simplified, but if you're interested enough in it, this is a great reference — an introduction to physics from Richard Feynman's point of view as delivered to the two-year introductory physics course he taught at CalTech. More interesting than the usual university introductory physics textbooks, even after all these years. Requires strong knowledge of high school math.

Richard Feynman (1918-1988) was awarded the Nobel Prize in Physics in 1965.

Greenberger, D. and Padesky, C.A. (1995) Mind Over Mood. New York: Guilford Press ISBN 978-0-89862-128-0. *At the time of this writing, this book is still available.*

Haruto is not the only one who suffers from an anxiety disorder — at the time of this writing, the NIMH (the USA National Institute of Mental Health) estimates that 18% of American adults, or 40 million people, suffer from these disorders, of which 2.7 million Americans have panic disorder and 2.2 million Americans have obsessive-compulsive disorder (OCD). This reference is a workbook that teaches the reader the basics of cognitive therapy (also known as 'CBT').

At the time of this reference, Dennis Greenberger is at the University of California at Irvine, and Christine A. Padesky is at the Center for Cognitive Therapy in Huntington Beach, California.

Gross, T. (2006, April 11) Football Killing Fields. National Review Online [Internet]
At the time of this writing old National Review articles are readily available on its web site. www.nationalreview.com/comment/gross200604111309.asp

In Smith's *I,robot* story, Israel has been singled out for boycotts and criticism throughout the world, which makes it harder to defend itself militarily as well as prevent the collapse of its economy. This premise is based on an extrapolation of the reality at the time of this writing. For example, the *nonpolitical* world soccer body FIFA has condemned Israel because it destroyed an empty soccer field used for terrorist training exercises, yet Gross points out FIFA has always looked the other way, eg, attack on Israeli soccer field which caused actual injuries a week before this article, Taliban using their soccer fields to slaughter violators of religious laws, etc. Gross states that this is another example of Israel being subjected to 'startling double standards' by the United Nations, the European Union, academic bodies, many Western media, etc, and now even by FIFA.

Harel, D. and Feldman, Y. (2004) Algorithmics: The Spirit of Computing,

<u>Third Edition</u>. Harlow, England: Pearson Education ISBN 978-0-321-11784-7.

The brains of the Alpha and Beta robots, in particular the Alpha Plus robots, did not just work on neural nets, but also used a variety of algorithms running on conventional processors. Algorithms are at the heart of what computers do. The electronic hardware of the computer and its software programs, both which come in a large assortment of specifications, are there to run algorithms. One of the best introductory books on the subject — suitable for a motivated high school student, yet offering even the experienced programmer some insights.

At the time of this reference, David Harel is at the Weizmann Institute in Rehovot, Israel, and Yishai Feldman is at the Interdisciplinary Centre in Herzliya, Israel.

Harris, W. (2005) <u>The New Face of Lebanon: History's Revenge</u>. Princeton, New Jersey: Markus Wiener Publishers ISBN 978-1-55876-392-0.

In Smith's *I,robot* rockets rain down on Israel from Lebanon. This reference provides some insight into the latter country. The reference starts noting that the French word for "Lebanonization" entered the French language officially in 1991 as meaning the fragmentation of a state due to the infighting between its different communities. Lebanon has communities of Maronite and other Christians, Sunni Muslims, Shi'i Muslims, Druze, and Palestinians, among others. Yet Beirut had become an important commercial and intellectual center for the Arab world, and political pluralism and stability is described during the first three decades of Lebanon's independence until about 1975, with the Maronites originally having "political supremacy." In 1932 Christians made up 51% of the population, Sunnis 22%, Shi'is 20% and Druze 7%. By the early 1980s the Christian population had dropped to 35%, while the Shi'is had risen to this level. In addition, Palestinians were using the country as a base for attacks on Israel. Israel's invasion of Southern Lebanon in 1982 where many Shi'is lived plus the extension of the Iranian revolution via *Hizballah* "Party of God" resulted in Islamic radicalism with terrorism against Western targets. Other changes in the country are discussed, and by the end of the eighties there was civil war between the Maronites, Sunnis, Shiites, Druze and Palestinians.

At the time of this reference, William Harris is at the University of Otago in New Zealand, but he previously taught at Haigazian University College in Lebanon and is married to a Lebanese Shi'ite.

Hillis, W.D. (1989, February) Richard Feynman and The Connection Machine. <u>Physics Today</u>. *This journal may be found at larger community libraries. Also, at the time of this writing, a copy of this article was available online at The Long Now Foundation website: www.longnow.org -> people -> Danny Hillis -> essays by Danny Hillis.*

Danny Hillis is the creator of the 1980s famous Connection Machine (and the company that produced it, Thinking Machines Corp.) which consists of thousands of processor/memory cells that can be linked together in the form of a Boolean n-cube, with routes for communication between the processors along this n-cube topology. Physicist Richard Feynman's son Carl was an undergraduate at MIT who helped Hillis with his thesis project on the Connection Machine. Richard

Feynman expressed an interest in the project, and the day after Thinking Machines was incorporated to everyone's surprise walked in and said he was reporting for duty. Among Feynman's many duties, which included buying office supplies, was to invite speakers, for a seminar series, who could appreciate the new machine. For the first seminar, he invited his friend from CalTech, John Hopfield. Although Hopfield is more famous now, and his networks serve as the basis of the Alpha and Beta robots of Smith's *I,robot*, in 1983 working on neural networks was not very fashionable. Feynman devised a method whereby he used one processor on the Connection Machine to simulate each of Hopfield's neurons in his neural nets.

At the time of this reference, W. Daniel Hillis is at Thinking Machines Corporation in Cambridge, Massachusetts. (Unfortunately Thinking Machines filed for bankruptcy in 1994. However, Hillis continued to work on his *Clock of the Long Now*. See the Lemley reference below.)

Hillis, W.D. (1999) Pattern on the Stone. New York: Basic Books ISBN 978-0-465-02596-1. *At the time of this writing, this book is still available.*
Hillis describes how computers work in simple language. Despite its age, an excellent starting point for the reader of Smith's *I,robot* interested in understanding computation.
At the time of this reference, W. Daniel Hillis is at Walt Disney Imagineering.

Hopfield, J.J. and Tank, D.W. (1986) Computing with Neural Circuits: A Model. Science, 233, 625-633. *This journal can be found in many community libraries.*
This reference may be of interest since the networks of Hopfield serve as the basis of the Alpha and Beta robots of Smith's *I,robot*. A hypothetical biological neural circuit consisting of a single axon input, four principal neurons with four output axons and three interneurons, is presented for analysis. It is then shown that a useful neural model is a network of nonlinear neurons symmetrically connected. Such a network will act so that its values, thought of a computational global energy, will move downhill until some local minimum is reached. It is noted that a number of biological neural circuits contain much symmetry.
At the time of this reference, J.J. Hopfield is at the California Institute of Technology in Pasadena, California and David W. Tank is at AT&T Bell Laboratories in Murray Hill, New Jersey.

Horn, B.K.P. (1986) Robot Vision. Cambridge, Mass.: MIT Press ISBN 0-262-08159-8.
Vision functions occupy a large part of our brains as well as a large number of processor cards in the Alpha and Beta robots. Processing visual images is hard work. This book is technical but provides an introduction to vision processing.
At the time of this reference, Berthold Klaus Paul Horn is at the Massachusetts Institute of Technology in Cambridge, Massachusetts.

Hornyak, T.N. (2006) Loving the Machine — The Art and Science of Japanese Robots. Tokyo/New York: Kodansha International ISBN 978-4770030122.
Smith's *I,robot* notes that Japan has a long history of developing robots. This illustrated book describes Japanese fascination with anthropomorphic robots, from the automatons of the Edo period to the advanced Honda humanoid robot *Asimo*

(an acronym, although many feel it actually is named in honor of Isaac Asimov). Japanese National Institute of Advanced Industrial Science and Technology and Kawada Industries' construction robot *HRP-2("Humanoid Robotics Project")* *Promet* shown at the beginning of Smith's *I,robot* is described at the end of this reference, noted to be able to perform real tasks such as serving drinks or installing wall paneling.

Hornyak, T.N. (2007, August) Playing It by Ear — A machine-listening system that understands three speakers at once. Scientific American, p. 28. *This journal can be found in many community libraries.*
The voice synthesis and recognition ability of the Alpha robots of Smith's *I,robot* story actually exists at the time of writing. **Hiroshi G. Okuno** of Kyoto University and colleagues have developed a voice recognition system that can comprehend multiple persons speaking at the same time. The system identifies different audio sources and then uses mathematical filters to separate out the sounds from the different speakers. The system then fills in missing sounds by comparing the sounds heard with a database of fifty million Japanese utterances.

Hyman, B.M. and Pedrick, C. (2005) The OCD Workbook: Your Guide to Breaking Free from Obsessive-Compulsive Disorder- 2nd edition. Oakland, California: New Harbinger ISBN 978-1-57224-422-1.
This workbook helps persons with obsessive-compulsive disorder (OCD) (what Haruto suffers from) to overcome some of their symptoms.
At the time of this reference, Bruce M. Hyman, Ph.D. is at the OCD Resource Center of Florida in Hollywood, Florida, and Cherry Pedrick, RN is in Las Vegas, Nevada.

Kurzweil, R. (2005) The Singularity is Near — When Humans Transcend Biology. New York: Viking ISBN 978-0-670-03384-3.
Exponential knowledge in genetics, nanotechnology and robotics will eventually result in a merging with human biology to create a new advanced species. This is not science fiction but solid work with Kurzweil making cogent arguments. This reference strongly supports the level of technology and artificial intelligence (AI) depicted in Smith's *I,robot*. Kurzweil points out ("AI Winter") that while there are those that feel that AI failed in the 1980s this is the normal "technology hype cycle for a paradigm shift". For example, with railroads, the Internet and AI (and other technologies) there is a start with unrealistic expectations which are not immediately met and disappointment occurs. However, AI is improving. Kurzweil quotes Rodney Brooks "...this stupid myth out there that AI has failed, but AI is everywhere around you every second of the day... fuel injection systems... land in an airplane... Microsoft software..."
At the time of this reference, Raymond Kurzweil is a futurist as well as an inventor in optical character recognition and other fields.

Kynge, J. (2006) China Shakes the World — The Rise of a Hungry Nation. London: Orion Publishing ISBN 978-0-7538-2155-8.
Early in this reference is the quotation attributed (possibly) to Napoleon, "Let China sleep, for when she wakes, she will shake the world." And indeed, China has now started to shake the world. Economically the Chinese 'appetite' is felt 'even in

the most remote corners of the Earth' — forests from Papua New Guinea, fish from all over the Pacific, iron ore from Australia, soybeans from the United States, oil from the Middle East, etc. In Smith's *I,robot* story, Japan creates the Alpha and Beta military robots to counter the threat from the millions and millions of soldiers in the Chinese army. Although Kynge's book does not consider the Chinese military in detail, he notes the hatred and vitriol of Chinese protestors against the Japanese surprised him.

At the time of this reference, James Kynge is an Edinburgh-educated journalist living in Beijing.

Lemley, B. (2005, November) Time Machine — The Clock of the Long Now. Discover Magazine. *This journal can be found in many community libraries.*

As Smith's *I,robot* story points out, there is the need for mankind to consider the long-term. The *Clock of the Long Now* project aims to produce a clock that will record time for some 10,000 years, as well as accompany a *10,000-Year Library* that is not vulnerable to the short lifespan of digital information. The clock is made of stone and steel. At the time of this reference, the clock's designer, **W. Daniel Hillis** had spent two decades already working on this project. Danny Hillis is also the creator of The Connection Machine, and the company that produced it, Thinking Machines Corp.

Lewis, B. (2006, Winter) The New Anti-Semitism. The American Scholar, 75, 25-36. *This journal can be found in larger public libraries.*

The setting of the end of Smith's *I,robot* in Israel involves hostility between Israelis (who are often, but not always, Jewish) and their neighbors. This article considers what is exactly Anti-Semitism. In Smith's *I,robot* Mara notes that her grandparents came from Jerusalem to Misgov, and indeed in this reference, Lewis notes that the ancient Jewish community in Jerusalem was evicted and destroyed when Israel formed (with Lewis complaining that while the Arab refugees received help from the United Nations the treatment of the Jewish refugees was very different).

At the time of this reference, Bernard Lewis is at Princeton University.

Lloyd, S. (1996) Universal Quantum Simulators. Science, 273, 1073-1078. *This journal can be found in many community libraries.*

In the last chapter of Smith's *I,robot* Isaac's father snaps open his chest plate to reveal quantum computers inside. Quantum computers can perform ordinary digital logic as well as quantum logic. In 1982 Feynman noted that simulating quantum systems on classical computers is difficult, and conjectured that quantum computers would be able to simulate such systems better than classical computers. Lloyd shows in this article that quantum computers can effectively be programmed to simulate local interaction quantum systems. A quantum computer can perform such simulations with tens of quantum bits and tens of steps versus a classical computer requiring trillions of gigabits and steps.

At the time of this reference, Seth Lloyd is at the Massachusetts Institute of Technology.

Lovelock, J. (1998) A Book For All Seasons. Science, 280, 832-833. *This journal can be found in many community libraries.*

In Smith's *I,robot* the civilization of humans passes to those of the robots, and so this reference may be of interest. Lovelock makes the argument that civilizations are ephemeral (for example, 30 some civilizations in the past 500 years). Unlike social insects, which carry instructions for nest building in their genes, humans have no permanent record of their civilization that could be used to rebuild it. It is argued that we need to have an up-to-date record of our civilization, including our science and technology, that is written on durable paper with long-lasting print in clear language that any intelligent person in the future could understand. It is noted that in the Dark Ages, the religious monasteries took care of our knowledge. However, at present, scientists are not guardians of our knowledge, and science does not possess the equivalent of the monasteries.

At the time of this reference, James Lovelock is an honorary visiting fellow at Oxford University in Oxford, U.K.

Madden, J.D. (2007) Mobile Robots: Motor Challenges and Material Solutions. Science, 318, 1094-1097. *This journal can be found in many community libraries.*

In humans or humanoids, you need about 200W (watts) mechanical power at the ankles during walking, rising to 700W for a sprint. This easy-to-read reference considers technologies that could enable agile robots. Gasoline engines give good power per mass but have transmission problems, eg, no force at zero speed, which makes use in robotics difficult, not to mention heat and noise issues. Hydraulics give lifelike movements but Madden unfortunately dismisses their potential due to wasted shunting of fluid for walking, and the noise and heat of an internal combustion engine powering the hydraulic pumps. He does not consider fully the use of, as shown in Smith's *I,robot*, hydraulics re-engineered lighter used with high power-to-mass fuel cells which also allow efficient bursts of power consumption. Although Smith's book glances over the technical details, the use of Smith's hydraulic spheres and the free running of cylinders unless they are needed, results in minimal fluid shunting. The reference then goes on to consider electric motors, which are considered more attractive. If a 150W/kg geared-electric motor is used, you need a 6.5kg motor in each ankle of the robot, although if you cool the motor and use a spring action in each step, lower weight and thus agility could significantly be improved. The promise of dielectric elastomers is then considered. However, as Seiko in Smith's *I,robot* says, "You probably expected to see electroactive polymer artificial muscles, but they weren't strong enough. Instead, we modified aviation hydraulics and made them lighter and faster." The reference then goes on to consider future technologies such as electrostatically stretching carbon nanotubes.

At the time of this reference, John D. Madden is at the University of British Columbia in Vancouver.

Matarić, M.J. (2007) The Robotics Primer. Cambridge, Mass.: MIT Press ISBN 978-0-262-63354-3.

Excellent introduction to the field of robotics without the necessity of a mathematics or engineering background. Covers the key concepts of the field — what is a robot, sensors and actuators, control architectures to learning. Control architectures are divided into *deliberative control, reactive control, hybrid control* and *behavior-based control*. Deliberative control works over a long timescale with planning into the future. Reactive control works on a shorter timescale and Rodney Brooks' subsumption architecture of Smith's *I,robot* is described here. (Of interest, Matarić was a PhD student of Brooks.) Hybrid control combines reactive and deliberative control, typically consisting of a reactive layer, a planner and a layer that links the previous two together. Behavior-based control emerged from reactive control with *behaviors* as basic control modules.

At the time of this reference, Maja J. Matarić is at the University of Southern California in Los Angeles.

Menzel, P. and D'Aluisio F. (2000) <u>Robo Sapiens — Evolution of a New Species</u>. Cambridge, Mass.: MIT Press ISBN 978-0-262-13382-1.

Wonderful photographs and stories of robotic research projects around the world. Includes **Hirochika Inoue** ("grand old man of Japanese robotics") of the University of Tokyo's project to develop a humanoid robot. Includes **Rodney Brooks** of MIT with, at the time, his latest version of *Cog*, a humanoid robot. Keep in mind that another decade of robotic development occurs before Smith's *I,robot* takes place.

Photojournalist Peter Menzel and writer Faith D'Aluisio are associated with Material World Books in California at the time of this reference.

Moravec, H. (1999, December) Rise of the Robots. <u>Scientific American</u>, pp. 124-135. *This journal can be found in many community libraries.*

It is predicted that by 2010 person-sized mobile robots with cognitive abilities similar to about a lizard (5000 million instructions per second, ie, 5000 MIPS) will be produced and used for chores such as delivering packages, followed by a second generation that has cognitive abilities of a mouse (100,000 MIPS) that could be more readily trainable, eventually followed by robots with the cognitive abilities of a monkey (5 million MIPS, ie, 5 million million instructions per second), and it is predicted that by 2050 robots with computer 'brains' executing 100 million MIPS will start to equal human intelligence. Given that edge or motion detection takes about 100 computer instructions, 1000 MIPS is required to allow performance equal to the human retina's 10 million detections per second. If the retina is .02 grams and requires 1000 MIPS for emulation, then it is extrapolated that the 1500-gram brain requires some 100 million MIPS.

For comparison (quick disclaimer: estimates dependent on exact machine configuration; ongoing controversy MIPS vs. FLOPs vs. other standards; trademarks property respective owners):

First Intel 8080 microprocessor PCs (1970s) — under 1 MIPS
Intel 386 (1988) — 9 MIPS
Intel 486 (1992) — 54 MIPS
Intel Pentium Pro (1996) — 540 MIPS
Intel Pentium III (1999) — 1350 MIPS
Intel Pentium 4 Extreme (2003) — 9700 MIPS (lizard brain?)
Intel Core 2 Extreme QX6700 (2006) — 57,000 MIPS (mouse brain?)
IBM Blue Gene (2007) — roughly 300 million MIPS (human brain?)

Although Blue Gene meets the MIPS estimate required to create the cognitive abilities of a person, at the time of this writing, no such AI system exists, because while such a system may require this processing power, a particular architecture as well as a software must be developed to make the system do the things desired. In Smith's *I,robot* adequate processing power for vision and cognition in the Alpha and Beta robots is assumed by use of multiple specialized processors, and of more importance, an approach to a type of architecture and software capable of allowing the desired cognitive abilities, is depicted.

At the time of this reference, Hans Moravec is at Carnegie Mellon University in Pittsburgh, Pennsylvania.

Morris, E. (Director) (2002) Fast, Cheap & Out of Control [DVD]. Sony Pictures, ASIN B00003CX9Z. *At the time of this writing, available from www. sonypictures.com as well as other online retailers.*

A documentary of four real persons trying to control their particular fields — including, of course, Rodney Brooks trying to tame his robots. In Smith's *I,robot* story the ideas of Rodney Brooks are used in constructing the brains of the Alpha and Beta robots. Brooks' ideas may not be the most powerful in the field of AI, but they are seductively useful for quickly creating functional robots. The term 'Fast, Cheap and Out of Control', for which this documentary film is named, comes from a paper which Brooks co-authored in 1989, where he argues that instead of launching expensive, complex space missions, it would be better to use large numbers of relatively small, simple, mass-produced autonomous robots. It is suggested that it may be possible to invade a planet with millions of these small robots. (The paper is in the 1989 vol. 42 issue of the *Journal of The British Interplanetary Society.*)

At the time of this writing, Errol Morris is a filmmaker living in Cambridge, Massachusetts.

Pyle, K.B. (2007) Japan Rising — The Resurgence of Japanese Power and Purpose. New York: Public Affairs ISBN 978-1-58648-417-0.

Since the end of World War II, Japan has been pacifistic and isolated with regard to global strategic issues. However, Pyle notes that Japan's strategic behavior has fluctuated wildly over the last few centuries, and that Japan is quietly awakening again. After the first Chinese nuclear weapons test in 1964, Prime Minister Satō wanted nuclear weapons for Japan too, but the majority of the Japanese population was very against this, and in 1967, he pledged that Japan would not manufacture or import nuclear weapons, rather, there would be reliance on the U.S defense/ nuclear umbrella. He would later receive the Nobel Peace Prize for this pledge, but in private, he did not accept the pledge, which he called "nonsense". Although many readers of Smith's *I,robot* would it assume that North Korea launching missiles (of any type) over Japan is simply a fictional element, indeed on August 31, 1998, for unknown reasons, North Korea launched a missile over Japan. The Japanese Yoshida Doctrine forbid the militarization of space, but in response to the Korean missile, Japan decided to go ahead and develop its own military surveillance satellites and work with the USA on ballistic missile defense. As well, Japan began to take notice of Chinese nuclear weapons and in 2002, senior politician Ozawa Ichiro on a visit to China stated that Japan had enough plutonium from its power plants to produce thousands of nuclear warheads if it wanted to. In 2003 the former vice minister of foreign affairs acknowledged that North Korea's nuclear weapons created a nightmare for Japan — becoming a victim for nuclear blackmail or else developing its own nuclear weapons which the population was against and would create 'internal turmoil'.

At the time of this reference, Kenneth B. Pyle is at the University of Washington in Seattle, Washington. A long-time expert on Japan, in 1999 he received the Order of the Rising Sun, an honorary civilian and military decoration.

Rumelhart, D.E. and Zipser, D. (1985) Feature Discovery by Competitive Learning. Cognitive Science, 9, 75-112, 1985. *This journal can be found in most university libraries.*

Learning is only briefly touched upon in Smith's *I,robot* story, but is an essential aspect of a neural net. An unsupervised competitive learning paradigm for a parallel network of neuron-like elements is presented in this reference. Although more recent textbooks on artificial neural networks may provide a better description of unsupervised competitive learning paradigms, this article is useful in presenting a history of the related concepts, from **Donald Hebb** (1949 - strengthen connections between elements of the network only when both the pre- and postsynaptic units were active simultaneously) to **Dean Edmonds** and **Marvin Minsky** (1951 - neural learning machine, where even if one of its hundreds tubes or thousands of soldered connections failed, it did not make much difference) to a high school classmate of Minsky, **Frank Rosenblatt** (1962 - perceptron learning theorem; spontaneous learning without a teacher) to Minsky and **Papert** (1969 - limitations of perceptrons), to various paradigms of learning (autoassociator - present partial pattern and system retrieves original pattern or one close to it; pattern associator - like autoassociator except pairs of patterns are initially presented to machine; classification paradigm - to classify patterns into fixed set of categories, as in the perceptron's operation; regularity detector - machine discovers statistically important features of the input, but unlike the classification paradigm there is no predetermined fixed set of categories).

At the time of this reference, David E. Rumelhart and David Zipser are at the University of California at San Diego.

Simon, H.A. (1996) The Sciences of the Artificial, Third Edition. Cambridge, Mass.: The MIT Press ISBN 978-0-262-69191-8.

Although the original edition (a reprint of a lecture given at MIT in 1968) is now some forty years old, this third edition is still available, and still relevant to those who want to understand the artificial sciences (such as building robots as in Smith's *I,robot* story).

Herbert A. Simon (1916- 2001) was awarded the 1978 Nobel Prize in Economics.

Smith, D.D. (2004, Spring) North Korea and the United States: A Strategic Profile. The Korean Journal of Defense Analysis, Vol. XVI, No. 1. At the time of this writing, a copy of this article is available at *www.derekdsmith.com* -> *Publications*

In Smith's *I,robot* Haruto worries about the nukes-for-robots dealing leading to nuclear proliferation. In this reference, it is argued that although to date North Korea has exported only its missile technology and not any weapons of mass destruction, due to North Korea's financial needs, reprocessed plutonium from its Yongbyon reactor may end up being sold to rogue states or terrorists.

At the time of this writing, Derek D. Smith is a lawyer in Washington, D.C.

Sofge, E. (2008, March) America's Robot Army: Ultimate Fighting Machines. Popular Mechanics, pp. 58-63. *This journal can be found in many community libraries.*

Easy-to-read article with color photos of first generation armed robots.

Tank, D.W. and Hopfield, J.J. (1987) Collective Computation in Neuronlike Circuits. Scientific American, pp. 104-114. *This journal can be found in many community libraries.*

Easy-to-understand article about neural networks similar to those used in the Alpha and Beta robots of Smith's *I,robot* story. A collective decision circuit where computing elements are richly interconnected, can be pictured as a landscape, with arrows moving from initial state to subsequent states, and with these arrows considered the 'computational energy' of the circuit. The circuit follows a path that decreases its computational energy to some minimum, just as gravity pulls an object down into some minimal well. Useful to create an association memory, ie, you present the memory not with a storage address but with details and patterns, and the associative memory returns the nearest match to these presented patterns.

At the time of this reference, David W. Tank is at AT&T Bell Laboratories, and John J. Hopfield is at California Institute of Technology in Pasadena, California.

University of Texas Libraries (2007) Perry-Castañeda Library Map Collection [Internet]. University Texas at Austin. *At the time of this writing, readily available to public at www.lib.utexas.edu/maps/*

Online map collection of thousands of high quality maps of most regions of the world that Smith's *I,robot* takes place in.

Watzman, H. (2007) Deep Divisions — Archaeologists are unearthing remarkable finds in Jerusalem. Nature, 447, 22-24. *This journal can be found in many community libraries.*

The chapter of Smith's *I,robot* that takes place in Jerusalem does not provide much background material about the location. The 'Old City' of Jerusalem, a walled area only about a kilometer squared, contains sacred Jewish, Islamic and Christian sites. The interpretation, strongly affected by politics, of archeological excavations of the 'City of David' just outside the Old City walls in the Palestinian village of Silwan, is discussed in this article. Homes stand on centuries of Muslim and Byzantine ruins, which in turn stand on top of the Jerusalem destroyed by the Babylonians in 586BC, which in turn stand on top of the Jerusalem destroyed by the Assyrians in 701BC, which in turn are expected to stand on top of David's and Solomon's Jerusalem, which in turn may stand on a Middle Bronze Age city dating from 1800BC. In 2005 pottery dated to around 1000BC, corresponding to the time of King David, was found. The group supporting the excavations believes that Jewish presence in the area is God's will and a precondition for the arrival of the Messiah.

Author's Note: Haiku by Ryōta

Ōshima Ryōta (1718-1787) studied under Rito (1680-1754). Ryōta became himself a popular teacher and competed with the Kikaku school.

Translations involve interpretation of Ryōta's intent. Translations in Smith's *I,robot* are creations of the author but are based on the scholarship provided by:

Blyth, R.H. (1950) Haiku- Volume II — Spring. Tokyo: Hokuseido Press.
Blyth, R.H. (1963) A History of Haiku — Volume One — From the Beginnings up to Issa. Tokyo: Hokuseido Press.
Bownas, G. and Thwaite, A. (1964) The Penguin Book of Japanese Verse. Middlesex, England: Penguin.
Guest, H. (2002) A Puzzling Harvest — Collected Poems 1955-2000. London: Anvil Press ISBN 085646354X.
Henderson, H.G. (1958) An Introduction to Haiku. Garden City, New York: Doubleday.

Author's Note: Acknowledgments

All mistakes in this book are those of the author, but he acknowledges the help provided by:

Dr. Peter W. Sloss of U.S. National Oceanic & Atmospheric Administration.
Dave King, master editor and patient teacher.
Kathy Harestad for the illustrations and design of this book, but even more so for her insight into its story.
My father, Gerald S., for vital suggestions to the full manuscript.
Kim Moritsugu, novelist and writing instructor.
John Kemeny, for a critical read of the manuscript.
Dr. Pierre-Yves F. Robin, geodynamicist, for tectonics advice.
Dalia Adler, for a critical read of the manuscript.
Dr. Atsuko Sakaki, Department of East Asian Studies, University of Toronto.

I, robot

Author's Note: Disclaimers

Kibbutz Misgov — Fictionally based on the real Kibbutz Misgav Am, but the people, events and details are all fictional.

al-Haleeb — Fictional organization. I did not use a real organization because there seems to be a new one dominant in the news every decade, and there already is enough hate and division in the world.

Governments and militaries of Japan, North Korea, South Korea, China, Israel, Lebanon, USA and other countries' governments, militaries, agencies — All depictions in this book are fictional.

Isaac Asimov, Ayu-sama, Rodney A. Brooks, Z. Gvirtzman and John J. Hopfield — Real persons whose works are mentioned in this book, but otherwise have no relationship whatsoever with this story. There is no depiction whatsoever that could reasonably be construed to be derogatory. Depiction of Gvirtzman's geological article as being useful for the testing of nuclear weapons is fictional — it simply is a scientific article in a reputable journal about the bathymetry in an area where the fictional testing occurs in this book. Full citation: Gvirtzman, Z., and R. J. Stern (2004), Bathymetry of Mariana trench-arc system and formation of the Challenger Deep as a consequence of weak plate coupling, *Tectonics*, *23*, TC2011, doi:10.1029/2003TC001581.

Aoyama Takamichi, Nelson Bennett, Mr. and Mrs. Coen, Policeman Fujii, Petty Officer Hirashi, Itou Yoshio, Kauma Ogawa, Major Kirzner, Menachem Levi, Machii Tomo, Secretary Matthews, Bob Murray, Nishimatsu Akihiro, Moshe Otzker, Eli Peletz, Seiko Haku, Suzuki Haruto, Takahashi Ichiro, Professor Tamaki, Colonel Tanaka, Toshifumi Haruka, Captain Watanabe and other characters not listed here – Completely fictional. Any resemblance to actual persons living or dead is completely coincidental.

Alev Electronics, Autonomous Products Division, Autonomous Products' Alpha and Beta Robots, Autonomous Products' Carl, Coen Electronics, Dragon Explorer, Haifa-Asia Car Corporation, Holland East, IT Chips, Kiiro Tower Hotel, Mikiyasu Industries, New Pacific Queen, Prime Clock, Silly Machines, Supermarket Turk and others not listed here – Completely fictional. Any resemblance to actual entities is completely coincidental.

Achzarit carriers, Aden Hotel, Airbus, Asahi Beer, Bank of Tokyo, BBC, C-130 Hercules, Caltech, Canon telephoto EF300, Capsule Inn, Carmel container terminal, Casio Pathfinder GPS watch, cheesies sounds similar to Cheezies registered by Hawkins, Clock of the Long Now Foundation, CNN, Coca-Cola, Coke, DHL, F-35, FIFA International Football Association, Ford, Fuji TV, Gakkan Mae Girls School, Gaviscon, General Motors, Geological Survey of Israel, Ginza Mitsukoshi Department Store, Glock, Google, google as a verb may also be a trademark of Google Inc., Hamamatsu, Heavenly Hash, Honda, Idaten, JAL, Jaffa oranges, Japan Kibbutz Association, Jewish Community Center Tokyo, Katyusha, Kawada

Industries, Kentucky Fried Chicken, Kurooshia, Lexus, M5, Manta Ray Restaurant, Mars Bars, Massachusetts Institute of Technology, MIT AI Lab, Memory Stick is a trademark of Sony, Mercedes, Merkava, Microsoft Windows, MIT, Minowamon Junmai Sake, Mitsubishi H-IIA rocket, Nissan, Noblesse, Ocean Bridge, Princeton, Rafael, Red Cross, Renault, Shiseido perfume, Technion, Thinking Machines, Time Magazine, Toyota, Trophy, Turkish Airlines, TV Tokyo, University of Tokyo, University of Tokyo AI Lab, Visa, Volkswagen, Windows, Yomiuri Shimbun and others not listed here – Real products, landmarks, companies, universities or organizations mentioned in the plot of this story but their depiction is completely fictional. There is no depiction whatsoever that could reasonably be construed to be derogatory. Any trademarks mentioned are property of their respective owners.

The use of general descriptive names, trade names, trademarks, websites and other descriptors, even if not particularly identified as such, is not be taken as an indication that such names or descriptors may be used freely. *Robot Binaries & Press* is a registered Canadian corporation. The website link *robotpress* is a contraction of the above name and is not associated with any other trade names related to *robot* or *press* or *robotpress* or *roboticpress*.

Author's Note: Permissions

All illustrations except as noted below are original creations of the author and artist. Assistance with rendering of the artwork by Kathy Harestad www.kathyart. com.

Cover artwork and photos: Kathy Harestad and Licensed © Alone of Graphics Manufacturer and Licensed © Philippe Schpilka.
Graphic: Cherry Trees. Licensed © Henning Stein.
Photograph: Clock of the Long Now Faceplate — Photo by Nicholas Rolfe Horn courtesy of The Long Now Foundation www.longnow.org.
Photograph: New Quarters Misgav Am — Photo by Chaver83. Released into the public domain worldwide August 26, 2007.
Graphics: North Korean nuclear explosion seismogram and xenon-133 dispersal — Courtesy of the CTBTO Preparatory Commission.
Photograph: Laws of Robotics Licensed © Sergey Tokarev.
Graphic: Howard Smith Licensed © Scott Maxwell.

About the Type

This book was set in Minion Pro, a type developed by Robert Slimbach, winner of the 1991 Prix Charles Peignot for Excellence in Type Design. Minion Pro derives from the late Renaissance typefaces that were both beautiful and readable. Minion Pro fonts are clear and balanced, and facilitate the read of a long novel.

Dr. Howard S. Smith is an MIT-trained engineer with an interest in artificial intelligence — the supermarket self-checkout machines are all based on his work — and natural intelligence — evolution of the brain.

The author welcomes your comments. Please write to him at: **authorhowardssmith@robotpress.net**

ForeWord
Reviews *of*
Good Books
Independently Published

Author Interview with ForeWord Magazine

When did you start reading, and what did you like to read as a kid?

I remember reading cereal boxes. Tony the Tiger sticks in my mind.

My next door neighbor and best friend Gary had a Childcraft Encyclopedia. Every day after playing with him I remember sitting on his floor and reading another few pages of the encyclopedia. In third grade our teacher told us about novels. The first one I ever read was *The Good Earth* by Pearl S. Buck. I remember thinking that novels had better endings than the encyclopedia.

There was a used bookstore next to the barbershop I went to as a kid. For a quarter, the store owner would go to one of his packed shelves, pull out a novel, blow off the dust, and tell me to come back when I finished it. With every haircut came another novel. The second book I ever read was *I, Robot* by Isaac Asimov.

When you were growing up did you have books in your home?

My mom kept buying me Dr. Seuss books. Aghh!! Yup, I know everyone loves them. I didn't. I thought they were stupid. My mother, however, had a membership in the Book of the Month Club. Her books were a lot better than mine.

In the fifth grade I switched to reading technical books. I took a correspondence course in electronics technology from the Cleveland Institute of Electronics that lasted a few years. I went to MIT after eleventh grade, and didn't really start reading fiction again for a while. I missed it. Television was okay, movies were better, but nothing satisfied like a good (fiction) book.

When did you think about becoming a writer? Was there someone who got you interested in writing?

Am I writer now? I never really thought about it like that. Writing *I,robot* (small 'r') was just something I wanted to do (okay… something I felt compelled to do), rather than being "a writer."

How do you write? Do you have a daily routine? What's good about it? What do you hate about it?

I try to write as quickly as possible. No agonizing about this or that word. Get the story out. Then once that's done, I can come back and agonize. So my daily writing routine is to snatch any period of free time and write.

I wrote *I,robot* in five drafts. In the first try, the story took place in the Far East. The intensity between the combatants was so high that World War III started off before there was enough time for the development and rise of the robots. No good. In the next draft I switched the setting to Iraq. Just the right amount of violence (… that sure seems like an oxymoron) but a confusing story. It never makes sense why people die, but in this case, it really didn't make sense. Into the trash bucket. I then moved the end of the story over to the Israeli-Lebanese border. It worked well. The fourth draft was a research exercise – what evidence supported every single line in the book. And then, in the fifth and final draft, I edited. One hundred fifty thousand words became one hundred ten, and the book read well.

Any particular story to tell concerning the writing of this book?

After seeing the 2004 *I, Robot* film I thought that someone should update this ridiculous notion of a "positronic brain" and update how the rise of the robots would really come about.

Asimov wrote his first robot stories in the early 1940s, shortly after the discovery of the positron (it's like an electron, but it has a positive charge). This was before even primitive computers such as the ENIAC existed. The trouble with positrons is that they immediately annihilate upon contact with electrons, so it really wouldn't make much sense to build a computer based on these particles.

However, Asimov's *I, Robot* is more about software than hardware – rules (Three Laws of Robotics) and how they affect the lives of robots and people. I have tried to be true to this principle in my rendition of Asimov's classic.